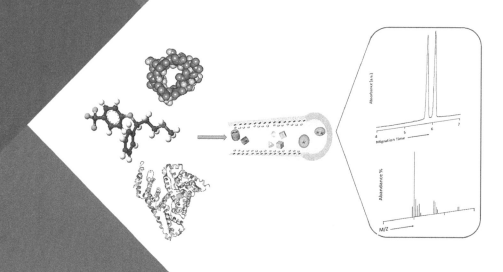

Capillary
Electrophoresis

Capillary Electrophoresis
Trends and Developments in Pharmaceutical Research

edited by

Suvardhan Kanchi
Salvador Sagrado
Myalowenkosi Sabela
Krishna Bisetty

Published by

Pan Stanford Publishing Pte. Ltd.
Penthouse Level, Suntec Tower 3
8 Temasek Boulevard
Singapore 038988

Email: editorial@panstanford.com
Web: www.panstanford.com

British Library Cataloguing-in-Publication Data
A catalogue record for this book is available from the British Library.

Capillary Electrophoresis: Trends and Developments in Pharmaceutical Research
Copyright © 2017 by Pan Stanford Publishing Pte. Ltd.
All rights reserved. This book, or parts thereof, may not be reproduced in any form or by any means, electronic or mechanical, including photocopying, recording or any information storage and retrieval system now known or to be invented, without written permission from the publisher.

For photocopying of material in this volume, please pay a copying fee through the Copyright Clearance Center, Inc., 222 Rosewood Drive, Danvers, MA 01923, USA. In this case permission to photocopy is not required from the publisher.

ISBN 978-981-4774-12-3 (Hardcover)
ISBN 978-1-315-22538-8 (eBook)

Printed in the USA

Contents

Preface	xiii

1. Capillary Electrophoresis: A Versatile Technique in Pharmaceutical Analysis — **1**

Imran Ali, Zeid A. Alothman, Abdulrahman Alwarthan, and Hassan Y. Aboul-Enein

1.1	Introduction		2
1.2	Theory of Capillary Electrophoresis		2
1.3	Types of Capillary Electrophoresis		5
1.4	Instrumentation		7
	1.4.1	Sample Introduction	8
	1.4.2	Capillary Electrophoresis Capillary	9
	1.4.3	Power Supply	10
	1.4.4	Background Electrolyte	11
	1.4.5	Detection	11
		1.4.5.1 UV/Vis absorbance detectors	12
		1.4.5.2 Conductivity detectors	13
		1.4.5.3 Amperometric detectors	14
		1.4.5.4 Thermooptical absorbance detectors	15
		1.4.5.5 Fluorescence detectors	15
		1.4.5.6 Atomic absorption spectroscopy detectors	16
		1.4.5.7 Inductively coupled plasma detectors	17
		1.4.5.8 Mass detectors	17
		1.4.5.9 Miscellaneous detectors	18
		1.4.5.10 Indirect detection	19
1.5	Data Integration		20
1.6	Sample Preparation		20
1.7	Method Development and Optimization		21
1.8	Validation of Methods		23
1.9	Applications		24
1.10	Conclusion		45

vi *Contents*

2. Recent Applications of Chiral Capillary Electrophoresis in Pharmaceutical Analysis   **71**

José María Saz and María Luisa Marina

2.1   Introduction   71
2.2   Chiral Selectors Added to BGE for Enantioseparation of Chiral Drugs by CCE   72
    2.2.1   Cyclodextrins   73
    2.2.2   Macrocyclic Antibiotics   81
    2.2.3   Polysaccharides   84
    2.2.4   Other Chiral Selectors   85
2.3   Applications of CCE in Pharmaceutical Analysis   87
    2.3.1   Analysis of Pharmaceutical Formulations   93
    2.3.2   Analysis of Illicit Drugs   96
    2.3.3   Analysis of Biological Samples   98

3. A Mini-Review on Enantiomeric Separation of Ofloxacin using Capillary Electrophoresis: Pharmaceutical Applications   **117**

Suvardhan Kanchi, Myalowenkosi Sabela, Deepali Sharma, and Krishna Bisetty

3.1   Introduction   117
3.2   Role of CE in Pharmaceutical Analysis   118
3.3   Challenges Involved in the Enantiomeric Separation of Ofloxacin   119
3.4   Developments in Enanatiomeric Separation of Ofloxacin   120
    3.4.1   Recent Developments in Enantiomeric Separation of Ofloxacin   121
3.5   Applications of Nanotechnology in Enantiomeric Separation of Ofloxacin   127
3.6   Role of Computational Techniques in Enantiomeric Separation of Ofloxacin   130
    3.6.1   Molecular Modeling   135
    3.6.2   Molecular Dynamic Simulations   137
3.7   Conclusion   138

4. Nano-Stationary Phases for Capillary Electrophoresis Techniques   **147**

Chaudhery Mustansar Hussain and Sagar Roy

4.1   Introduction   148

4.2	Different Types of Nanoparticles/ Nanomaterials		149
4.3	Synthetic Strategies for Immobilization for Nano-Stationary Phases		154
4.4	Evaluation of Nano-Stationary Phases		157
4.5	Conclusion		159

5. Capillary Electrophoresis Coupled to Mass Spectrometry for Enantiomeric Drugs Analysis **165**

Vítězslav Maier, Martin Švidrnoch, and Jan Petr

5.1	Enantioseparation of Pharmaceutical Products		166
5.2	Basic Concepts of Enantioseparation by Capillary Electrophoresis		167
5.3	Capillary Electrophoresis with Mass Spectrometry		169
	5.3.1	Electrospray Ionization in Hyphenation of CE with MS	170
5.4	Interfacing of CE-ESI-MS		172
	5.4.1	Sheathless Interface	173
	5.4.2	Sheath Liquid Interface	174
	5.4.3	Liquid-Junction Interface	177
5.5	Compatibility of Capillary Electrophoresis and Mass Spectrometry		177
	5.5.1	Basic Principles of Partial Filling Technique	179
5.6	Cyclodextrin as CS for CE-MS Enantioseparation		183
	5.6.1	Experimental Setup without Restrictions of the CD Entering MS Ion Source	184
	5.6.2	Counter-Current Mode for CD-Mediated Enantioseparation by CE-ESI-MS	186
	5.6.3	Partial Filling Technique Employing Neutral Derivatives of CDs	187
	5.6.4	Combination of PFT and Counter-Current Migration for CD-Mediated CE-ESI-MS Enantioseparation	189
5.7	Enantioseparation using MEKC-MS		193
5.8	Crown Ethers Mediated Enantioseparations by CE-MS		200

viii | Contents

5.9	Macrocyclic Antibiotics as CS for CE-MS	201
5.10	Non-Aqueous Capillary Electrophoresis-Mass Spectrometry for Enantioseparation	204
5.11	Enantioseparation by CEC-MS	206
5.12	Specifics of Quantitative Analysis of Enantiomers by CE-MS	208

6. Enantioselective Drug–Plasma Protein-Binding Studies by Capillary Electrophoresis **225**

Laura Escuder-Gilabert, Yolanda Martín-Biosca, Salvador Sagrado, and María José Medina-Hernández

6.1	Introduction	225
6.2	Plasma Proteins	227
6.3	Capillary Electrophoresis for Enantioselective Protein-Binding Experiments	229
	6.3.1 Combination of CE with other Separation Techniques	231
	6.3.2 CE for Direct Enantioselective Protein-Binding Evaluation	233
6.4	Experimental Design and Mathematical Models in Protein-Binding Studies	235
	6.4.1 Mathematical Models and Deficiencies	235
	6.4.2 Experimental Design and Verification of the Assumptions	237
	6.4.3 Examples Illustrating the Direct Approach Strategy	240
6.5	Studies on the Application of Capillary Electrophoresis for Evaluating Enantioselective Plasma Protein Binding of Chiral Drugs	242
6.6	Conclusion	251

7. Clinical Use of Capillary Zone Electrophoresis: New Insights into Parkinson's Disease **259**

Pedro Rada, Luis Betancourt, Sergio Sacchettoni, Juan Félix del Corral, Hilarión Araujo, and Luis Hernández

7.1	Introduction	260
	7.1.1 Parkinson's Disease	260
	7.1.2 Basal Ganglia Circuitry	261

	7.1.3	In Vivo Monitoring of Brain Molecules	263
7.2		Capillary Zone Electrophoresis	264
	7.2.1	GABA in the Ventrolateral Nucleus of the Thalamus in Two PD Patients	266
	7.2.2	Sample Preparation	267
	7.2.3	Running Conditions	267
7.3		Result	267
	7.3.1	Polyamine Putrescine and Parkinson's Disease	268
	7.3.2	Blood Sampling	269
	7.3.3	Running Conditions	269
	7.3.4	Putrescine Levels in RBCs and Plasma of PD Patients	271

8. Electrophoretically Mediated Microanalysis for Evaluation of Enantioselective Drug Metabolism — **277**

Yolanda Martín-Biosca, Laura Escuder-Gilabert, Salvador Sagrado, and María José Medina-Hernández

8.1		Introduction	277
8.2		Enzymatic Reactions in CE: Electrophoretically Mediated Microanalysis	279
8.3		Evaluation of Enantioselective Metabolism by EMMA	281
	8.3.1	Study of Stereoselectivity of Flavin-Containing Monooxygenase Isoforms using Cimetine as Substrate	285
	8.3.2	Characterization of the Enantioselective CYP3A4 Catalyzed N-demethylation of Ketamine	287
	8.3.3	Evaluation of Enantioselective Metabolism of Verapamil and Fluoxetine by the at-Inlet EMMA Mode	292
8.4		Conclusion	299

9. Capillary Electrophoresis for the Quality Control of Intact Therapeutic Monoclonal Antibodies — **305**

Anne-Lise Marie, Grégory Rouby, Emmanuel Jaccoulet, Claire Smadja, Nguyet Thuy Tran, and Myriam Taverna

9.1	Introduction	305
9.2	Impurities and Drug-Related Substances	308

		9.2.1	Process-Related Impurities	308
		9.2.2	Product-Related Substances	310
			9.2.2.1 Chemical degradations	310
			9.2.2.2 Biochemical degradations	315
			9.2.2.3 Physical degradations	316
	9.3	Identity and Heterogeneity of mAbs		318

9.3 Identity and Heterogeneity of mAbs 318
9.3.1 Source of Natural Heterogeneity of Proteins:
Post-Translational Modifications 318
9.3.2 Identity Control Issues after mAb Compounding 320
9.4 Quality Control of mAbs at Different Stages of Their Production 322
9.4.1 In-process Controls 322
9.4.2 Control of Drug-Related Substances 324
9.4.2.1 Capillary zone electrophoresis 324
9.4.2.2 Capillary gel electrophoresis 329
9.4.2.3 Capillary isoelectric focusing 336
9.4.3 Control of Compounded mAb before Patient Administration 343
9.5 Conclusion 346

10. Molecular Simulation of Chiral Selector–Enantiomer Interactions through Docking: Antimalarial Drugs as Case Study 363

Myalowenkosi Sabela, Suvardhan Kanchi, Deepali Sharma, and Krishna Bisetty

10.1 Introduction 363
10.2 Procedure 368
10.2.1 Experimental Protocol 368
10.2.2 Simulation Protocol 368
10.2.2.1 Ligand and receptor preparation 368
10.2.2.2 Minimization 369
10.2.2.3 Molecular properties 370
10.2.3.4 HOMO-LUMO calculations 370
10.2.3.5 Docking 371
10.2.3.6 Score ligand poses and analysis 371
10.3 Results and Discussion 372

	10.3.1 Experimental Enantioseparation	372
	10.3.2 Score Ligand Poses and Analysis	374
	10.3.3 Analysis of Enantioselective Interaction	376
	10.3.4 Enantioresolution	378
10.4	Conclusion and Future Prospect	380

Index 385

Preface

Capillary electrophoresis (CE) has recently received widespread recognition as an analytical technique of choice and has become an established method that can be used in numerous analytical laboratories, including industrial and academic sectors, and in pharmaceutical and biochemical research and quality control, which are the most important fields of CE applications.

Considering the rapid growth of CE, this book seeks to broaden the understanding of modern CE applications, developments, and prospects focused on molecules of pharmaceutical interest. Accordingly, this book describes recent developments and applications that are related to compounds of pharmaceutical interest, such as drugs, natural products, metabolites, and impurities in formulations. It also discusses about the latest pharmaceutical applications in quality control, drug and disease monitoring, drug metabolism, estimation of physicochemical properties, etc. It describes how various methods can be developed for sample preparation (e.g., in-capillary pre-concentration techniques), capillary coating, stationary phases, (enantio-) separations (e.g., use of nanoparticles and bio-macromolecules as chiral selectors), and miniaturization among others. The book also attempts to offer new trends involving the use of molecular simulations, currently used to broaden the understanding of pseudo separation mechanisms.

The book is aimed at beginners in this field as some fundamentals of CE technique have been introduced. It clearly outlines the procedures that can be used to hurdle over several barriers in a range of analytical problems. Some of these barriers include detection limits, signal detection, changing capillary environment, and reproducible and improvement on resolution separations of analytes and hyphenation of mass spectrometry with CE. Each chapter outlines a specific electrophoretic variant with detailed instructions and some standard operating procedures. The reported works have been summarized in the form of tables wherever necessary. Overall, the book provides a comparative assessment of related techniques on mode selection, methods development, detection, and quantitative analysis and estimation of pharmacokinetic parameters. In that respect, we are confident that this book will meet its desired goal to render assistance to lovers of electrophoresis.

Suvardhan Kanchi
Salvador Sagrado
Myalowenkosi Sabela
Krishna Bisetty
Spring 2017

Chapter 1

Capillary Electrophoresis: A Versatile Technique in Pharmaceutical Analysis

Imran Ali,[a] Zeid A. Alothman,[b] Abdulrahman Alwarthan,[b] and Hassan Y. Aboul-Enein[c]

[a]*Department of Chemistry, Jamia Millia Islamia (A Central University), New Delhi 110025, India*
[b]*Department of Chemistry, College of Science, King Saud University, Riyadh 11451, Kingdom of Saudi Arabia*
[c]*Department of Medicinal and Pharmaceutical Chemistry, Pharmaceutical and Drug Industries Research Division, National Research Centre, Dokki, Cairo 12311, Egypt*
drimran_ali@yahoo.com

Capillary electrophoresis is a resourceful technique of high speed, sensitivity, low limit of detection, and inexpensive running cost. It is a remarkable development separation science. Various publications have made available the literature on capillary electrophoresis for the analyses of diverse analytes. The present chapter describes introduction of capillary electrophoresis, theory, instrumentation, sample injection, separation capillary, high-voltage power supply, background electrolyte, direct detection, indirect detection,

---

Capillary Electrophoresis: Trends and Developments in Pharmaceutical Research
Edited by Suvardhan Kanchi, Salvador Sagrado, Myalowenkosi Sabela, and Krishna Bisetty
Copyright © 2017 Pan Stanford Publishing Pte. Ltd.
ISBN 978-981-4774-12-3 (Hardcover), 978-1-315-22538-8 (eBook)
www.panstanford.com

Capillary Electrophoresis

data integration, sample preparation, optimization, validation, applications, and conclusion. The main emphasis is given to pharmaceutical analyses.

1.1 Introduction

The expression of electrophoresis relates to the movement of ions or charged species in the influence of an electric field. Basically, it is a separation technique based on electrical driven forces. The analytes are separated based on their different electrophoretic mobility under applied voltage. The electrophoretic mobility depends on charge of analytes and viscosity of background electrolyte (BGE). The movement of the analytes is directly proportional to the applied potential. Capillary electrophoresis is referred to the electrophoresis carried out in capillary. It is a technique of high speed, good sensitivity, low limit of detection, and quantification with economic experimental price. That is why many papers have appeared in the literature [1–5]. Small quantities of material can be separated due to low detection limit with capillary electrophoresis. The first capillary electrophoresis equipment was fabricated by Hjertén in 1967 [6]. The present status of capillary electrophoresis is related to publications of Jorgenson and Lukacs [7–10]. These publications discussed instrumentation, viz., silica capillary, electrode, buffer reservoirs, power supply, and detector. It is important to point out here that capillary electrophoresis got appreciation in 1989 after the first international conference on high-performance capillary electrophoresis, which was held in Boston, USA [7]. The different aspects of capillary electrophoresis are discussed in this chapter.

1.2 Theory of Capillary Electrophoresis

The discussion of the principle of capillary electrophoresis is important for readers to handle a capillary electrophoresis instrument properly. Basically, capillary electrophoresis is based on the divergence in the electrophoretic and electroosmotic mobilities of molecules. These mobilities are the basis for the separation of analytes. The dissimilar migration velocities of the analytes are because of diverse charge–size ratios. The larger the charge–size ratio, the higher the mobility and, consequently, the lower migration time. The diverse sizes of the analytes are also accountable for their diverse migrations (steric affect). The schematic representation of the separation in capillary electrophoresis is shown in Fig. 1.1. Basically, the mixture of the analytes moves in the capillary in the form of different zones and separation occurs as different zones during the migration time. Figure 1.1a depicts the loading stage of the mixture (say, a, b, and c) on to capillary, while Fig. 1.1b indicates a partial separation of these three components. Contrarily, Fig. 1.1c represents the clear separation of analytes a, b, and c, which are detected by the detector and printed by the recorder in the form of peaks (Fig. 1.1d).

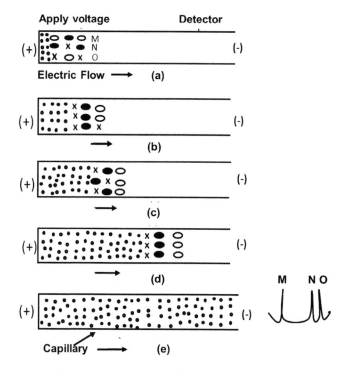

Figure 1.1 Schematic representation of separation in capillary electrophoresis.

Mathematically, migration of the analytes is restricted by the sum of intrinsic electrophoretic mobility (μ_{ep}) and electroosmotic mobility (μ_{eo}) because of the action of electroosmotic flow (EOF).

The experiential mobility (μ_{obs}) of the analyte is related to μ_{eo} and μ_{ep} as given by the following equation:

$$\mu_{obs} = E \left(\mu_{eo} + \mu_{ep} \right) \tag{1.1}$$

where E is the applied voltage (kV).

Capillary electrophoresis separations are measured by the migration times (t), electrophoretic mobility (μ_{ep}), separation (α), and resolution factors (R_s). The values of these parameters can be estimated by the following equations:

$$\alpha = \frac{\mu_{ep1}}{\mu_{ep2}} \text{ or } \frac{t_2}{t_1} \tag{1.2}$$

$$R_s = \frac{2\Delta t}{W_1 + W_2} \tag{1.3}$$

where μ_{ep1} and μ_{ep2} are the electrophoretic mobilities of the analytes 1 and 2, respectively. Likewise, t_2 and t_1 are the migration times of the analytes 1 and 2, correspondingly. Also Δt is the difference in the migration times of the two peaks, and W_1 and W_2 are the base widths of peak 1 and peak 2, correspondingly. If the values of α and R_s are one or greater, the separation is measured to be complete. If the individual values of these parameters are lower than one, the separation is measured to be partial or incomplete.

The simplest approach to describe the separation of two components is resolution factor (R_s). The value of separation factor may be connected with μ_{app} and μ_{ave} by the following equation:

$$R_s = \frac{1}{4} x \frac{\Delta \mu_{app}}{\mu_{ave}} N^{1/2} \tag{1.4}$$

where μ_{app} is the apparent mobility of two analytes and μ_{ave} is the average mobility of the two molecules. The usefulness of Eq. (1.4) is to permit independent assessment of the two factors that affect separation, selectivity, and efficiency. The selectivity is reflected in the mobility of the analytes, while the efficiency of the separation process is measured by N (theoretical plate number). A different term for N is derived from the following equation:

$$N = 5.54 \left(\frac{L}{W_{1/2}} \right)^2 \tag{1.5}$$

where L and $W_{1/2}$ are capillary length and peak width at half-height, correspondingly.

It is important to point out here that it is ambiguous to argue theoretical plates in capillary electrophoresis, while it is a carryover from chromatographic speculation. The separation is controlled by the comparative mobilities of the analytes in the used electric field, which is a function of their charge, mass, and shape. Theoretical plate in capillary electrophoresis is just a convenient idea to explain the analyte peaks' profile and to evaluate the factors that influence the separation. The effectiveness of the separation on column is articulated by N, but it is tricky to employ the factors that affect efficiency. This is owing to the performance of a particular component throughout the separation progression. It is not appropriate to explain the separation in capillary electrophoresis. Nevertheless, an extra functional parameter is the height equivalent of a theoretical plate (HETP). It is given by the following equation:

$$\text{HETP} = \frac{L}{N} = \frac{\sigma^2_{\text{tot}}}{L} \tag{1.6}$$

The HETP may be measured as the function of the capillary occupied by the analytes and additional useful to assess separation effectiveness in contrast to N. σ^2_{tot} is affected by diffusion, differences in the mobilities, Joule heating effect, and interaction of the analytes with the capillary wall. Consequently, σ^2_{tot} is expressed by the following equation:

$$\sigma^2_{\text{tot}} = \sigma^2_{\text{diff}} + \sigma^2_{T} + \sigma^2_{\text{inj}} + \sigma^2_{\text{wall}} + \sigma^2_{\text{electos}} + \sigma^2_{\text{electmig}} + \sigma^2_{\text{sorp}} + \sigma^2_{\text{oth}} \tag{1.7}$$

where σ^2_{tot}, σ^2_{diff}, σ^2_{T}, σ^2_{inj}, σ^2_{wall}, $\sigma^2_{\text{electos}}$, $\sigma^2_{\text{electmig}}$, σ^2_{Sorp} and σ^2_{oth} are the square roots of standard deviations of total, diffusion, transport, injection, wall, electroosmosis, electromigration, sorption and other phenomenon, correspondingly.

1.3 Types of Capillary Electrophoresis

Throughout the time, diverse changes have been made in capillary electrophoresis leading to different modes of electrophoresis. The most significant types are capillary zone electrophoresis (CZE), capillary gel electrophoresis (CGE), capillary isoelectric focusing

(CIEF), capillary isotachphoresis (CITP), micellar electrokinetic chromatography (MEKC), affinity capillary electrophoresis (ACE), capillary electrokinetic chromatography (CEKC), and capillary electrochromatography (CEC). These capillary electrophoresis modes vary in the working principles and, consequently, can be applied for a variety of applications. Among different sorts of capillary electrophoresis, CZE is the most accepted because of its extensive range of applications [1–5, 11]. The analytes in CZE are separated according to the type of zones and, consequently, called capillary zone electrophoresis.

CITP is a stirring border-line capillary electrophoretic method involving amalgamation of two buffers to generate a state that separates zones of all that move at equal velocity. The zones remain sandwiched between the so-called leading and terminating electrolytes. The stable state velocity in CITP is found as the electric field changes in each zone, and, consequently, fine pointed boundaries of the zones appear. CGE relates to the allocation of analytes according to their charge and size in a carrier electrolyte. A gel-forming medium is supplemented, and, consequently, it is an adjoining technique to slab electrophoresis. This method is appropriate for big molecules such as proteins and nucleic acids. In CIEF, a pH gradient is created inside the capillary by means of compounds that have both acidic and basic groups with pI ranging from 3 to 9. A procedure named as focusing containing basic and acidic solutions at cathode and anode, results in the separation of analytes. This form of capillary electrophoresis is applied for the separation of immunoglobulins and hemoglobulins and for quantifying pI of proteins. ACE is applied for the separation of some biologically connected molecules. The separation in this modality is based on the same types of specific, reversible interactions, which are available in biological systems, viz., binding of an enzyme with a substrate or an antibody with antigen. It is also an important means for the separation of halocarbons in the environmental and biological matrices. MEKC allows the separation of uncharged species through the smart utilization of charged micelles as a pseudo-phase in which partition of the analyte occurs. Occasionally, a surfactant molecule (at a concentration above its critical micellar concentration) is added for the optimization of the separation in capillary electrophoresis. The separation mechanisms shifted toward chromatographic principle; therefore, the method is named micellar electrokinetic chromatography. This method was developed by Terabe et al. [12] in 1984. The best surfactant for MEKC shows fine solubility in buffer solution, uniform micellar solution with compatibility with detector and with low viscosity. On the whole, the micelle and buffer lean to travel toward the positive and negative ends, correspondingly. The progression of the buffer is stronger than the micelle movement, and, consequently, the buffers go toward the negative end. The separation in this form of capillary electrophoresis depends on the allocation of the analytes between micelle and the aqueous phases.

CEC is a hybrid method between HPLC and CE, which was explored in 1990 [13]. It is anticipated to coalesce high peak efficiency, which is a feature of electrically driven separations, with high separation selectivity in HPLC. The chromatographic band expansion mechanisms are pretty dissimilar in both the modes. The chromatographic and electrophoretic mechanisms work concurrently in CEC, and numerous combinations are achievable. The separation occurs on the mobile/stationary phases boundary, and the swap kinetics between the mobile and stationary phases is imperative. The separation principle is a combination of liquid chromatographic and capillary electrophoresis theories. CEKC was introduced by Terabe in 1984 [12], and the analytes are separated by electrophoretic migration and chromatographic principles both. The main frequently used pseudo-phases in CEKC are artificial and natural micelles and microemulsions, proteins, peptide, charged linear, charged macromolecules, and cyclic oligosaccharides. It is applied for the separation of neutral molecules, which are unfeasible in normal capillary electrophoresis.

1.4 Instrumentation

The schematic diagram of a capillary electrophoresis setup is shown in Fig. 1.2, which consists of injector, capillary, electrodes, BGE vessels, high-voltage power supply, detector, and recorder. The sample is loaded into the capillary at the injection end, frequently by siphoning and/or electromigration. The motivating electric current is supplied by a high-voltage power supply. Migrating zones of the sample are detected at the detection end of the capillary by a detector. The ease of this experimental setup makes an apparatus available to most of the laboratories. Nevertheless, for custom performance, supplementary refined, and preferably automated, instrumentation is needed. The diverse parts of capillary electrophoresis are discussed in detail in the following sub-sections.

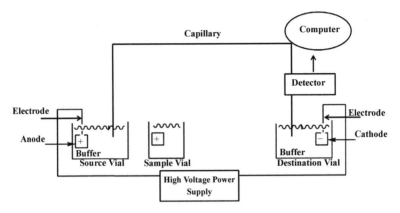

Figure 1.2 The schematic diagram of a capillary electrophoresis setup.

1.4.1 Sample Introduction

The art of sample injecting is significant for reproducible results in capillary electrophoresis. The effects of sample volume on separation effectiveness have been described by several researchers [12, 14–16]. Consequently, a variety of methods of sample loading have been projected from time. These include electroosmotic–electromigration technique [7], rotary-type injector [17], miniaturized sampling valve [18], electric sample splitter [19], and electrokinetic [7, 8, 20] and hydrostatic [21] techniques. Among these sample injection modes, electrokinetic and hydrostatic methods are most accepted because of their reproducible outcome and ease of operation. Consequently, these are used in all viable available capillary electrophoresis equipment.

The electrokinetic sample method is based on electrophoretic migration and electroosmotic flow of electrolyte in the capillary of capillary electrophoresis. The sampling end of capillary and platinum electrode is permitted to dip into the sample vial. An electric current is supplied for a small period of time for 10 s. The electroosmotic flow is produced throughout these sampling pumps in the capillary. After the sampling is finished, the end of the capillary is stirred reverse into the buffer container following capillary electrophoresis analysis. Nevertheless, this form of sampling restricts its utilization, especially when the sample solution contains more than one component (as the amount of each component introduced into the capillary) and is selectively dogged by its electrophoretic mobility. If the electroosmotic flow is toward the negative electrode, the positively charged solutes are injected selectively to a greater extent in comparison to neutral and anionic molecules. Furthermore, anions with electrophoretic velocities greater than electroosmosis do not enter the capillary. Hence, the selection of sampling depends on the type of species to be analyzed.

Sample injection through the hydrostatic method (also known as hydrodynamic or siphoning method) is the simplest, general, and often used practice in most of capillary electrophoresis equipment. In this sort of sampling, one end of the capillary is permitted to dip into the sample solution for a period of 1–30 s. The hydrodynamic flow generated by hydrostatic pressure introduces a little volume of the sample into capillary. Subsequently, the end of the capillary is returned back to the buffer container following the analysis. The hydrostatic method is predictable to initiate a real delegative aliquot of sample in contrast to electrokinetic modality. The reproducibility of sampling is meager if it is carried out by hand. For this reason, it is suitable to use fully automated siphonic samplers.

1.4.2 Capillary Electrophoresis Capillary

The capillary is the most significant element of the capillary electrophoresis equipment since separation depends on the form and character of the material. The main features for a good class of capillaries are mechanical and chemical stability and good releasing of Joule heat. A fine clearness to UV radiation is mainly significant for on-column ocular detection. Fused silica capillaries of diverse lengths and diameters are available, meeting all requirements. Open tubular fused silica capillaries have low price and are readily available in a wide variety of inner diameters. Capillaries made of other materials such as glass or Teflon are rarely used. Capillaries of rectangular cross sections having theoretically the best cooling properties and improved path length for absorbance detection are seldom applied [22, 23].

Principally, the selection of the diameter and length of capillaries depends on separation conditions. Usually, longer and narrower capillaries acquiesce the maximum separation effectiveness. In general, capillaries with an inner diameter of 0.2–0.005 mm and a length of 30–100 cm are ideal. The separation efficiencies differ from thousands to millions of theoretical plates. Narrow bore capillaries are beneficial owing to low Joule heat generation. Fast analyses can be done by decreasing the length of the capillary. Besides, a low amount of sample should be injected to achieve fine separation efficiency. The necessary capillary length can be achieved by cutting with a ceramic cleaving stone or a diamond cutter. The detection window on capillary can be released by burning the covered polyimide coating. Two aluminum foils can be applied to shield the polyimide layer around the detection window for cutting an exact and sharp window. A detection window (0.5 mm) can be simply organized in this way. The window can also be designed using a drop of 96–98% sulfuric acid followed by heating up to 130°C [24]. The window prepared in this way is not fragile as in the case of the burning method. Additionally, the danger of capillary injury is low in the acid drop process in contrast to the burning method. Likewise, the window can also be designed by concentrated KOH. Also, the window can be designed by directly scratching the polyamide layer, but it may injure the capillary and is, therefore, not advisable. As discussed, fused silica capillaries have a broad range of applications. However, coating of the inner face of capillaries has been described to achieve the requirements of particular applications [1]. Customized capillaries are hydrophilic, hydrophobic, and gel filled in character.

1.4.3 Power Supply

A high electric field strength is necessary to assist separation in capillary electrophoresis. It ranges from 10 to 100 kV depending on the internal diameter of the capillary, composition of the BGE, and character of analytes. The resultant dynamic current seldom exceeds 100 µA, or else capillary overheating would be anticipated. A reversible polarity, high-voltage power supply is advised where the electrode can be grounded, preferably the one close to the detector. Generally, this array eliminates an augmentation in the sound of the detector signal owing to high electrostatic tension between the detection end and the detector electronics. A fine high-voltage power supply is assumed to have a voltage ranging from 1 to 20 kV with stability better than 1%, current range of 0–200 μA, the option to act at either constant voltage or constant current, reversible polarity, and interlock for the worker protection.

1.4.4 Background Electrolyte

Background electrolyte is used to keep up a high-voltage gradient across the sample-containing solution in the capillary. It requires a higher conductivity of the electrolyte than the sample. Hence, buffers are used as BGEs in most capillary electrophoresis analyses. Also, to control the pH of BGE, the use of buffers is important. The electrolyte uniqueness and concentration must be selected cautiously for the best separation of the species. The choice of BGEs depends on their conductivity and the sort of species to be separated. The comparative conductivities of dissimilar electrolytes can be anticipated from their condosities. The condosities are defined as the concentration of NaOH, which has the same electrical conductance as the molecules under investigation [25]. A broad range of electrolytes can be applied to design buffers for capillary electrophoresis. Low UV-absorbing components are needed for the preparation of buffers in case the UV detector is the device of detection. Also, volatile chemicals are needed if MS or ICP are the detection devices. Nevertheless, these circumstances considerably edge the option to a reasonable number of electrolytes. The pH of BGE is also an additional issue that decides the option of buffers. For low-pH buffers, phosphate and citrate have normally been used, even the absorbing character of the later is strong at wavelengths <260 nm. The basic buffers used are borate, TRIS, CAPS, etc. Practical buffers with their pHs and wavelengths are given in Table 1.1 [1].

1.4.5 Detection

Detection is significant to any analytical equipment, which controls the application range of capillary electrophoresis. Similar to liquid chromatography, EC is also coupled with several detectors. These include electrochemical, conductivity, photochemical, fluorescence, atomic absorption, mass, inductively coupled plasma, etc. The booming coupling of these detection strategies made capillary electrophoresis competent to detect species in the range of 10^{-8} to 10^{-9} M. A number of significant detectors used in capillary electrophoresis are discussed as follows.

1.4.5.1 UV/Vis absorbance detectors

It is well known that most of the molecules take up UV light; therefore, more than 50% analyses are achieved using photometric detectors in capillary electrophoresis. These detectors are based on the absorbance of UV or visible light. The first online UV detection in capillary electrophoresis was developed by Hjertén [6]. The detection gap for light absorbance is designed in capillary as described earlier. Offline detection systems were also used in capillary electrophoresis. The separated zones leave capillary and are forced to detection cell of UV photometer [26, 27]. The function of offline detector is inadequate due to dilution and mutual mixing of previously separated zones.

In these detectors, a narrow ray of light from lamp passes via capillary window, which is detected by a photosensor, i.e., photomultiplier, photodiode. The nonlinearity is important for wide bore capillaries or concentrated solutions of strong absorbing substances. The recorded peak of a detected zone does not symbolize the factual concentration outline in the zone. The slit width can alter the detector reply, which is significant for stray light refusal and augmentation of the linearity of the detection reply. The longitudinal tallness of the slit is significant with respect to the efficiency of the separation. In general, the light source in the UV detector is deuterium (D_2) lamps. Nevertheless, cadmium (229, 326 nm), zinc (214 nm), mercury (254 nm), and iodine (206, 270 nm) may also be applied. A tungsten lamp is used for radiation in the visible region. Generally, light strength reaching to the photosensor in capillary electrophoresis is faintly concentrated in contrast to HPLC detection cell. Therefore, small fused silica planoconvex lenses or ball lenses are applied to avoid this difficulty.

The photodiode array detectors offer the scanning arrangement detection [28–30] pattern, which brings numerous advantages over the fixed wavelength detection. All the compounds are detected at a fixed wavelength in normal photometric detectors, which may have different λ_{max} values. Therefore, the detection limits of the detector cannot be determined precisely. Obstinately, in photodiode array detector, the scanning spectra of all the compounds are recorded by the detector to determine λ_{max} values for each analyte. The detection is individual, which results in maximum detection of all the compounds analyzed. In the photodiode array detector, radiation from lamp is focused onto window on the separation capillary. It passes to a holographic grating polychromator via slit. The entire wavelength range in UV and visible regions is scanned by a linear array of photodiodes. The signal is evaluated and recorded by a computer, after electronic processing. The scan rate (the number of spectra taken during 1 s interval) depends on the spectral resolution of 2–20 nm.

1.4.5.2 Conductivity detectors

Conductivity detection (CD) acts by measuring the conductance or a voltage drop in BGE. The conductivity detectors are common and non-selective, which may detect a zone of any compound with an effectual mobility diverse from BGE. The conductivity of a moving zone differs from BGE, which is measured by the detector. If the conductivities of BGE ions and the analyte are alike, it is not possible to detect the analytes properly. The potential gradient detector (PGD) also acts on a similar principle and measures voltage across two sensing microelectrodes. The net detection signal is proportional to the difference between electric field strength in BGE and the migrating zone. The surfaces of both CD and PGD cells should be as minute as possible to decrease the effects of the electrode reaction, i.e., bubble development throughout analysis. Usually, wires of platinum or platinum–iridium alloy, less than 0.05 mm diameter, are used as the electrodes. The fabrication of the detector cell is essential in capillary electrophoresis. It is fabricated by molding expertise to produce a detection cell in a block of polyester resin [31] or by laser drilling method to create an on-column detection cell straight in a silica capillary [32]. The detection in these detectors depends on both the effective mobility of the analytes and BGE, providing a practical and corresponding information to that provided by a selective detector. Thus, these can also be used to recognize the zones in capillary electrophoresis.

Occasionally, a tiny potential drop across the sensing electrodes and any current leakage may produce unwanted signals due to the electrochemical reactions with noisy and drifting baseline. In such cases, these signals can be concealed by the accumulation of a non-ionic detergent in BGE, which forms a defensive film on the sensing electrodes. Hydroxypropylcellulose (0.2%) or Triton X-100 can be applied for this reason. Huang et al. [32] introduced a new conductivity and amperometric detection in capillary electrophoresis. It worked by measuring conductivity between the outlet of the separation capillary and the grounded electrode of the power supply. The electrode was made of 50 μm platinum wire fixed in a fused silica capillary. This arrangement avoided the opportunity of any electrochemical reaction and enhanced the constancy of the detection signal.

1.4.5.3 Amperometric detectors

In amperometric detection, the redox potential of an analyte is detected and recorded at the electrode surface. In this method, the detected ions are chemically oxidized or reduced throughout their passage through the detection cell. The practical array of potential, which can be exploited in practice, depends on the solvent and material of the electrodes. The practical range of potential of glassy carbon is from −0.8 to +1.1 V versus Ag/AgCl reference electrode. At a more negative potential, an augmentation of the background current is obtained, due to the decrease in oxygen dissolved in the BGE and hydrogen overvoltage. The positive potential limit is connected to the oxidation of water and the electrode material both. When an electrochemically active solute (oxidizable) is introduced into the electrochemical cell, the electric current is produced and a characteristic potential ($E_{1/2}$) is obtained, which depends on the character of the diffusion transport of the solute to the surface of the working electrode. Therefore, further augmentation of the potential will not boost the detection signal.

In capillary electrophoresis, the separation is obtained by high electric field strength; therefore, the location of the amperometric detection cell in the capillary is not probable. Consequently, the separated zones from the outlet of the separation capillary were primarily detected by offline amperometric detectors [33]. Later, superior designs of amperometric detection were developed [34–37]. The separated zones were transported into the detector associated at the end of a small fused silica detection capillary, which was linked to the separation capillary by a specially developed electrically conductive joint made of porous glass. The sensitivity of an amperometric detection is high, which can be achieved up to nanomole to attomole levels.

1.4.5.4 Thermooptical absorbance detectors

Thermooptical absorbance is a laser-based detection device used in capillary electrophoresis [38, 39]. In this detection method, a beam of high-energy laser is focused onto the separation capillary and the analytes are excited. Heat is generated in the separated analytes, during non-radiative relaxation of excited states. The temperature rise is proportional to the absorbance of analytes and the laser intensity. The refractive index of BGE changes with temperature, and the heated area acts as a thermal lens. The alteration in refractive index is calculated by another laser beam, which is directed across the capillary in the direction perpendicular to the probe beam. The deflection of this is measured by a photodiode. These sorts of detectors are applied only for species that powerfully absorb the radiation emitted by the laser. Hence, the species should be tagged with an appropriate derivatizing agent. The detection limits of these types of detectors can be obtained up to attomole levels.

1.4.5.5 Fluorescence detectors

In this form of detection, the separated species absorbs a photon and reaches the excited state following return to the ground sate by emitting a photon. This phenomenon is called fluorescence. The excited radiation from the elevated energy source is focused onto the detection pan of capillary. The emitted fluorescence is measured accordingly. Molecules of good fluorescence quantum yield, high absorptivity, and photostability are measured exactly with low limit of detection. In this way, fluorescence detection is more sensitive than light absorbance but more selective. In the simplest arrangement, a high-pressure mercury lamp emitting at 365 nm is used as the excitation source. But in stylish detectors, xenon arc lamps, giving a broad spectrum in the entire UV–Vis region, are used as the excitation sources. The fluorescence intensity is directly proportional to the intensity of the excitation radiation. Therefore, lasers should be used as the excitation sources. Nevertheless, laser-based fluorescence detectors have some drawbacks in comparison to lamp-based detectors. The chief drawback is no possibility to tune the wavelength of the laser emission to the absorbance maximum of a fluorescing compound. Therefore, a limited number of compounds can be detected directly. Hence, derivatization (pre- or post-column) of the compounds, to convert into fluorescent sensitive, is required before their detection by this method.

The most important derivatizing reagents are naphthalene dialdehyde, dansylchloride, fluoresceinisothiocyanate, fluores-camine, phenylthiohydantoin derivatives, fluorescein derivatives, 2-aminopyridine derivatives, 4-chloro-7-nitrobenzofurazan, and 3-(4-carboxybenzoyl)-2-quinolinecarboxyaldehyde [2]. A fine derivatizing reagent should show fast reaction rate, good excitation and emission maxima of fluorescent, high quantum yield, and only one derivatizing product. In both pre- and post-column detections, the optimization of the detection is based on a suitable assortment of flow rate of the derivatizing reagent. The detection limit by these detectors varies from 10^{-8} to 10^{-9} mol/L, but the reported detection limit was 1.2×10^{-20} mol for fluorescein isothiocyanate. However, further improvement of sensitivity in fluorescent detector is probable in the near future with the use of new detection approaches such as two-dimensional charge-coupled device [40].

1.4.5.6 Atomic absorption spectroscopy detectors

Atomic absorption spectrometry (AAS) was introduced by Alan Walsh in 1950. It is being used as a potent instrument in the quantitative analysis of trace elements. AAS has been hyphenated at several occasions with capillary electrophoresis, especially in the application of metal ion analysis and compounds having metal ions such as hemoglobin and cyanocobalamine. This detection method gives total concentrations of metal ions and is independent of molecular form of metal ion in the sample. The impurities present in the sample do not change the detection. Therefore, it is very discriminating in character. The absorption energy by ground-state atoms in the gaseous state forms the basis of AAS. Metal ion atoms are vaporized by flame in AAS. The atoms in gaseous form absorb energy from the energy source (AAS lamp). The absorption of energy is directly proportional to the number of atoms in the separated zone. The different sorts of nebulizers have been developed to couple AAS with capillary electrophoresis. This type of coupling resulted in the trace analysis of many metal ions [40–43]. The detection limit of this detection has been reported from mg/L to µg/L level.

1.4.5.7 Inductively coupled plasma detectors

Inductively coupled plasma (ICP) spectrometry is a capable emission spectroscopic method. Commercial ICP systems became available in 1974. In this method, the carrier gas (argon) is heated at a high temperature (9000–10,000 K). The plasma is developed in which the excitation of atomic electrons occurs simply and exactly. Therefore, ICP is considered an improved detector than AAS. The metal ions enter the plasma in the form of aerosol. The droplets are dried and dissolved, and the matrix is decomposed in the plasma. In the high-temperature region of the plasma, atomic and ionic species (in various energy states) are formed. This method can be coupled itself with mass spectrometer (MS), which is considered to be a better detector with the detection of multi-elemental capabilities and wide linear dynamic range. Quadrupole mass filters are the most general mass analyzers as these are rather low-cost. Double focusing magnetic/electrostatic sector equipment and time-of-flight mass analyzers are also used [44]. Many interfaces have been described for coupling ICP to capillary electrophoresis. These hyphenations resulted in the efficient separations of numerous metal ions [42, 45, 46] with the detection limit ranging from mg/L to ng/L.

1.4.5.8 Mass detectors

In mass detectors, the molecules under examination are bombarded with a beam of electrons, which generate ionic fragments of the original species. The resulting collection of charged particles is separated as per their masses and charge ratios. The spectrum generated is called mass spectrum, which is a record of information regarding different masses generated and their relative abundances.

Mass detector has been hyphenated with capillary electrophoresis effectively for several applications.

In spite of its huge accomplishment, it should be realized that in its normal configuration, i.e., equipped with a pneumatic nebulizer for sample introduction and with a quadrupole filter, MS also indicates a number of significant restrictions. The occurrence of spectral interferences may hamper exact determination of trace elements. The application of a double focusing sector field mass spectrometer in MS instrumentation gives a higher mass resolution. In addition, photons are proficiently eliminated from the ion beam, resulting in very small background intensities, which makes it suitable for intense trace analysis. Hence, the hyphenation of mass spectrometer to capillary electrophoresis is a significant position to be measured. It shows a new separation dimension. Some articles have appeared in the literature on this subject with numerous designs [47–50]. The viable interfaces for the hyphenation of capillary to mass spectrometer are accessible in the market [51]. The first CE–MS coupling was described by Smith and Udseth [47] with an electrospray ionization interface for the connection of the separation capillary to a quadrupole mass spectrometer. The arrangement of CE–MS may be applied for quaternary ammonium salts, vitamins, azo dyes, peptides, amino acids, and proteins, with good sensitivities [47, 48, 52]. CE–MS and CE-tandem-MS have been described to have detection limit up to femtomole of peptides, digests, and proteins. The coupling of capillary electrophoresis with MS shows one of the mainly fascinating instrumental developments in separation science. Additionally, widespread study is necessary to discover all the chattels of this coupling in capillary electrophoresis.

1.4.5.9 Miscellaneous detectors

In the premature stage of capillary electrophoresis growth, Lukacs [53] attempted to employ refractive index for detection, but owing to difficulty in signal and its noise and drift, it could not be applied fruitfully. Later, Chen et al. [54] and Demana et al. [55] introduced refractive index detector in capillary electrophoresis. Kaniansky [56] enhanced the design of this type of detector. The use of refractive index detector in capillary electrophoresis is not general due to its partial sensitivity. One more expansion is the use of radio detectors in which the species are detected by determining their radioactivity. Therefore, it is a very choosy and susceptible method to recognize merely radioactive species [57, 58]. Geiger–Müller and other photomultipliers are used for the detection of the separated species. A number of designs of radio detectors hyphenated with capillary electrophoresis have been reported in the literature [59–61]. Radioactive emission is a haphazard procedure, and signal augmentation was obtained by declining the driving current at the moment of the detection. Therefore, the detection was enhanced by the extended transition time of the detected substance [57, 59, 62]. The detection was based on Raman spectroscopy and is due to the irradiation of a high-energy monochromatic light on the species and its measurement after scattering by those substances. Raman bands are measures of rotational and vibrational bonds in polarizable molecules. Therefore, Raman spectrum balances the conventional infrared spectrum. Hence, additional powerful Stokes band at frequency lower than the event laser radiation is appropriate for the detection. Chen and Morris [63] developed the first detector in capillary electrophoresis based on Raman spectroscopy. Afterward, Christensen and Yeung [64] modernized this design for multichannel Raman spectroscopic detection. The described sensitivity was 500 attomoles of methyl orange equivalent to a concentration of 10^{-7} M. In addition, optical detector is the mere option in separation science to decide the optical clarity of the species. This detector operates on the principle of rotation of polarized light by the species.

1.4.5.10 Indirect detection

Besides the above-cited detectors, indirect detection has also been adopted in capillary electrophoresis owing to its worldwide use. In indirect detection, various properties of BGE constituents such as UV absorbance, fluorescence, electrochemistry, or transformation are considered. The detection signal is obtained from a decline in the background signal due to the migration of substance zone, which replaces the detection-active components of BGE. Originally, Hjertén [6] attempted to employ the indirect method of detection in capillary electrophoresis, but afterward the utilization of indirect detection was accepted. The applications of indirect detection in capillary electrophoresis are the use of UV photometry [65, 66], fluorescence [67, 68], and amperometry [69, 70]. In brief, most of the detection methods can be used in the indirect mode. In the indirect mode of detection, the selection of BGE is significant as this should be employed with co-ion or counter ion. To evade any zone broadening, sample concentration should be lower than BGE (about 100 times). The application range of indirect detection is limitless, ranging from simple inorganic ions to macromolecules with detection limit up to the femtomole level [71–73]. Despite the widespread character of indirect detection, it suffers from some disadvantages. For example, samples with ions having strong direct signals may pose detection difficulty due to the due to overlapping of ionic zones, and this difficulty may increases when complex actual samples are to be analyzed.

1.5 Data Integration

Data integration is the final and imperative part of capillary electrophoresis equipment, which provides direct information about the analyses. Previously, the detectors were linked to labor-intensive integrators or recorders to obtain electropherograms. But these days, computer-based software are accessible, with the aid of which integration of the analysis results can be carried out exactly and precisely. Electropherograms can be obtained at diverse values of data collection, which provides the exact information about the analysis. The different manufactures who provide sophisticated data integration units with a capillary electrophoresis machine are Bio-Rad, Lab Alliance, Waters, Groton Technology Inc., and Agilent Technologies, United States; LUMEX, Russia; Picometrics, France; and Beckman Coulter Ltd., United Kingdom. Similar to chromatography, the results of the analysis are described in terms of migration times, electrophoretic mobilities, and separation and resolution factors.

1.6 Sample Preparation

In the analysis of unknown samples of biological or environmental origin, sample handling becomes a significant matter in capillary electrophoresis. Despite thousands of applications, little awareness has been generated for sample-handling methods.

Samples having high ionic matrix may cause difficulty in capillary electrophoresis. The EOF in the capillary can be changed with the influence of the sample matrix resulting in reduced resolution. In addition, the detector baseline is usually disturbed when the pH of the sample differs very much from the pH of the BGE. Samples having UV-absorbing materials are also tricky in the detection of the unidentified matrix. Due to these difficulties, a number of researchers have recommended sample cleanup methods concerning diverse sample-handling methods [74–77]. The actual samples frequently need the applications of simple methods such as filtration, extraction, and dilution. The electromigration of sample cleanup suffers from strict medium reliance effects; even then it has been used for pre-concentration in inorganic analysis. The sample treatment techniques have been described in a number of reviews [76, 78–80]. In addition, numerous papers are also available on dialysis and electrodialysis of sample cleanup preceding to capillary electrophoresis injections [76, 78–80]. Numerous online sample preparation methods have been developed and used in capillary electrophoresis. Whang and Pawliszyn [81] considered an interface that enables the solid phase microextraction (SPME) fiber to be inserted straight into the injection end of capillary electrophoresis. The authors arranged a "semi-custome made" polyacrylate fiber to attain the SPME-CE interface. Similarly, Kuban and Karlberg [82] also described an online dialysis/FIA sample cleanup procedure in capillary electrophoresis. As well, several reviews are available in the literature on sample preparation before going to capillary electrophoresis [74, 77, 83, 84]. The readers are advised to read these papers.

1.7   Method Development and Optimization

Method development in capillary electrophoresis begins with the choice of a suitable BGE, applied voltage, detector, etc. Primarily, the structures of the molecules to be analyzed should be explored, and a detection mode should be finalized. If the species are UV responsive, a value of λ_{max} should be dogged and applied throughout the experiments. The choice of the BGE should be carried out keeping in mind the wavelength cut of the buffers (Table 1.1).

22 | *Capillary Electrophoresis*

Table 1.1 Commonly used buffers with their suitable pHs and wavelengths for chiral separations in CE

Buffers	pH	Wavelength (nm)
Phosphate	1.14–3.14	195
Citrate	3.06–5.40	260
Acetate	3.76–5.76	220
MES	5.15–7.15	230
PIPES	5.80–7.80	215
Phosphate	6.20–8.20	195
HEPES	6.55–8.55	230
Tricine	7.15–9.15	230
Tris	7.30–9.30	220
Borate	8.14–10.14	180
CHES	9.50	< 190

Source: Reproduced from Ref. [1], Copyright 1967, with permission from Elsevier. CHES: 2-(*N*-Cyclohexylamino) ethanesulphonic acid; MES: Morpholino ethanesulfonic acid; PIPES: Piperazine-*N,N'*-bis(2-ethanesulfonic acid); HEPES: *N*-2-Hydroxyethylpiperazine-*N'*-2-ethanesulfonic acid

Generally, 10 kV is the starting value of the applied voltage following its optimization as per the obligation. The separation in capillary electrophoresis is optimized by calculating a variety of experimental variables. These factors may be divided into two categories: independent and dependent. The independent variables are under the direct control of the machinist. These parameters include the option of the buffer, pH of the buffer, ionic strength of the buffer, voltage applied temperature of the capillary, dimension of the capillary, and background additives. Obstinately, the dependent variables are directly exaggerated by the independent variables and are not under the direct control of the machinist. These variables are EOF, field strength (V/m), BGE viscosity, sample mobility, sample diffusion, sample charge, sample size and shape, Joule heat, sample interaction with capillary and BGE, and molar absorptivity. In this way, the necessary separation of the species is achieved. A concise experimental procedure for the separation optimization in capillary electrophoresis is specified in Scheme 1.1.

1.8 Validation of Methods

Capillary electrophoresis is rising as a speedy, proficient, and adaptable method in separation science. But still it needs improvements in operational conditions to attain reproducibility. To the best of our knowledge, merely a few papers are accessible in the literature on the corroboration of methods connected to the determination of the molecules by capillary electrophoresis [77–80]. So in these days, extra importance is needed on method justification. For usual investigation, it is necessary to keep the migration times steady to permit automatic peak recognition by means of viable data analysis software. Automatic peak recognition and quantification are merely promising if the relative standard deviation of migration times is less than 0.5% [85]. Numerous reviews are available in the literature, which explain the development in the reproducibility of capillary electrophoresis analyses [86–92]. A further approach for the qualitative and quantitative reproducibility is the alteration of the total time x-scale of electrophoretic data to the equivalent effective mobility scale (μ-scale) [78, 93]. The exchange leads to an improved explanation of the obtained electropherograms in terms of separation. It enables enhanced straight assessment of the electropherograms and easier peak tracking when trying to recognize solitary components from complex matrix, especially when UV–visible signatures of the species are also presented [78]. A quantitative perfection was obtained in the μ-scale with considerably improved peak area precision, which equates to enhanced precision in quantitative analysis in contrast to the primary timescale integration. Nevertheless, electrophoretic data processing capillary electrophoresis software is required to handle electrophoretic data directly. Hence, it may be understood that the selectivity of various compounds by capillary electrophoresis is fairly fine. However, the reproducibility is still a trouble. Hence, several efforts have been made to resolve this difficulty. Organic solvents are mixed with

BGE to enhance reproducibility. It has also been described that the organic solvents improve the solubility of hydrophobic complexes. They reduce their sorption onto the capillary wall and adjust the allocation of the molecules between aqueous phase and micellar phase. They also adjust the viscosity of the separation medium, thus achieving enhancement in reproducibility. The stable pH of the BGE is extremely significant from the viewpoints of selectivity and reproducibility. pH controls the performance of EOF, acid/base dissociation equilibria of complexes, and state of obtainable complex. Hence, the selectivity and reproducibility may be enhanced by adjusting the pH of the BGE.

1.9 Applications

Capillary electrophoresis has obtained a status of elevated separation efficiency, high sensitivity, small analysis time, and short running cost methods. It has an extremely wide range of applications from small ions to macromolecules. It has been adopted to separate biomolecules, xenobiotics, and chiral compounds. In biological science, its main applications include separation of amino acids, proteins, peptides, DNA, enzymes, antibiotics, vitamins, carbohydrates, profens, amines, β-blockers, toxins, alcohols, organic acids, dyes, and a number of other drugs and pharmaceuticals. In these days, capillary electrophoresis has emerged as one of the most significant separation techniques in chiral separation of drugs, pharmaceuticals, pollutants, and other compounds. Unlike in chromatography, the chiral selectors in capillary electrophoresis are mixed in BGE. Therefore, these are called BGE additives. Many applications on the analysis of a variety of compounds using capillary electrophoresis are available in the literature. Hence, many reviews, monographs, and books have been published [1–3, 11, 94–99]. Therefore, it is not possible to quote them all in this chapter. However, efforts have been made to include the most important papers in this chapter, especially on pharmaceutical analyses. The various applications of capillary electrophoresis are given in Table 1.2.

Table 1.2 Applications of capillary electrophoresis and related technologies

Compounds	Sample Matrix	Electrolytes	Detection	Reference
Capillary Electrophoresis				
Analysis of Biomolecules				
Simple Mixture Analysis				
Amines				
Tetramethyl ammonium bromide, tetramethyl ammonium perchlorate, tetrapropyl ammonium hydroxide, tetrabutyl ammonium hydroxide and trimethylphenyl ammonium iodide	—	0.1 mM KCl in 50% methanol	MS	[103]
Polyamines	—	Formic acid, 0.01% ED, 5% EG	Fluorescence	[104]
Alkylamines	—	0.5 M PB	Fluorescence	[7]
Histamine	Wine	0.1 M BB, 0.2 M KCl, pH 9.5	Post-column fluorescence	[105]
Putrescine, cadaverine, spermidine, and spermine	—	0.005 M BB with 0.1% ED, 2% SDS, and 5% EG	Post-column fluorescence	[106]
Biogenic amines	Oysters	50 mM phosphate buffer with $Ru(bpy)_3{}^{2+}$	Electrochemiluminescence	[107]

(*Continued*)

Table 1.2 (*Continued*)

Compounds	Sample Matrix	Electrolytes	Detection	Reference
Polyamines and acetylpolyamines	Human urine	20 mM borate buffer at pH 7.4	Fluorescence	[108]
Amitriptyline, nortriptyline, imipramine, desipramine, doxepin, and nordoxepin	Human serum	50 mM CAPSO (pH 9.54) in methanol/water with KCl	—	[109]
Indoleamines	Urine and serum	100 mM tris-borate buffer (pH 9.0)	Fluorescence	[110]
Acids				
Formic, acetic, propionic, and butyric	—	0.025 M Na veronal, pH 8.6	Indirect UV, 225 nm	[19]
Malonic, lactic, aspartic, glutamic, glucornic, hydrochloric, and phosphoric	—	0.02 m benzoic acid, histidine, pH 6.2, 0.1% triton	Indirect UV, 254 nm	[111]
Benzoic, benzilic, and naphthoic	—	0.1 M tris-acetic acid, pH 8.6, 20% dextran	UV 205 nm	[112]
Picric, cinnamic, and sorbic	—	0.01 M KCl, pH 5.6, ME: 0.01 M HCl	UV 254 nm	[113]
Hippuric and gibberilic	—	0.1 M ammonium acetate, MeCN (1:9)	—	[114]

Compounds	Sample Matrix	Electrolytes	Detection	Reference
Succinic, citric, salicylic, malic, benzoic, sorbic, ascorbic, and tartaric acid	Blueberry juices	Ammonium acetate buffer of different concentration and pH	ES-IMS	[115]
Oleanolic acid, ursolic acid, quercetin, and apigenin	Swertia mussotii Franch by capillary	50×10^{-3} M borate-phosphate buffer (pH 9.5) with 5.0×10^{-3} M β-cyclodextrin	UV 250 nm	[116]
Uric acid	Plasma	75 mm glycylglycine solution titrated with NaOH 5 m (pH 9.0)	UV 292 nm	[117]
Carbohydrates				
Neutral	—	0.2 M boric acid, KOH, pH 5.0	UV 240 nm	[118]
Oligosaccharides	—	0.01 M Na_2HPO_4, 0.01 M $Na_2B_4O_7$, pH 9.4	Fluorescence	[119]
Cyclodextrins	—	0.03 M benzoic acid, tris, pH 6.2	Indirect UV 254 nm	[65]
Glucose	Serum	25 mM sodium hydrogen phosphate (pH 7.0)	LIF	[120]
Glucose, sucrose, and fructose	Fruits	0.5% potassium sorbate, 0.62% cetyltrimethylammonium bromide, and 0.02% potassium hydroxide	UV 254 nm	[121]

(*Continued*)

Table 1.2 (*Continued*)

Compounds	Sample Matrix	Electrolytes	Detection	Reference
Oligosaccharides	Milk	55% aqueous (200 mM NaH_2PO_4, pH 7.05, containing 100 mM SDS) and 45% methanol	UV 200 nm	[122]
Amino Acids				
Amino acids	—	1.0 mM PB, pH 5.31 or 7.12	—	[37]
Amino acids	—	0.2 mM Na salicylate, 0.04 mM Na_2CO_3, NaOH, pH 9.7	Indirect fluorescence	[66]
Debsyl derivatives	Urine	0.02 M PB, pH 7.0, MeCN (1:1)	—	[123]
Fluorescamine derivatives	—	10% propanol, pH 10.16	Fluorescence	[124]
2,4-Dinitrophenol derivatives	—	LE: 0.02 M HCl, his, pH 5.5, 0.1% HEC, TE: 0.01 M MES, his, pH 5.3	Visible 405 nm	[125]
NDA derivatives	—	0.01 M boric acid, 0.02 M KCl, pH 9.5	Fluorescence	[126]
FITC derivatives	—	0.05 M BB, pH 9.0	Fluorescence	[127]
	—	0.2 M BB, pH 7.8	Fluorescence	[128]
Dansyl derivatives	—	0.05 M PB, pH 7.0	Fluorescence	[7, 20, 129]

Compounds	Sample Matrix	Electrolytes	Detection	Reference
Aromatic amino acids	Human blood plasma	Tis phosphate 80 mM at pH 1.4	UV 200 nm	[130]
Analysis of aromatic acids	Water samples	1-Ethyl-3-methylimidazolium chloride and 1-ethyl-3-methylimidazolium hydrogen sulfate	UV 199 nm	[131]
Amino acids	African gourd seed milks	30 mM boric acid buffer adjusted to pH 9.3 with 12 mM SDS and 5% ethylene glycol	Fluorescence	[132]
Amino acids	Tea	1.5 formic acid (pH 1.5)	ELSD	[133]
Amino acids	Dried blood spots	mM NH_4Ac, adjusted to a pH of 10.7 with NH_4OH	MS	[134]
Amino acids	Physiological fluids	Disodium monophosphate (10 mM at pH 2.90)	Indirect absorbance	[135]
Proline and pipemidic acid	Human urine	5 mM $Ru(bpy)_3^{+2}$ and 50 mM phosphate buffer	Electrochemiluminescence	[136]
Peptides and proteins				
Dipeptides	—	0.15 M H_3PO_4, pH 1.5	UV 190 nm	[137]
Tri- and pentapeptides	—	5 mM ammonium acetate, NaOH, pH 8	MS	[138]

(Continued)

Table 1.2   (*Continued*)

Compounds	Sample Matrix	Electrolytes	Detection	Reference
β-Lactoglulin	—	0.02 M citrate buffer, pH 2.5	UV 200 nm	[139, 140]
Ovalbumin and oligopeptide	—	00125 M PB, pH 6.86	Fluorescence	[141]
Digest of β-casein	—	0.25 M ammonium acetate, pH 7.2	UV 210 nm	[142]
Bovine	—	0.1 M formate buffer, pH 4.0	UV 230 nm	[53, 143]
Mycoglobin, cytochrome C, and lysozyme	—	0.03 M KH_2PO_4 buffer, pH 3.8	UV 205 nm	[144]
Ribonuclease A, B_1, and B_2	—	0.02 M CAPS buffer, pH 11.0	UV 200 nm	[140]
Different proteins	Human serum	0.05 M Na borate, pH 10.0	UV 200 nm	[145]
HGH	—	0.1 M PB, pH 2.56	UV 200 nm	[146]
Insulin	—	0.09 M tris-NaH_2PO_4, pH 8.6 + 8.0 M urea, 0.1% SDS	UV	[147]
Different proteins	Human urine	0.1 M tris-acetic acid, pH 8.6	UV 280 nm	[148]
Acidic and basic proteins	—	30 mM NaH_2PO_4 buffer (pH 5.0) containing 4 mM C18-4-$C_{18}P$	UV 214 nm	[149]
Nucleic Acids				
E. Coli t-RNA	*E. Coli*	0.1 M tris-acetic acid, pH 8.6	UV 280 nm	[150]

Compounds	Sample Matrix	Electrolytes	Detection	Reference
P(dA)$_{40\text{-}60}$	—	0.1 M tris, 0.25 M borate, 7 M urea, pH 7.6	UV 260	[151]
SV40 DNA	—	9 mM tris-9 mM boric acid, pH 8.0, 2 mM EDTA	Fluorescence	[152]
Plasmid DNA	Plasmid	89 mM tris, 89 mM boric acid, 2 mM EDTA	UV-induced fluorescence	[153]
DNA, size standards	—	89 mM tris, 89 mM boric acid, 2.5 mM EDTA	UV 260 nm	[24]
RNA in hydroxyethylcellulose polymer	—	0.5× TBE, 1× SYBR Green II and 4.0 M urea	Fluorescence	[154]
miRNAs	Human serum	25 mM ammonium acetate (pH 6.0)	MS	[155]
Adenine nucleotides	Saccharomyces cerevisiae	Borate buffer (range 30–60 mM)	UV 210 nm	[156]
Nucleotide monophosphates	Human milk	30 mM ammonium formate-ammonia medium, (9.6 pH)	ESI-MS	[157]
Alkaloids				
(−)-Ephedrine, (+)-pseudoephedrine, (−)-*N*-methylephedrine, and (+)-*N*-methylpseudoephedrine	—	0.2 M borate buffer (pH 9.5)	Amperometric	[158]

(*Continued*)

Table 1.2 *(Continued)*

Compounds	Sample Matrix	Electrolytes	Detection	Reference
Harmala alkaloids	—	25 mM ammonium acetate (pH 7.8) and 10% (v/v) methanol	MS	[159]
Ergot	Cereal	—	UV	[160]
Isoquinoline alkaloids	Rhizoma Coptidis	20 mM sodium acetate in methanol-acetonitrile (80:20, v/v)	UV 190 nm	[161]
Curine, sinomenine, and magnoflorine	Sinomenium acutum	60 mM acetate buffer (pH 5.20)–15% (v/v) methanol	Chemiluminescence	[162]
Tetrandrine, fangchinoline, and sinomenine	—	80 mM ammonium acetate, in a mixture of 70% methanol, 20% ACN, and 10% water, which also contained 1% acetic acid	MS	[163]
Brucine, strychnine, atropine sulfate, anisodamine hydrobromide, scopolamine hydrobromide, and anisodine hydrobromide	Human plasma and urine	Ammonium acetate buffer of different concentrations and pHs	MS	[164]
Epinephrine and norepinephrine	Rat brain	50 mM boric acid, 40 mM borax (pH 8.9)	LIF	[165]
Quinine	Beverages	—	UV 200–600 nm	[166]

Compounds	Sample Matrix	Electrolytes	Detection	Reference
Aconite	Aconite roots	200 mm tris, 150 mm perchloric acid, and 40% 1,4-dioxane (pH 7.8)	UV 214 nm	[167]
Flavonoids				
Kaempferol, quercetin, and luteolin	Semen Plantaginis	Different concentrations of 1-ethyl-3-methylimidazolium hydrogen sulfate and 1-ethyl-3-methylimidazolium tetrafluoroborate	UV 254 nm	[168]
Quercetin, kaempferol, and flavopiridol	—	Tris/phosphate/MgCl$_2$ (pH 7.2, 120 mM ionic strength)	UV 254 nm	[169]
Drugs				
Profen group	—	PB, pH 6.1–8.4	UV 215 nm	[170]
Sulfa groups	—	0.03 M, PB, pH 7.0	UV 215 nm	[171]
Alkaoids	—	0.05 M PB, pH 7.0	UV 245 nm	[172]
Toxins	—	PB, pH 8.7	Fluorescence	[173]
Antibiotics	Human serum	BB, pH 10.0, 100 mM SDS	—	[174]
Vitamins	—	0.05 M tris-borate buffer, pH 8.4, 0.05 SDS	UV 200	[175]

(*Continued*)

Table 1.2 (*Continued*)

Compounds	Sample Matrix	Electrolytes	Detection	Reference
β-Blockers	—	$Na_2B_4O_7$-H_3BO_3 (50 mM), pH 9.0	UV	[176]
Histamines	—	BB, pH 9.0, 100 mM SDS	UV 210 nm	[177]
Analgesics	Human urine	PB, 60% EG, pH 7.9	MS	[178]
Anti-allergic	—	PB	UV	[179]
Anti-diabetic	—	50 mM borate buffer (pH 9.0)	UV 210 nm	[180]
Non-steroidal anti-inflammatory drugs	Urine	40 mM anhydride sodium tetra borate (9.2 pH)	UV 214 nm	[181]
Non-steroidal anti-inflammatory drugs	Milk and dairy products	30 mM acetate buffer at pH 4.0 containing 25% ACN, v/v	UV 210 nm	[182]
Steroids	—	PB	UV	[183]
Valsartan, amlodipine besylate, and hydrochlorothiazide	Tablets	40 mM phosphate buffer at pH 7.5 with	UV 230 nm	[184]
Cisatracurium besylate and its degradation	Pharmaceutical preparations	Phosphate buffer solution (PBS) of 350 μL (100 mM, pH 8.5) containing $Ru(bpy)_3^{2+}$ (2 mM)	Electrochemiluminescence detection	[185]

Applications | 35

Compounds	Sample Matrix	Electrolytes	Detection	Reference
Piroxicam	Tablets	10 mM borate buffer (pH 9.0) containing 10% (v/v) methanol	UV 204 nm	[186]
Fesoterodine	—	10 mM sodium phosphate buffer at pH 6.5	UV 208 nm	[187]
Nitrate and hydrocortisone acetate	Cream formulation	Phosphate buffer (50 mM, pH 4)	UV 230 nm	[188]
Metformin and cyanoguanidine	—	40 mM citrate buffer (pH 6.7)	UV 214 nm	[189]
Omeprazole and their main metabolites in human urine	Urine	60% phosphate buffer (pH 12, 25 mM) and 40% acetonitrile	ESI-MS	[190]
Loratadine, desloratadine, and cetirizine	—	25 mM phosphate buffer, pH 2.5	UV 240 nm	[191]
Cathartics and appetite pharmaceuticals	Dietary supplements	20 mM ammonium formate in 20% v/v acetonitrile-water (pH 8.0)	MS	[192]
Methylphenidate and a metabolite	Drosophila melanogaster	50 mM citric acid electrolyte (pH ≈ 2.1)	MS	[193]
1-(2-Chlorophenyl)piperazine, 1-(3-chlorophenyl)piperazine and 1-(4-chlorophenyl) piperazine	Confiscated pills	20 mM phosphoric acid adjusted to pH 2.5 with triethylamine and 10 mM α-cyclodextrin	UV 236 nm	[194]

(Continued)

Table 1.2 (*Continued*)

Compounds	Sample Matrix	Electrolytes	Detection	Reference
Aspartame, cyclamate, saccharin, and acesulfame K	—	150 mM 2-(cyclohexylamino) ethanesulfonic acid and 400 mM tris(hydroxymethyl) aminomethane at pH 9.1	Contactless conductivity detection	[195]
Fatty acids	—	10 mM DOC in methanol	Contactless conductivity	[196]
Trimethoprim and sulfamethoxazole	—	10 mM lithium phosphate buffer (pH 7.1)	Contactless conductivity	[197]
Chlorpromazine and promethazine and their main metabolites	—	40 mM phosphate buffer solution (pH 6.5) containing 5 mM tris(2,2'-bipyridyl) ruthenium	Electrochemiluminescence	[198]
Sitagliptin and metformin	Pharmaceutical preparations	60 mM phosphate buffer at pH 4.0 with	UV 203 nm	[199]
Lappaconitine hydrobromide and isopropiram fumarate	Rabbit plasma	20 mM phosphate buffer (pH 8.5) containing 5% (v/v) ACN and 0.17 M SDS	Electrochemiluminescence	[200]
Sotalol, alprenolol, and atenolol		Phosphate buffer (pH 8.5)	Amperometric	[201]
Metoprolol and hydrochlorothiazide	Dosages	50 mM phosphate at pH 9.5	UV 240 and 214 nm	[202]

Compounds	Sample Matrix	Electrolytes	Detection	Reference
Zopiclone and its impurities	Tablets	80 mM sodium phosphate buffer pH 2.5 and 5 mM carboxymethyl-β-cyclodextrin	UV 305 and 200 nm	[203]
Posaconazole	Patient plasma	1.25 M formic acid	UV 195 nm	[204]
Amphetamine and related drugs	Equine plasma	25 mM ammonium formate in acetonitrile/methanol (20: 80, v/v) plus 1 M formic acid	MS	[205]
Glutathione	—	5.0 mM NDA, 20 mM borate buffer (pH 9.2)	Fluorescence	[206]
Verticine and verticinone	Bulbus Fritillariae	40 mM 1-butyl-3-methylimidazolium tetrafluoroborate (BMImBF(4)) IL-8 mM phosphate buffer	Electrochemiluminescence	[207]
Norfloxacin, ciprofloxacin, and ofloxacin	Urine	70 mM Na_2B4O_7–NaH_2PO_4 (pH 8.5)	Chemiluminescence	[208]
Chelerythrine and sanguinarine	Chinese herbal medicines	2.0 mL of 200 mM ammonium acetate, 0.5 mL acetic acid and 5.0 mL ACN	Laser-induced native fluorescence	[209]
Mefenacet		5 mM $Ru(bpy)_3{}^{2+}$ solution (pH 7.38)	Electrochemiluminescence	[210]

(Continued)

Table 1.2 (*Continued*)

Compounds	Sample Matrix	Electrolytes	Detection	Reference
Tamoxifen, imipramine, and their main metabolites	Urine	17 mM ammonium acetate and 1.25% acetic acid in 80:20 (v:v) methanol-acetonitrile	Diode array	[211]
Sodium tripolyphosphate	Meat samples	10 mM succinic acid + 15 mM BALA + 0.1% HEC	Conductometric	[212]
Clenbuterol	—	Phosphate buffer	Chemiluminescence	[213]
Chiral Mixture Analysis				
Amine drugs	—	0.1 M BB, pH 9.5, β-CD	UV	[214]
Amino acids	—	PB with β-CD	UV	[215]
Amino acids	Hydrolyzed protein fertilizers		MS	[216]
Amino acids	—	100 mM tris-borate buffer (pH 10.0) containing 2 mM β-CD and 10 mM hexamethylenediamine	Fluorescence	[217]
AA-dansyl derivatives	—	PB with β-CD	UV	[218]

Compounds	Sample Matrix	Electrolytes	Detection	Reference
AA-2,4-Dinitrophenyl derivatives		PB with 6-amino-6-deoxy-β-CD and its N-hexyl derivatives	UV	[219]
AA-AQC derivatives	—	PB with vancomycin	UV	[220]
AA-FMOC derivatives	—	PB with vancomycin	UV	[220]
AA-PTH derivatives	—	PB with vancomycin		[220]
AA-AEOC derivatives	—	PB with γ-CD	UV	[221]
β-Blockers	—	PB with TM-β-CD	UV	[222]
Profens	—	PB with β-CD	UV	[223]
Dipeptides	—	PB with (+)-18-crown-6-tetracarboxylic acid	UV	[224]
Antifungals	—	PB with CDs	UV	[225]
Alkaloids	—	PB with CDs	UV	[226]
Benzoin	—	PB with HSA	UV	[227]
Anti-allergic	—	PB with CDs	UV	[228]
Diltiazem	—	PB with dextran sulfide	UV	[229]
Baclofen	—	PB with β-CDs	UV	[230]
Thalidomide	—	PB with β-CDs	UV	[231]
Warfarin	—	PB with saccharides	UV	[232]

(Continued)

Table 1.2 *(Continued)*

Compounds	Sample Matrix	Electrolytes	Detection	Reference
Ketamine and metabolites	Plasma	3.0 Phosphate buffer containing 0.66 % of highly sulfated γ-cyclodextrin	UV 200 nm	[233]
Tapentadol	—	100 mM sodium borate buffer (pH 9.5) with non-charged hydroxypropylated CDs (2-hydroxypropyl-β-CD, 2-hydroxypropyl-γ-CD	Array detector	[234]
Ketoconazole	—	10 mM phosphate buffer at pH 2.5 containing 20 mM TMβCD, 5 mM SDS, and 1.0% (v/v) with heptakis (2, 3, 6-tri-O-methyl)-β-cyclodextrin	Array detector	[235]
Dansyl amino acids and dipeptides	—	100.0 mM boric acid, 5.0 mM ammonium acetate, 3.0 mM Zn(II), 6.0 mM L-hydroxyproline, and 4.0 mM γ-CD at pH 8.2	UV 254 nm	[236]

Compounds	Sample Matrix	Electrolytes	Detection	Reference
Alprenolol, atenolol, metoprolol, clenbuterol, methoxyphenamine, pindolol, propranolol, sotalol, synephrine, labetalol, and fenoterol	—	70 mM TBA with 60 mM H_3BO_3	UV 225 nm	[237]
Meptazinol and its three intermediate	—	Mono-6-deoxy-6-piperdine-β-cyclodextrin	UV 271 nm	[238]
Phenylalanine and tryptophan	—	15 mM sodium tetraborate, 5 mm β-CD, and 4 mm chiral ionic liquid at pH 9.5	UV 214 nm	[239]
Amlodipine	Tablets	Phosphate buffer (100 mM, pH 4) containing 10% w/v maltodextrin	UV 214 nm	[240]
Indapamide	—	Phosphate buffer (pH 7.0) with β-CD	UV 242 nm	[241]
Mirtazapine, N-demethylmirtazapine, 8-hydroxymirtazapine, and mirtazapine-N-oxide	—	6.25 mM borate-25 mM phosphate solution at pH 2.8 containing 5.5 mg/mL carboxymethyl-β-cyclodextrin	UV 200 nm	[242]

(Continued)

Table 1.2 (*Continued*)

Compounds	Sample Matrix	Electrolytes	Detection	Reference
A diaryl-pyrazole sulfonamide derivative	—	67 mM phosphate buffer at pH 7.4 with amino-β-CD and β-CD	Photo diode array	[243]
	—	Water:methanol (90:10, v/v) and consisting of 10.7 or 16.1 mM with penicillin	UV 264 nm	[244]
Meptazinol and its three intermediates	—	Phosphate buffer (pH 6.0) with CM-β-CD	UV 237nm	[245]
Amphetamine-like designer drugs	—	63.5 mM phosphate buffer and 46.9 mM NaOH with sulfated-β-CD, caroboxymethyl-β-CD, dimethyl-β-CD	UV 200–400 nm	[246]
Tolterodine and methoxytolterodine	Commercial pills	Tris/phosphate buffer (2.5 pH) with α-, β-CD and phosphated-γ-CD	UV 200 nm	[247]
Gossypol	Cotton flower petals and seed	Borate buffer (pH 9.3)	Matrix diode array	[248]

Compounds	Sample Matrix	Electrolytes	Detection	Reference
Clenbuterol	—	Phosphate buffer (50 mM, pH 3.5) containing 10 mM CM-β-CD	NMR	[249]
Lipoic acid	Dietary supplements	100 mM phosphate buffer (pH 7.0) containing 8 mM trimethyl-β-cyclodextrin	UV 200 nm	[250]
Norephedrine	—	Native α-CD, β-CD, heptakis(2,3-di-O-acetyl-6-O-sulfo)-β-CD (HDAS-β-CD), and heptakis(2,3-di-O-methyl-6-O-sulfo)-β-CD	NMR	[251]
Amphetamine, methamphetamine, ephedrine, pseudoephedrine, norephedrine, and norpseudoephedrine	—	Acetic acid (pH 2.5 and 2.8) with carboxymethyl-β-cyclodextrin and heptakis(2,6-di-O-methyl)-β-cyclodextrin (DMBCD) and chiral crown ether (+)-(18-crown-6)-2,3,11,12-tetracarboxylic acid	Contactless conductivity	[252]

(Continued)

Table 1.2 (Continued)

Compounds	Sample Matrix	Electrolytes	Detection	Reference
Threo-methylphenidate in human plasma	—	Phosphate buffer (50 mM, pH 3.0) containing 20 mM HP-β-CD and 30 mM triethanolamine	UV 200 nm	[253]
Ketamine	—	Phosphate buffer pH 2.5 buffer comprising 50 mM Tris and phosphoric acid together with either multiple isomer sulfated β-cyclodextrin (10 mg/mL) or highly sulfated γ-cyclodextrin (2%, w/v)	UV 195 nm	[254]
Camphorquinone	—	25 mM borate buffer at pH 9.0 with 20 mM α-CD, 10 mM carboxymethyl-β-CD	Laser-induced phosphorescence	[255]
Palonosetron hydrochloride	—	30 mM NaH_2PO_4 (pH 3.0) containing 150 mM β-CD and 10% (v/v) methanol	UV 214 nm	[256]

Compounds	Sample Matrix	Electrolytes	Detection	Reference
Levocetirizine	Pharmaceuticals	Anionic cyclodextrin	Photo diode	[257]
Amphetamine	Human urine	—	UV 200 nm	[258]
Pheniramine (PHM), dimethindene (DIM), dioxopromethazine	Biological matrices	A negatively charged carboxyethyl-beta-cyclodextrin	UV 240 nm	[259]

AEOC: 2-(9-Anthryl)-ethylchloroformate; AQC: 6-aminopropylhydroxy succinimidyl carbamate; dAMP: 2′-deoxyadenosine-5′-monophosphate; BB: borate buffer; CAPS: 3-[cyclohexylamino]-1-propane-sulfonic acid; CD: cyclodextrin; CHES: 2-(N-cyclohexylamino)ethanesulphonic acid; CDTA: cyclohexane-1,2-diaminetetraacetic acid; dCMP: 2′-deoxycytidine-5′-monophosphate; DHBP: 1,1′-di-n-heptyl-4,4′ bipyridinium hydroxide; DNA: deoxyribonucleic acid; DOSS: sodium dioctyl sulfosuccinate; DTPA: diethylenetriaminepentaacetic acid; ED: ethylene cyanohydrin; EDTA: ethylenediaminetetraacetic acid; EG: ethylene glycol; FITC: fluorescein isothiocyanate; FMOC: 9-fluorenylmethyl chloroformate; dGMP: 2′-HDB: hexadimethrine bromide, oxyguanosine-5′-monophosphate; HEC: hydroxyethyl cellulose; HIBA: α-hydroxyisobutyric acid; HQS: 8-hydroxyquinoline-5-sulfonic acid; HS: human albumin serum; ME: modified electrolyte; MeCN: acetonitrile; MES: morpholinoethanesulfonic acid; MS: mass spectrometry; NDA: naphthalene-2,3-dicarboxyaldehyde; o-OPA: o-phthaldialdehyde; PAHs: polynuclear aromatic hydrocarbons; PAPS: 3′-phosphoadenosine-5′-phosphosulfate; PB: phosphate buffer; PCBs: polychlorinated biphenyls; PTH: phenylthiohydantion; t-RNA: t-ribonucleic acid; SDS: sodium dodecylsulphate; TBABr: tetrabutylammonium bromide; TE: terminating electrolyte; TEA: triethanolamine; TM-β-CDs: trimethyl-β-cyclodextrins; dTMP: 2′-deoxythymidine-5′-monophosphate; TTAOH: tetradecyl-trimethylammonium hydroxide; TTHA: triethylenetetraminehexa acetic acid; UV: ultraviolet

1.10 Conclusion

Analysis at trace level is an extremely significant and challenging issue, especially in unidentified matrix. Capillary electrophoresis has been used often for this reason. But unfortunately, capillary electrophoresis could not attain a reputable position in the practice analysis owing to reduced reproducibility. Therefore, many researchers have recommended diverse modifications to build capillary electrophoresis as a method of selection. The detection limit may be enhanced by using fluorescent and radioactive complexing agents since detection via fluorescent and radioactive detectors is extra-sensitive and reproducible with the small limit of detection. To build capillary electrophoresis applications more reproducible, the BGE should be developed in such a manner so that its physical and chemical properties stay unchanged during the experimental run. The choice of the capillary wall chemistry, pH and ionic strength of BGE, detectors and optimization of BGE are discussed in many papers [79, 80, 100–102]. Apart from the above aspects described for the perfection of capillary electrophoresis, some other issues should also be touched so that capillary electrophoresis may be adopted as the everyday method in analytical science. Basically, the non-reproducibility of capillary electrophoresis may be due to the heating of BGE following extensive use of the capillary electrophoresis equipment. Hence, to maintain steady temperature during experiments, a cooling appliance should be incorporated in the capillary electrophoresis equipment. There are very few research papers describing method validation. To make the developed method more appropriate, the validation of the methodology should be determined. A good hyphenation of capillary electrophoresis equipment with AAS, ICP, mass, etc. devices should be developed, which may result into superior reproducibility and small limits of detection. In a nutshell, all the capabilities and potential of capillary electrophoresis as a separation technique have been explored and are in progress. Nevertheless, capillary electrophoresis will be realized as an extensively familiar technique of preference in separation science. In brief, a lot has to be developed for the improvement of capillary electrophoresis. Certainly, it will be one of the best analytical techniques in the coming years. Let us hope for the best for capillary electrophoresis.

Acknowledgment

The authors are thankful to King Saud University, Riyadh, Saudi Arabia, for the International Scientific Partnership Program. The authors extend their appreciation to the International Scientific Partnership Program ISPP at King Saud University for funding this research work through ISPP# 0037.

References

1. Hjertén, S. (1967). Free zone electrophoresis. *Chromatogr. Rev.*, **9**, 122–219.
2. Jorgenson, J.W., and Lukacs, K.D. (1981). Zone electrophoresis in open-tubular glass capillaries. *Anal. Chem.*, **53**, 1298–1302.
3. Jorgenson, J.W., and Lukacs, K.D. (1981). High-resolution separations based on electrophoresis and electroosmosis. *J. Chromatogr.*, **218**, 209–216.
4. Jorgenson, J.W., and Lukacs, K.D. (1983). Capillary zone electrophoresis. *Science*, **222**, 4621.
5. Wehlr, T., and Zhu, M. (1993). Capillary electrophoresis: Historical perspectives, in J.P. Landers (Ed.), *Handbook of Capillary Electrophoresis* (CRC Press, Boca Raton).
6. Landers, J.P. (1993). *Handbook of Capillary Electrophoresis* (CRC Press, Boca Raton).
7. Foret, F., Krivankova, L., and Bocek, P. (1993). *Capillary Zone Electrophoresis* (VCH Publ, Weinheim).
8. Wehr, T., Rodriguez-Diaz, R., and Zhu, M. (1998). *Capillary Electrophoresis of Proteins*, Vol. 80 (Marcel Dekker, Inc., New York).
9. Khaledi, M.G. (1998). *High Performance Capillary Electrophoresis: Theory, Techniques and Applications* (John Wiley and Sons, New York).
10. Lunn, G. (2000). *Capillary Electrophoresis Methods for Pharmaceutical Analysis* (John Wiley and Sons, New York).
11. Chankvetadze, B. (1997). *Capillary Electrophoresis in Chiral Analysis* (John Wiley and Sons, New York).
12. Terabe, S., Otsuka, K., Ichikawa, A., and Ando, T. (1984). Electrokinetic separations with micellar solutions and open-tubular capillaries. *Anal. Chem.*, **56**, 111–113.
13. Mayer, S., and Schurig, V. (1992). Enantiomer separation by electrochromatography on capillaries coated with chirasil-dex. *J. High Resolut. Chromatogr.*, **15**, 129–131.

14. Lukacs, K.D., and Jorgenson, J.W. (1985). Capillary zone electrophoresis: Effect of physical parameters on separation efficiency and quantitation. *J. High Resolut. Chromatogr.*, **8**, 407–411.

15. Foret, F., Deml, M., and Bocek, P. (1988). Capillary zone electrophoresis: Quantitative study of the effects of some dispersive processes on the separation efficiency. *J. Chromatogr.*, **452**, 601–613.

16. Hjertén, S. (1990). Zone broadening in electrophoresis with special reference to high performance electrophoresis in capillaries: An interplay between theory and practice. *Electrophoresis*, **11**, 665–690.

17. Tsuda, T., Mizuno, T., and Akiyama, J. (1981). Rotary-type injector for capillary zone electrophoresis. *Anal. Chem.*, **59**, 799–800.

18. Mikkers, F.E.P., Everaerts, F.M., and Verheggen, T.P.E.M. (1979). High-performance zone electrophoresis. *J. Chromatogr.*, **169**, 11–20.

19. Deml, M., Foret, F., and Bocek, P. (1985). Electric sample splitter for capillary zone electrophoresis. *J. Chromatogr.*, **320**, 159–165.

20. Jorgenson, J.W., and Lukacs, K.D. (1981). Free-zone electrophoresis in glass capillaries. *Clin. Chem.*, **27**, 1551–1553.

21. Terabe, S., Otsuka, K., Ichikawa, K., Tsuchiya, A., and Ando, T. (1984). Electrokinetic separations with micellar solutions and open-tubular capillaries. *Anal. Chem.*, **56**, 111–113.

22. Tsuda, T., Sweedler, J.V., and Zare, R.N. (1990). Rectangular capillaries for capillary zone electrophoresis. *Anal. Chem.*, **622**, 2149–2152.

23. Thormann, W., Arn, D., and Schumacher, E. (1984). Detection of transient and steady states in electrophoresis: Description and applications of a new apparatus with 255 potential gradient detectors along the separation trough. *Electrophoresis*, **5**, 323–337.

24. Bocek, P., and Chramback, A. (1991). Capillary electrophoresis of DNA in agarose solutions at 40°C. *Electrophoresis*, **12**, 1059–1061.

25. Wolf, A.V., Morden, G.B., and Phoebe, P.G. (1987). In R.C. Weast, M.J. Astle, and W.H. Beyer (Eds.), *CRC Handbook of Chemistry and Physics*, 68th ed. (CRC Press, Boca Raton, USA).

26. Hjertén, S., and Zhu, M. (1987). *Physical Chemistry of Colloids and Macromolecules IUPAC* (Blackwell Scientific Publishing, London).

27. Tsuda, T., Nakagawa, G., Sato, M., and Yagi, K. (1983). Separation of nucleotides by high-voltage capillary electrophoresis. *J. Appl. Biochem*, **5**, 330–336.

28. Hjertén, S., and Zhu, M. (1985). Micropreparative version of high-performance electrophoresis: The electrophoretic counterpart of

narrow-bore high-performance liquid chromatography. *J. Chromatogr.*, **327**, 157–164.

29. Vindevogel, J., Sandra, P., and Verhagen, L.C. (1990). Separation of hop bitter acids by capillary zone electrophoresis and micellar electrokinetic chromatography with UV-diode array detection. *J. High Resolut. Chromatogr.*, **13**, 295–298.

30. Kobayashi, S., Ueda, T., and Kikumoto, M. (1989). Photodiode array detection in high-performance capillary electrophoresis. *J. Chromatogr.*, **480**, 179–184.

31. Foret, F., Deml, M., Kahle, V., and Bocek, P. (1986). Online fiber optic UV detection cell and conductivity cell for capillary zone electrophoresis. *Electrophoresis*, **7**, 430–432.

32. Huang, X., Zare, R.N., Sloss, S., and Ewing, A.G. (1991). End-column detection for capillary zone electrophoresis. *Anal. Chem.*, **63**, 189–192.

33. Wallingford, R.A., and Ewing, A.G. (1987). Capillary zone electrophoresis with electrochemical detection. *Anal. Chem.*, **59**, 1762–1766.

34. Wallingford, R.A., and Ewing, A.G. (1988). Retention of ionic and non-ionic catechols in capillary zone electrophoresis with micellar solutions. *J. Chromatogr.*, **441**, 299–309.

35. Wallingford, R.A., and Ewing, A.G. (1989). Separation of serotonin from catechols by capillary zone electrophoresis with electrochemical detection. *Anal. Chem.*, **61**, 98–100.

36. Gaitonde, C.D., and Pathak, P.V. (1990). Capillary zone electrophoretic separation of chlorophenols in industrial waste water with on-column electrochemical detection. *J. Chromatogr.*, **514**, 389–393.

37. Enggstrom-Silverman, C.E., and Ewing, A.G. (1991). Copper wire amperometric detector for capillary electrophoresis. *J. Microcol. Sep.*, **3**, 141.

38. Bornhop, D.J., and Dovichi, N.J. (1986). Simple nanoliter refractive index detector. *Anal. Chem.*, **58**, 504–505.

39. Yu, M., and Dovichi, N.J. (1989). Attomole amino acid analysis: Capillary-zone electrophoresis with laser-based thermo-optical detection. *Appl. Spectros.*, **43**, 196–201.

40. Dovichi, N.J., Nolan, T.G., and Weimer, A.W. (1984). Theory for laser-induced photothermal refraction. *Anal. Chem.*, **56**, 1700–1704.

41. Liu, W., and Lee, H.K. (1999). Chemical modification of analytes in speciation analysis by capillary electrophoresis, liquid chromatography and gas chromatography. *J. Chromatogr. A*, **834**, 45–63.

42. Michalke, B., and Schramel, P. (1997). Coupling of capillary electrophoresis with ICP-MS for speciation investigations. *Fresenius J. Anal. Chem.*, **357**, 594–599.

43. Ergolic, K.J., Stockton, R.A., and Chakarborti, D. (1983). In W.H. Lederer and R.J. Fensterheim (Eds.), *Arsenic: Industrial, Biochemical and Environmental Prospectives* (Vnostrand Reinhold, New York).

44. Medina, I., Rubi, E., Mejuto, M.C., and Cela, R. (1993). Speciation of organomercurials in marine samples using capillary electrophoresis, *Talanta*, **40**, 1631–1636.

45. Magnuson, M.L., Creed, J.T., and Brockhoff, C.A. (1997). Speciation of selenium and arsenic compounds by capillary electrophoresis with hydrodynamically modified electroosmotic flow and on-line reduction of selenium (VI) to selenium (IV) with hydride generation inductively coupled plasma mass spectrometric detection. *Analyst*, **122**, 1057–1067.

46. Michalke, B., and Schramel, P. (1998). Capillary electrophoresis interfaced to inductively coupled plasma mass spectrometry for element selective detection in arsenic speciation. *Electrophoresis*, **19**, 2220–2025.

47. Smith, R.D., and Udseth, H.R. (1988). Capillary zone electrophoresis-MS, *Nature*, **331**, 639.

48. Udseth, H.R., Loo, J.A., and Smith, R.D. (1989). Capillary isotachophoresis/mass spectrometry, *Anal. Chem.*, **61**, 228–232.

49. Moseley, M.A., Deterding, L.J., Tomer, K.B., and Jorgenson, J.W. (1990). Capillary zone electrophoresis-mass spectrometry using a coaxial continuous-flow fast atom bombardment interface, *J. Chromatogr.*, **516**, 167–173.

50. Moseley, M.A., Deterding, L.J., Tomer, K.B., and Jorgenson, J.W. (1991). Determination of bioactive peptides using capillary zone electrophoresis/mass spectrometry, *Anal. Chem.*, **63**, 109–114.

51. Pawliszyn, J. (1988). Nanoliter volume sequential differential concentration gradient detector. *Anal. Chem.*, **60**, 2796–2801.

52. Loo, J.A., Udseth, H.R., and Smith, R.D. (1989). Peptide and protein analysis by electrospray ionization-mass spectrometry and capillary electrophoresis-mass spectrometry, *Anal. Biochem.*, **179**, 404–412.

53. Lukacs, K.D. (1983). *Theory, Instrumentation and Application of Capillary Zone Electrophoresis*, Ph.D. Thesis, University of North Carolina at Chapel Hill, NC, USA.

54. Chen, C., Demana, T., Huang, S., and Morris, M.D. (1989). Capillary zone electrophoresis with analyte velocity modulation. Application to refractive index detection, *Anal. Chem.*, **61**, 1590–1593.

55. Demana, T., Chen, C., and Moris, M.D. (1990). Non-ideal behavior in analyte velocity modulation capillary electrophoresis, *J. High Resolut. Chromatogr.*, **13**, 587–589.

56. Kaniansky, D. (1981). Ph.D. Thesis, Komensky University, Bratislava.

57. Pentoney, S.L., Zare, R.N., and Quint, J.F. (1989). On-line radioisotope detection for capillary electrophoresis, *Anal. Chem.*, **61**, 1642–1647.

58. Petru, A., Rajec, P., Cech, R., and Kunc, J. (1989). Determination of radiolytic products of two-phase tributylphosphate water systems by capillary isotachophoresis, *J. Radioanal. Nucl. Chem.*, **129**, 229–232.

59. Pentoney, S.L., Zare, R.N., and Quint, J.F. (1989). Semiconductor radioisotope detector for capillary electrophoresis, *J. Chromatogr.*, **480**, 259–270.

60. Altria, K.D., Simpson, C.F., Bharij, A.K., and Theobald, A.E. (1990). A gamma-ray detector for capillary zone electrophoresis and its use in the analysis of some radiopharmaceuticals, *Electrophoresis*, **11**, 732–734.

61. Chen, C.Y., and Morris, M.D. (1988). Raman spectroscopic detection system for capillary zone electrophoresis, *J. Appl. Spectroscop.*, **42**, 515–518.

62. Kaniansky, D., Rajec, P., Svec, A., Marak, J., Koval, M., Lucka, M., Franko, S., and Sabanos, G. (1989). On-column radiometric detector for capillary isotachophoresis, *J. Radioanal. Nucl. Chem.*, **129**, 305–325.

63. Chen, C., and Morris, M.D. (1991). On-line multichannel Raman spectroscopic detection system for capillary zone electrophoresis, *J. Chromatogr.*, **540**, 355–363.

64. Christensen, P., and Yeung, E.S. (1989). Fluorescence-detected circular dichroism for on-column detection in capillary electrophoresis, *Anal. Chem.*, **61**, 1344–1347.

65. Nardi, A., Fanali, S., and Foret, F. (1990). Capillary zone electrophoretic separation of cyclodextrins with indirect UV photometric detection, *Electrophoresis*, **11**, 774–776.

66. Kuhr, W.G., and Yeung, E.S. (1988). Indirect fluorescence detection of native amino acids in capillary zone electrophoresis, *Anal. Chem.*, **60**, 1832–1834.

52 | *Capillary Electrophoresis*

67. Gross, L., and Yeung, E.S. (1989). Indirect fluorimetric detection and quantification in capillary zone electrophoresis of inorganic anions and necleotides, *J. Chromatogr.*, **480**, 169–178.

68. Yeung, E.S., and Kuhr, W.G. (1991). Indirect detection methods for capillary separations, *Anal. Chem.*, **63**, 275A–282A.

69. Dovichi, N.J., Martin, J.C., Jeff, J.H., and Keller, R.A. (1983). Attogram detection limit for aqueous dye samples by laser-induced fluorescence, *Science*, **219**, 845–847.

70. Olefirowicz, T.M., and Ewing, A.G. (1990). Capillary electrophoresis with indirect amperometric detection, *J. Chromatogr.*, **499**, 713–719.

71. Foret, F., Fanali, S., Nardi, A., and Bocek, P. (1990). Capillary zone electrophoresis of rare earth metals with indirect UV absorbance detection, *Electrophoresis*, **11**, 780–783.

72. Ali, I., and Aboul-Enein, H.Y. (2002). Speciation of metal ions by capillary electrophoresis, *Critical Rev. Anal. Chem.*, **32**, 337–350.

73. Ali, I., and Aboul-Enein, H.Y. (2002). Determination of metal ions in water, soil, and sediment by capillary electrophoresis, *Anal. Lett.*, **35**, 2053–2076.

74. Martinez, D., Cugat, M.J., Borrull, F., and Calull, M. (2000). Solid-phase extraction coupling to capillary electrophoresis with emphasis on environmental analysis, *J. Chromatogr. A*, **902**, 65–89.

75. Malik, A.K., and Faubel, W. (2001). A review of analysis of pesticides using capillary electrophoresis, *Crit. Rev. Anal. Chem.*, **31**, 223–279.

76. Haddad, P.R., Doble, P., and Macka, M. (1999). Developments in sample preparation and separation techniques for the determination of inorganic ions by ion chromatography and capillary electrophoresis, *J. Chromatogr. A*, **856**, 145–177.

77. Dabek-Zlotorzynska, E., Aranda-Rodriguez, R., and Keppel-Jones, K. (2001). Recent advances in capillary electrophoresis and capillary electrochromatography of pollutants, *Electrophoresis*, **22**, 4262–4220.

78. Pacakova, V., Coufal, P., and Stulik, K. (1999). Capillary electrophoresis of inorganic cations, *J. Chromatogr. A*, **834**, 257–275.

79. Liu, B.F., Liu, B.L., and Cheng, J.K. (1999). Analysis of inorganic cations as their complexes by capillary electrophoresis, *J. Chromatogr. A*, **834**, 277–308.

80. Valsecchi, S.M., and Polesello, S. (1999). Analysis of inorganic species in environmental samples by capillary electrophoresis, *J. Chromatogr. A*, **834**, 363–385.

81. Whang, C., and Pawliszyn, J. (1998). Solid phase microextraction coupled to capillary electrophoresis, *Anal. Commun.*, **35**, 353–356.

82. Kuban, P., and Karlberg, B. (1997). On-line dialysis coupled to a capillary electrophoresis system for determination of small anions. *Anal. Chem.*, **69** (1999), 1169–1173.

83. Ezzel, J.L., Richter, B.E., Felix, W.D., Black, S.R., and Meikle, J.E. (1995). A comparison of accelerated solvent extraction with conventional solvent extraction of organophosphorus pesticides and herbicides, *LC-GC*, **13**, 390–398.

84. Majors, R.E. (1995). Sample preparation and handling for environmental and biological analysis, *LC-GC*, **3**, 542–555.

85. Raber, G., and Greschonig, H. (2000). New preconditioning strategy for the determination of inorganic anions with capillary zone electrophoresis using indirect UV detection, *J. Chromatogr. A*, **890**, 355–361.

86. Faller, A., and Engelhardt, H. (1999). How to achieve higher repeatability and reproducibility in capillary electrophoresis, *J. Chromatogr. A*, **853**, 83–94.

87. Dabek-Zlotorzynska, E., Piechowski, M., McGrath, M., and Lai, E.P.C. (2001). Determination of low-molecular-mass carboxylic acids in atmospheric aerosol and vehicle emission samples by capillary electrophoresis, *J. Chromatogr. A*, **910**, 331–345.

88. Macka, M., Johns, C., Doble, P., and Haddad, P.R. (1999). Indirect photometric detection in CE using buffered electrolytes-Part I, principles, *LC-GC*, **19**, 38–47.

89. Timerbaev, A.R. (1997). Strategies for selectivity control in capillary electrophoresis of metal species, *J. Chromatogr. A*, **792**, 495–518.

90. Doble, P., and Haddad, P.R. (1999). Indirect photometric detection of anions in capillary electrophoresis, *J. Chromatogr. A*, **834**, 189–212.

91. Dabek-Zlotorzynskaa, E., Lai, E.P.C., and Timerbaev, A.R. (1998). Capillary electrophoresis: The state-of-the-art in metal speciation studies. *Anal. Chim. Acta*, **359**, 1–26.

92. Ikuta, N., Yamada, Y., Yoshiyama, T., and Hirokawa, T. (2000). New method for standardization of electropherograms obtained in capillary zone electrophoresis. *J. Chromatogr. A*, **894**, 11–17.

93. Schmitt-Kopplin, P., Garmash, A.V., Kudryavtsev, A.V., Menzinger, P., Perminova, I.V., Hertkorn, N., Freitag, D., Petrosyan, V.S., and Kettrup, A. (2001). Quantitative and qualitative precision improvements

by effective mobility-scale data transformation in capillary electrophoresis analysis. *Electrophoresis*, **22**, 77–87.

94. Rathore, A.S., and Guttman, A. (2003). *Electrokinetic Phenomena* (Marcel Dekker Inc., New York).

95. Neubert, R., and Rüttinger, H.H. (2003). *Affinity Capillary Electrophoresis in Pharmaceutics and Biopharmaceutics* (Marcel Dekker Inc., New York).

96. El Rassi, Z., and Giese, R.W. (Eds.) (1997). *Selectivity and Optimization in Capillary Electrophoresis* (Elsevier, Amsterdam).

97. El-Rassi, Z. (2002). *CE and CEC Reviews: Advances in the Practice and Application of Capillary Electrophoresis and Capillary Electrochromatography* (John Wiley and Sons, New York).

98. Ali, I., Gupta, V.K., and Aboul-Enein, H.Y. (2003). Chiral resolution of some environmental pollutants by chirality electrophoresis. *Electrophoresis*, **24**, 1360–1374.

99. Krull, I.S., Stevenson, R.L., Mistry, K., and Swartz, M.E. (2000). *Capillary Electrochromaography and Pressurized Flow Capillary Electrochromatography* (HNB Publishing, New York).

100. Timerbaev, A.R., and Buchberger, W. (1999). Prospects for detection and sensitivity enhancement of inorganic ions in capillary electrophoresis. *J. Chromatogr. A*, **834**, 117–132.

101. Horvath, J., and Dolnike, V. (2001). Polymer wall coatings for capillary electrophoresis. *Electrophoresis*, **22**, 644–655.

102. Mayer, B.X. (2001). How to increase precision in capillary electrophoresis. *J. Chromatogr. A*, **907**, 21–37.

103. Olivers, J.A., Nguyen, N.T., Yonker, C.R., and Smith, R.D. (1987). Online mass spectrometric detection for capillary zone electrophoresis. *Anal. Chem.*, **59**, 1230–1232.

104. Tsuda, T., Kobayashi, Y., Hori, A., Matsumoto, T., and Suzuki, O. (1988). Post-column detection for capillary zone electrophoresis. *J. Chromatogr.*, **456**, 375–381.

105. Rose, D.J., and Jorgenson, J.W. (1988). Post-capillary fluorescence detection in capillary zone electrophoresis using o-phthaldialdehyde. *J. Chromatogr.*, **447**, 117–131.

106. Tsuda, T., Kobayashi, Y., Hori, A., Matsumoto, T., and Suzuki, O. (1990). Separation of polyamines in rat tissues by capillary electrophoresis. *J. Microcol. Sepn*, **2**, 21–25.

107. An, D., Chen, Z., Zheng, J., Chen, S., Wang, L., Huang, Z., and Weng, L. (2015). Determination of biogenic amines in oysters by capillary electrophoresis coupled with electrochemiluminescence. *Food Chem.*, **168**, 1–6.

108. Elbashir, A.A., Krieger, S., and Schmitz, O.J. (2014). Simultaneous determination of polyamines and acetylpolyamines in human urine by capillary electrophoresis with fluorescence detection. *Electrophoresis*, **35**, 570–576.

109. Madej, K., Woźniakiewicz, M., and Karabinowska, K. (2012). Capillary electrophoresis screening method for six tricyclic antidepressants in human serum. *Acta. Pol. Pharm.*, **69**, 1023–1029.

110. Li, M.D., Tseng, W.L., and Cheng, T.L. (2009). Ultrasensitive detection of indoleamines by combination of nanoparticle-based extraction with capillary electrophoresis/laser-induced native fluorescence. *J. Chromatogr. A*, **1216**, 6451–6458.

111. Foret, F., Fanali, S., Ossicini, L., and Bocek, P. (1989). Indirect photometric detection in capillary zone electrophoresis. *J. Chromatogr.*, **470**, 299–308.

112. Hjertén, S., Valtcheva, L., Elenbring, K., and Eaker, D. (1989). High-performance electrophoresis of acidic and basic low-molecular-weight compounds and of proteins in the presence of polymers and neutral surfactants. *J. Liq. Chromatogr.*, **12**, 2471–2499.

113. Bocek, P., Deml, M., and Pospichal, J. (1990). New option in capillary zone electrophoresis: Use of a transient ionic matrix (dynamic pulse). *J. Chromatogr.*, **500**, 673–680.

114. Lee, E.D., Mück, W., Henion, J.D., and Covey, T.R. (1989). Liquid junction coupling for capillary zone electrophoresis/ion spray mass spectrometry. *Biomed. Environ. Mass Spectrom.*, **18**, 844–850.

115. Li, B., Yongku, L., Wang, X., Wang, F., Wang, X., Wang, Y., and Meng, X. (2015). Simultaneous separation and determination of organic acids in blueberry juices by capillary electrophoresis-electrospray ionization mass spectrometry. *J. Food Sci. Technol.*, **52**, 5228–5235.

116. Gao, R., Wang, L., Yang, Y., Ni, J., Zhao, L., Dong, S., and Guo, M. (2015). Simultaneous determination of oleanolic acid, ursolic acid, quercetin and apigenin in Swertia mussotii Franch by capillary zone electrophoresis with running buffer modifier. *Biomed. Chromatogr.*, **29**, 402–409.

117. Sotgia, S., Carru, C., Scanu, B., Pisanu, E., Sanna, M., Pinna, G.A., Gaspa, L., Deiana, L., and Zinellu, A. (2009). Improved rapid assay of plasma

uric acid by short-end injection capillary zone electrophoresis. *Anal. Bioanal. Chem.*, **395**, 2577.

118. Honda, S., Iwase, S., Makino, A., and Fujiwara, S. (1989). Simultaneous determination of reducing monosaccharides by capillary zone electrophoresis as the borate complexes of N-2-pyridylglycamines. *Anal. Biochem.*, **176**, 72–77.

119. Liu, J., Shirota, O., Wiesler, D., and Novotny, M. (1991). Ultrasensitive fluorometric detection of carbohydrates as derivatives in mixtures separated by capillary electrophoresis. *Proc. Natl. Acad. Sci., USA*, **88**, 2302–2306.

120. Guan, Y., and Zhou, G. (2016). Ultrasensitive analysis of glucose in serum by capillary electrophoresis with LIF detection in combination with signal amplification strategies and on-column enzymatic assay, *Electrophoresis*, **37**, In Press. doi: 10.1002/elps.201500395.

121. Yakuba, Y.F., and Markovsky, M.G. (2015). The determination of glucose, sucrose and fructose by the method of capillary electrophoresis. *Vopr. Pitan.*, **84**, 89–94.

122. Monti, L., Cattaneo, T.M., Orlandi, M., and Curadi, M.C. (2015). Capillary electrophoresis of sialylated oligosaccharides in milk from different species. *J. Chromatogr. A*, **1409**, 288–291.

123. Yu, M., and Dovichi, N.J. (1989). Attomole amino acid determination by capillary zone electrophoresis with thermooptical absorbance detection. *Anal. Chem.*, **61**, 37–40.

124. Wallingford, R.A., and Ewing, A.G. (1987). Characterization of a microinjector for capillary zone electrophoresis. *Anal. Chem.*, **59**, 678–681.

125. Kaniansky, D., and Marak, J. (1990). On-line coupling of capillary isotachophoresis with capillary zone electrophoresis. *J. Chromatogr.*, **498**, 191–204.

126. Nickerson, B., and Jorgenson, J.W. (1988). High-speed capillary zone electrophoresis with laser-induced fluorescence detection. *J. High Resolut. Chromatogr.*, **11**, 533–534.

127. Sweedler, J.V., Shear, J.B., Fishman, H.A., Zare, R.N., and Scheller, R.H. (1991). Fluorescence detection in capillary zone electrophoresis using a charge-coupled device with time-delayed integration. *Anal. Chem.*, **63**, 496–502.

128. Pentoney, S.L., Huang, X., Burgi, D.S., Zare, R.N. (1988). On-line connector for microcolumns: application to the on-column o-phthaldialdehyde derivatization of amino acids separated by capillary zone electrophoresis. *Anal. Chem.*, **60**, 2625–2629.

129. Jorgenson, J.W., and Lukacs, K.D. (1981). Zone electrophoresis in open-tubular glass capillaries: Preliminary data on performance. *J. High Resolut. Chromatogr.*, **4**, 230–231.

130. Forteschi, M., Sotgia, S., Assaretti, S., Arru, D., Cambedda, D., Sotgiu, E., Zinellu, A., and Carru, C. (2015). Simultaneous determination of aromatic amino acids in human blood plasma by capillary electrophoresis with UV-absorption detection. *J. Sep. Sci*, **38**, 1794–1799.

131. Lu, Y., Wang, D., Kong, C., Zhong, H., and Breadmore, M.C. (2014). Analysis of aromatic acids by nonaqueous capillary electrophoresis with ionic-liquid electrolytes. *Electrophoresis*, **35**, 3310–3316.

132. Enzonga, J., Ong-Meang, V., Couderc, F., Boutonnet, A., Poinsot, V., Tsieri, M.M., Silou, T., and Bouajila, J. (2013). Determination of free amino acids in African gourd seed milks by capillary electrophoresis with light-emitting diode induced fluorescence and laser-induced fluorescence detection. *Electrophoresis,* **34**, 2632–2638.

133. Bouri, M., Salghi, R., Zougagh, M., and Ríos, A. (2013). Capillary electrophoresis coupled to evaporative light scattering detection for direct determination of underivatized amino acids: Application to tea samples using carboxyled single-walled carbon nanotubes for sample preparation. *Electrophoresis,* **34**, 2623–2631.

134. Jeong, J.S., Kim, S.K., and Park, S.R. (2013). Amino acid analysis of dried blood spots for diagnosis of phenylketonuria using capillary electrophoresis-mass spectrometry equipped with a sheathless electrospray ionization interface. *Anal. Bioanal. Chem.*, **405**, 8063–8072.

135. Zunić, G.D., Spasić, S., and Jelić-Ivanović, Z. (2012). Capillary electrophoresis of free amino acids in physiological fluids without derivatization employing direct or indirect absorbance detection. *Methods Mol. Biol.*, **828**, 243–254.

136. Sun, H., Li, L., and Su, M. (2010). Simultaneous determination of proline and pipemidic acid in human urine by capillary electrophoresis with electrochemiluminescence detection. *J. Clin. Lab. Anal.*, **24**, 327–333.

137. McCormick, R.M. (1988). Capillary zone electrophoretic separation of peptides and proteins using low pH buffers in modified silica capillaries. *Anal. Chem.*, **60**, 2322–2328.

138. Moseley, M.A., Deterding, L.J., Tomer, K.B., and Jorgenson, J.W. (1989). Capillary-zone electrophoresis/fast-atom bombardment mass spectrometry: Design of an on-line coaxial continuous-flow interface. *Rapid Commun. Mass Spectrom.*, **3**, 87–93.

58 | *Capillary Electrophoresis*

139. Grossman, P.D., Wilson, K.J., Petrie, G., and Lauer, H.H. (1988). Effect of buffer pH and peptide composition on the selectivity of peptide separations by capillary zone electrophoresis. *Anal. Biochem.*, **173**, 265–270.

140. Grossman, P.D., Colburn, J.C., Lauer, H.H., Nielsen, R.G., Riggin, R.M., Sittampalam, G.S., and Rickard, E.C. (1989). Application of free-solution capillary electrophoresis to the analytical scale separation of proteins and peptides. *Anal. Chem.*, **61**, 1186–1194.

141. Bushey, M.M., and Jorgenson, J.W. (1990). Automated instrumentation for comprehensive two-dimensional high-performance liquid chromatography/capillary zone electrophoresis. *Anal. Chem.*, **62**, 978–984.

142. Tehrani, J., Macomber, R., and Day, L. (1991). Capillary electrophoresis: An integrated system with a unique split-flow sample introduction mechanism. *J. High Resolut. Chromatogr.*, **14**, 10–14.

143. Jorgenson, J.W. (1984). Zone electrophoresis in open-tubular capillaries. *Trends Anal. Chem.*, **3**, 51–54.

144. Bruin, G.J.M., Chang, J.P., Kuhlman, R.H., Zegers, K., Kraak, J.C., and Poppe, H. (1989). Capillary zone electrophoretic separations of proteins in polyethylene glycol-modified capillaries. *J. Chromatogr.*, **471**, 429–436.

145. Gordon, M.J., Lee, K.J., Arias, A.A., and Zare, R.N. (1991). Protocol for resolving protein mixtures in capillary zone electrophoresis. *Anal. Chem*, **63**, 69–72.

146. Frenz, J., Wu, S.L., and Hancock, W.S. (1989). Characterization of human growth hormone by capillary electrophoresis. *J. Chromatogr.*, **480**, 379–391.

147. Cohen, A.S., and Karger, B.L. (1987). High-performance sodium dodecyl sulfate polyacrylamide gel capillary electrophoresis of peptides and proteins. *J. Chromatogr*, **397**, 409–417.

148. Hjertén, S. (1983). High-performance electrophoresis: The electrophoretic counterpart of high-performance liquid chromatography. *J. Chromatogr. Rev.*, **270**, 1–6.

149. Tian, Y., Li, Y., Mei, J., Cai, B., Dong, J., Shi, Z., and Xiao, Y. (2015). Simultaneous separation of acidic and basic proteins using gemini pyrrolidinium surfactants and hexafluoroisopropanol as dynamic coating additives in capillary electrophoresis. *J. Chromatogr. A*, **1412**, 151–158.

150. Hjertén, S., and Hirai, H. (Eds.) (1984). *Electrophoresis* (Walter de Gruyter, Berlin).

151. Guttman, A., Cohen, A.S., Heiger, D.N., and Karger, B.L. (1990). Analytical and micropreparative ultrahigh resolution of oligonucleotides by polyacrylamide gel high-performance capillary electrophoresis. *Anal. Chem.*, **62**, 137–141.

152. Kasper, T.J., Melera, M., Gozel, P., and Brownlee, R.G. (1988). Separation and detection of DNA by capillary electrophoresis. *J. Chromatogr.*, **458**, 303–312.

153. Zhu, M., Hansen, D.L., Burd, S., and Gannon, F. (1989). Factors affecting free zone electrophoresis and isoelectric focusing in capillary electrophoresis. *J. Chromatogr.*, **480**, 311–319.

154. Li, Z., Liu, C., Zhang, D., Luo, S., and Yamaguchi, Y. (2016). Capillary electrophoresis of RNA in hydroxyethylcellulose polymer with various molecular weights. *J. Chromatogr. B Anal. Technol. Biomed., Life Sci.*, **1011**, 114–120.

155. Khan, N., Mironov, G., and Berezovski, M.V. (2016). Direct detection of endogenous MicroRNAs and their post-transcriptional modifications in cancer serum bycapillary electrophoresis-mass spectrometry, *Anal. Bioanal. Chem.*, In Press. doi:10.1007/s00216-015-9277-y

156. Zhu, P., Wang, S., Wang, J., Zhou, L., and Shi, P. (2016). A capillary zone electrophoresis method for adenine nucleotides analysis in Saccharomyces cerevisiae. *J. Chromatogr. B Anal. Technol. Biomed. Life Sci.*, In Press. 10.1016/j.jchromb.2015.11.040

157. Mateos-Vivas, M., Rodríguez-Gonzalo, E., Domínguez-Álvarez, J., García-Gómez, D., Ramírez-Bernabé, R., and Carabias-Martínez, R. (2015). Analysis of free nucleotide monophosphates in human milk and effect of pasteurisation or high-pressure processing on their contents by capillary electrophoresis coupled to mass spectrometry. *Food Chem.*, **174**, 348–355.

158. Chen, X., Tang, Y., Wang, S., Song, Y., Tang, F., and Wu, X. (2015). Field-amplified sample injection in capillary electrophoresis with amperometric detection for the ultratrace analysis of diastereomeric ephedrine alkaloids. *Electrophoresis*, **36**, 1953–1961.

159. Tascon, M., Benavente, F., Sanz-Nebot, V.M., and Gagliardi, L.G. (2015). Fast determination of harmala alkaloids in edible algae by capillary electrophoresis mass spectrometry. *Anal. Bioanal. Chem*, **407**, 3637–3645.

160. Felici, E., Wang, C.C., Fernández, L.P., and Gomez, M.R. (2015). Simultaneous separation of ergot alkaloids by capillary electrophoresis after cloud point extraction from cereal samples. *Electrophoresis*, **36**, 341–347.

161. Hou, J., Li, G., Wei, Y., Lu, H., Jiang, C., Zhou, X., Meng, F., Cao, J., and Liu, J. (2014). Analysis of five alkaloids using surfactant-coated multi-walled carbon nanotubes as the pseudostationary phase in nonaqueous capillary electrophoresis. *J. Chromatogr. A*, **1343**, 174–181.

162. Shi, G., Li, J., Yin, Y., Xu, X., and Chen, G. (2013). Determination of alkaloids in Sinomenium acutum by field-amplified sample stacking in capillary electrophoresis with chemiluminescene detection. *Luminescence*, **28**, 468–473.

163. Chen, Q., Zhang, J., Zhang, W., and Chen, Z. (2013). Analysis of active alkaloids in the Menispermaceae family by nonaqueous capillary electrophoresis-ion trap mass spectrometry. *J. Sep. Sci*, **36**, 341–349.

164. Yu, Z., Wu, Z., Gong, F., Wong, R., Liang, C., Zhang, Y., and Yu, Y. (2012). Simultaneous determination of six toxic alkaloids in human plasma and urine using capillary zone electrophoresis coupled to time-of-flight mass spectrometry. *J. Sep. Sci.*, **35**, 2773–2780.

165. Liu, W.L., Hsu, Y.F., Liu, Y.W., Singco, B., Chen, S.W., Huang, H.Y., and Chin, T.Y. (2012). Capillary electrophoresis-laser-induced fluorescence detection of rat brain catecholamines with microwave-assisted derivatization. *Electrophoresis*, **33**, 3008–3011.

166. Mikuš, P., Maráková, K., Veizerová, L., and Piešt'anský, J. (2011). Determination of quinine in beverages by online coupling capillary isotachophoresis to capillary zone electrophoresis with UV spectrophotometric detection. *J. Sep. Sci.*, **34**, 3392–3398.

167. Song, J.Z., Han, Q.B., Qiao, C.F., But, P.P., and Xu, H.X. (2010). Development and validation of a rapid capillary zone electrophoresis method for the determination of aconite alkaloids in aconite roots. *Phytochem. Anal.*, **21**, 137–143.

168. Lu, Y., Jia, C., Yao, Q., Zhong, H., and Breadmore, M.C. (2013). Analysis of flavonoids by non-aqueous capillary electrophoresis with 1-ethyl-3-methylimidazolium ionic-liquids as background electrolytes. *J. Chromatogr. A*, **1319**, 160–165.

169. Nehmé, R., Nehmé, H., Roux, G., Destandau, E., Claude, B., and Morin, P. (2013). Capillary electrophoresis as a novel technique for screening natural flavonoids as kinase inhibitors. *J. Chromatogr. A*, **1318**, 257–264.

170. Wainright, A. (1990). Capillary electrophoresis applied to the analysis of pharmaceutical compounds. *J. Microcol. Sepn.*, **2**, 166–175.

171. Lux, J.A., Yin, H.F., and Schomburg, G. (1990). Construction, evaluation and analytical operation of a modular capillary electrophoresis instrument. *Chromatographia*, **30**, 7–15.

172. Debets, A.J.J., Hupe, K.P., Brinkman, U.A.Th., and Kok, W.T. (1990). A new valve for zone-electrophoretic sample treatment coupled on-line with high performance liquid chromatography. *Chromatographia*, **29**, 217–222.

173. Wright, B.W., Ross, G.A., and Smith, R.D.S. (1989). Capillary zone electrophoresis with laser fluorescence detection of marine toxins. *J. Microcol. Sepn.*, **1**, 85–89.

174. Kitihashi, T., and Fruta, I. (2001). Determination of vancomycin in human serum by micellar electrokinetic capillary chromatography with direct sample injection. *Clin. Chem. Acta.*, **312**, 221–225.

175. Fujiwara, S., Iwase, S., and Honda, S. (1988). Analysis of water-soluble vitamins by micellar electrokinetic capillary chromatography. *J. Chromatogr.*, **447**, 133–140.

176. Maguregui, M.I., Jimenez, R.M., Alonso, R.M., and Akesolo, U. (2002). Quantitative determination of oxprenolol and timolol in urine by capillary zone electrophoresis. *J. Chromatogr. A*, **949**, 91–97.

177. Nishiwaki, F., Kuroda, K., Inoue, Y., and Endo, G. (2000). Determination of histamine, 1-methylhistamine and N-methylhistamine by capillary electrophoresis with micelles. *Biomed. Chromatogr*, **14**, 184–187.

178. Wey, A.B., and Thorman, W. (2002). Capillary electrophoresis and capillary electrophoresis–ion trap multiple-stage mass spectrometry for the differentiation and identification of oxycodone and its major metabolites in human urine. *J. Chromatogr. B*, **770**, 191–205.

179. Bernal, J.L., del Nozal, M.J., Martin, M.T., Diez-Masa, J.C., and Cifuentes, A. (1998). Quantitation of active ingredients and excipients in nasal sprays by high-performance liquid chromatography, capillary electrophoresis and UV spectroscopy. *J. Chromatogr. A*, **823**, 423–431.

180. Doomkaew, A., Prapatpong, P., Buranphalin, S., van der Heyden, Y., and Suntornsuk, L. (2015). Fast and simultaneous analysis of combined anti-diabetic drugs by capillary zone electrophoresis. *J. Chromatogr. Sci.*, **53**, 993–999.

181. García-Vázquez, A., Borrull, F., Calull, M., and Aguilar, C. (2016). Single-drop microextraction combined in-line with capillary electrophoresis

for the determination of nonsteroidal anti-inflammatory drugs in urine samples. *Electrophoresis*, **37**, 274–281.

182. Alshana, U., Göğer, N.G., and Ertaş, N. (2013). Dispersive liquid–liquid microextraction combined with field-amplified sample stacking in capillary electrophoresis for the determination of non-steroidal anti-inflammatory drugs in milk and dairy products. *Food Chem.*, **138**, 890–897.

183. Vogel, H., Ehrat, M., and Bruin, G.J.M. (2002). Immunoaffinity screening with capillary electrochromatography. *Electrophoresis*, **23**, 1255–1262.

184. Ebeid, W., Salim, M., Elkady, E., Elzahr, A., El-Bagary, R., and Patonay, G. (2015). Simultaneous determination of valsartan, amlodipine besylate and hydrochlorothiazide using capillary zone electrophoresis (CZE). *Pharmazie*, **70**, 368–373.

185. Zuo, M., Gao, J., Zhang, X., Cui, Y., Fan, Z., and Ding, M. (2015). Capillary electrophoresis with electrochemiluminescence detection for the simultaneous determination of cisatracurium besylate and its degradation products in pharmaceutical preparations, *J. Sep. Sci.*, **38**, 2332–2339.

186. Dal, A.G., Oktayer, Z., and Doğrukol-Ak, D. (2014). Validated method for the determination of piroxicam by capillary zone electrophoresis and its application to tablets. *J. Anal. Methods Chem.*, **2014**, 1–7.

187. Sangoi, M.S., Todeschini, V., and Steppe, M. (2013). Determination of fesoterodine in a pharmaceutical preparation by a stability-indicating capillary zone electrophoresis method. *J. AOAC Int.*, **96**, 1308–1314.

188. Korany, M.A., Maher, H.M., Galal, S.M., and Ragab, M.A. (2013). Development and optimization of a capillary zone electrophoresis technique for simultaneous determination of miconazole nitrate and hydrocortisone acetate in a cream pharmaceutical formulation. *J. AOAC Int.*, **96**, 1295–1301.

189. Doomkaew, A., Prutthiwanasan, B., and Suntornsuk, L. (2014). Simultaneous analysis of metformin and cyanoguanidine by capillary zone electrophoresis and its application in a stability study. *J. Sep. Sci.*, **37**, 1687–1693.

190. Nevado, J.J., Peñalvo, G.C., Dorado, R.M., and Robledo, V.R. (2014). Simultaneous determination of omeprazole and their main metabolites in human urine samples by capillary electrophoresis using electrospray ionization-mass spectrometry detection. *J. Pharm. Biomed. Anal.*, **92**, 211–219.

191. Hancu, G., Campian, C., Rusu, A., Mircia, E., Kelemen, H. (2014). Chiral separation of indapamide enantiomers by capillary electrophoresis, *Adv. Pharm. Bull.*, **4**, 161–172.

192. Akamatsu, S., and Mitsuhashi, T. (2014). Simultaneous determination of pharmaceutical components in dietary supplements for weight loss by capillary electrophoresis tandem mass spectrometry. *Drug Test. Anal.*, **6**, 426–433.

193. Phan, N.T., Hanrieder, J., Berglund, E.C., and Ewing, A.G. (2013). Capillary electrophoresis–mass spectrometry-based detection of drugs and neurotransmitters in drosophila brain. *Anal. Chem.*, **85**, 8448–8454.

194. Siroká, J., Polesel, D.N., Costa, J.L., Lanaro, R., Tavares, M.F., and Polášek, M. (2013). Separation and determination of chlorophenylpiperazine isomers in confiscated pills by capillary electrophoresis. *J. Pharm. Biomed. Anal.*, **84**, 140–147.

195. Stojkovic, M., Mai, T.D., and Hauser, P.C. (2013). Determination of artificial sweeteners by capillary electrophoresis with contactless conductivity detection optimized by hydrodynamic pumping. *Anal. Chim. Acta.*, **787**, 254–259.

196. Buglione, L., See, H.H., and Hauser, P.C. (2013). Rapid separation of fatty acids using a poly(vinyl alcohol) coated capillary in nonaqueous capillary electrophoresis with contactless conductivity detection. *Electrophoresis*, **34**, 2072–2077.

197. da Silva, I.S., Vidal, D.T., do Lago, C.L., and Angnes, L. (2013). Fast simultaneous determination of trimethoprim and sulfamethoxazole by capillary zone electrophoresis with capacitively coupled contactless conductivity detection. *J. Sep. Sci.*, **36**, 1405–1409.

198. Li, X., Yang, Y., and Zhou, K. (2012). Simultaneous determination of chlorpromazine and promethazine and their main metabolites by capillary electrophoresis with electrochemiluminescence. *Se Pu.*, **30**, 938–942.

199. Salim, M., El-Enany, N., Belal, F., Walash, M., and Patonay, G. (2012). Simultaneous determination of sitagliptin and metformin in pharmaceutical preparations by capillary zone electrophoresis and its application to human plasma analysis. *Anal. Chem. Insights,* **7**, 31–46.

200. Zhou, M., Li, Y., Liu, C., Ma, Y., Mi, J., and Wang, S. (2012). Simultaneous determination of lappaconitine hydrobromide and isopropiram fumarate in rabbit plasma by capillary electrophoresis with electrochemiluminescence detection. *Electrophoresis,* **33**, 2577–2583.

201. Xu, L., Guo, Q., Yu, H., Huang, J., and You, T. (2012). Simultaneous determination of three β-blockers at a carbon nanofiber paste electrode by capillary electrophoresis coupled with amperometric detection. *Talanta,* **97**, 462–467.

202. Alnajjar, A.O., Idris, A.M., Attimarad, M.V., Aldughaish, A.M., and Elgorashe, R.E. (2013). Capillary electrophoresis assay method for metoprolol and hydrochlorothiazide in their combined dosage form with multivariate optimization. *J. Chromatogr. Sci.,* **51**, 92–97.

203. Tonon, M.A., and Bonato, P.S. (2012). Capillary electrophoretic enantioselective determination of zopiclone and its impurities. *Electrophoresis,* **33**, 1606–1612.

204. Liao, H.W., Lin, S.W., Wu, U.I., and Kuo, C.H. (2012). Rapid and sensitive determination of posaconazole in patient plasma by capillary electrophoresis with field-amplified sample stacking. *J. Chromatogr. A,* **1226**, 48–54.

205. Li, X.Q., Uboh, C.E., Soma, L.R., Guan, F.Y., You, Y.W., Kahler, M.C., Judy, J.A., Liu, Y., and Chen, J.W. (2010). Simultaneous separation and confirmation of amphetamine and related drugs in equine plasma by non-aqueous capillary-electrophoresis-tandem mass spectrometry. *Drug Test. Anal.,* **2**, 70–81.

206. Zhang, L.Y., and Sun, M.X. (2009). Fast determination of glutathione by capillary electrophoresis with fluorescence detection using β-cyclodextrin as modifier. *J. Chromatogr. B Analyt. Technol. Biomed. Life Sci.,* **877**, 4051–4054.

207. Gao, Y., Xu, Y., Han, B., Li, J., and Xiang, Q. (2009). Sensitive determination of verticine and verticinone in Bulbus Fritillariae by ionic liquid assisted capillary electrophoresis–electrochemiluminescence system. *Talanta,* **80**, 448–453.

208. Liu, Y.M., Jia, Y.X., and Tian, W. (2008). Determination of quinolone antibiotics in urine by capillary electrophoresis with chemiluminescence detection. *J. Sep. Sci,* **31**, 3765–3771.

209. Liu, Q., Liu, Y., Guo, M., Luo, X., and Yao, S. (2006). A simple and sensitive method of nonaqueous capillary electrophoresis with laser-induced native fluorescence detection for the analysis of chelerythrine and sanguinarine in Chinese herbal medicines. *Talanta,* **70**, 202–207.

210. Liu, S., Liu, Y., Li, J., Guo, M., Pan, W., and Yao, S. (2006). Determination of mefenacet by capillary electrophoresis with electrochemilumines-cence detection. *Talanta,* **69**, 154–159.

211. Flores, J.R., Nevado, J.J., Salcedo, A.M., and Díaz, M.P. (2005). Nonaqueous capillary electrophoresis method for the analysis of tamoxifen, imipramine and their main metabolites in urine. *Talanta,* **65**, 155–162.

212. Jastrzebska, A. (2006). Determination of sodium tripolyphosphate in meat samples by capillary zone electrophoresis with on-line isotachophoretic sample pre-treatment. *Talanta,* **69**, 1018–1024.

213. Ji, X., He, Z., Ai, X., Yang, H., and Xu, C. (2006). Determination of clenbuterol by capillary electrophoresis immunoassay with chemiluminescence detection. *Talanta,* **70**, 353–357.

214. Leroy, P., Belluci, L., and Nicolas, A. (1995). Chiral derivatization for separation of racemic amino and thiol drugs by liquid chromatography and capillary electrophoresis. *Chirality,* **7**, 235–242.

215. Yoshinaga, M., and Tanaka, M. (1994). Use of selectively methylated β-cyclodextrin derivatives in chiral separation of dansylamino acids by capillary zone electrophoresis. *J. Chromatogr. A,* **679**, 359–365.

216. Sánchez-Hernández, L., Serra, N.S., Marina, M.L., and Crego, A.L. (2013). Enantiomeric separation of free l- and d-amino acids in hydrolyzed protein fertilizers by capillary electrophoresis tandem mass spectrometry. *J. Agric. Food Chem.,* **61**, 5022–5030.

217. Liu, K., and Wang, L. (2013). Enantioseparations of amino acids by capillary array electrophoresis with 532 nm laser induced fluorescence detection. *J. Chromatogr. A,* **1295**, 142–146.

218. Schmitt, T., and Engelhardt, H. (1995). Optimization of enantiomeric separations in capillary electrophoresis by reversal of the migration order and using different derivatized cyclodextrins. *J. Chromatogr. A,* **697**, 561–570.

219. Egashira, N., Mutoh, O., Kurauchi, Y., and Ogla, K. (1996). Notes chiral separation of α-amino acid derivatives by capillary electrophoresis using 6-amino-6-deoxy-β-cyclodextrin and its n-hexyl derivative as chiral selectors. *Anal. Sci.,* **12**, 503–505.

220. Grasper, M.P., Berthod, A., Nair, U.B., and Armstrong, D.W. (1996). Comparison and modeling study of vancomycin, ristocetin A, and teicoplanin for CE enantioseparations. *Anal. Chem.,* **68**, 2501–2514.

221. Wan, H., Engstrom, A., and Blomberg, L.G. (1996). Direct chiral separation of amino acids derivatized with 2-(9-anthryl)ethyl chloroformate by capillary electrophoresis using cyclodextrins as chiral selectors. Effect of organic modifiers on resolution and enantiomeric elution order. *J. Chromatogr. A,* **731**, 283–292.

222. Wren, S.A.C. (1993). Theory of chiral separation in capillary electrophoresis. *J. Chromatogr. A*, **636**, 57–62.

223. Lelievre, F., and Gariel, P. (1996). Chiral separations of underivatized arylpropionic acids by capillary zone electrophoresis with various cyclodextrins. Acidity and inclusion constant determinations. *J. Chromatogr. A*, **735**, 311–320.

224. Schmidt, M.G., and Gübitz, G. (1995). Capillary zone electrophoretic separation of the enantiomers of dipeptides based on host–guest complexation with a chiral crown ether. *J. Chromatogr. A*, **709**, 81–88.

225. Chankvetadze, B., Endresz, G., and Blaschke, G. (1995). Enantiomeric resolution of chiral imidazole derivatives using capillary electrophoresis with cyclodextrin-type buffer modifiers. *J. Chromatogr. A*, **700**, 43–49.

226. Swartz, M.E. (1991). Method development and selectivity control for small molecule pharmaceutical separations by capillary electrophoresis. *J. Liq. Chromatogr.*, **14**, 923–938.

227. Ahmed, A., Ibrahim, H., Pastore, F., and Lloyd, D.K. (1996). Relationship between retention and effective selector concentration in affinity capillary electrophoresis and high-performance liquid chromatography. *Anal. Chem.*, **68**, 3270–3273.

228. Quang, C., and Khaledi, M. (1995). Extending the scope of chiral separation of basic compounds by cyclodextrin-mediated capillary zone electrophoresis, *J. Chromatogr. A*, **692**, 253–265.

229. Nishi, H., Nakamura, K., Nakai, H., and Sato, T. (1995). Enantiomeric separation of drugs by mucopolysaccharide-mediated electrokinetic chromatography. *Anal. Chem.*, **67**, 2334–2341.

230. Ali, I., and Aboul-Enein, H.Y. (2003). Optimization of the chiral resolution of baclofen by capillary electrophoresis using β-cyclodextrin as the chiral selector. *Electrophoresis*, **24**, 2064–2069.

231. Aumatell, A., Wells, R.J., and Wong, D.K.Y. (1994). Enantiomeric differentiation of a wide range of pharmacologically active substances by capillary electrophoresis using modified β-cyclodextrins. *J. Chromatogr. A*, **686**, 293–307.

232. Hulst, A. D., and Uerbeke, N. (1992). Chiral separation by capillary electrophoresis with oligosaccharides. *J. Chromatogr.*, **608**, 275–287.

233. Theurillat, R., Sandbaumhüter, F. A., Wolfensberger, R. B., and Thormann, W. (2016). Microassay for ketamine and metabolites in plasma and serum based on enantioselective capillary electrophoresis with highly sulfated γ-cyclodextrin and electrokinetic analyte injection. *Electrophoresis*, **37**, 1129–1138. doi: 10.1002/elps.201500468

234. Znaleziona, J., Fejős, I., Ševčík, J., Douša, M., Béni, S., and Maier, V. (2015). Enantiomeric separation of tapentadol by capillary electrophoresis: Study of chiral selectivity manipulation by various types of cyclodextrins. *J. Pharm. Biomed. Anal.*, **105**, 10–16.

235. Wan Ibrahim, W.A., Arsad, S.R., Maarof, H., Sanagi, M.M., and Aboul-Enein, H.Y. (2015). Chiral separation of four stereoisomers of ketoconazole drugs using capillary electrophoresis. *Chirality*, **27**, 223–227.

236. Mu, X., Qi, L., Qiao, J., Yang, X., and Ma, H. (2014). Enantioseparation of dansyl amino acids and dipeptides by chiral ligand exchange capillary electrophoresis based on Zn(II)-L-hydroxyproline complexes coordinating with γ-cyclodextrins, *Anal. Chim. Acta.*, **10**, 68–74.

237. Lebedeva, M.V., Prokhorova, A.F., Shapovalova, E.N., and Shpigun, O.A. (2014). Clarithromycin as a chiral selector for enantioseparation of basic compounds in nonaqueous capillary electrophoresis. *Electrophoresis,* **35**, 2759–2764.

238. Yu, J., Zhao, Y., Song, J., and Guo, X. (2014). Enantioseparation of meptazinol and its three intermediate enantiomers by capillary electrophoresis using a new cationic β-cyclodextrin derivative in single and dual cyclodextrin systems. *Biomed. Chromatogr.*, **28**, 868–874.

239. Yujiao, W., Guoyan, W., Wenyan, Z., Hongfen, Z., Huanwang, J., and Anjia, C. (2014). Chiral separation of phenylalanine and tryptophan by capillary electrophoresis using a mixture of β-CD and chiral ionic liquid ([TBA] [l-ASP]) as selectors. *Biomed. Chromatogr.*, **28**, 610–614.

240. Nojavan, S., Pourmoslemi, S., Behdad, H., Fakhari, A.R., and Mohammadi, A. (2014). Application of maltodextrin as chiral selector in capillary electrophoresis for quantification of amlodipine enantiomers in commercial tablets. *Chirality*, **26**, 394–399.

241. Tero-Vescan, A., Hancu, G., Oroian, M., and Cârje, A. (2014). Chiral separation of indapamide enantiomers by capillary electrophoresis. *Adv. Pharm. Bull.*, **4**, 267–272.

242. Wen, J., Zhang, W.T., Cao, W.Q., Li, J., Gao, F.Y., Yang, N., and Fan, G.R. (2014). Enantioselective separation of mirtazapine and its metabolites by capillary electrophoresis with acetonitrile field-amplified sample stacking and its application. *Molecules,* **19**, 4907–4923.

243. Rogez-Florent, T., Foulon, C., Six, P., Goossens, L., Danel, C., and Goossens, J.F. (2014). Optimization of the enantioseparation of a diaryl-pyrazole sulfonamide derivative by capillary electrophoresis in a dual CD mode using experimental design. *Electrophoresis*, **35**, 2765–2771.

68 | *Capillary Electrophoresis*

244. Dixit, S., and Park, J.H. (2014). Penicillin G as a novel chiral selector in capillary electrophoresis. *J. Chromatogr. A*, **1326**, 134–138.

245. Yu, J., Jiang, Z., Sun, T., Ji, F., Xu, S., Wei, L., and Guo, X. (2014). Enantiomeric separation of meptazinol and its three intermediate enantiomers by capillary electrophoresis: Quantitative analysis of meptazinol in pharmaceutical formulations. *Biomed. Chromatogr.*, **28**, 135–141.

246. Burrai, L., Nieddu, M., Pirisi, M.A., Carta, A., Briguglio, I., and Boatto, G. (2013). Enantiomeric separation of 13 new amphetamine-like designer drugs by capillary electrophoresis, using modified-B-cyclodextrins. *Chirality*, **25**, 617–621.

247. Lehnert, P., Přibylka, A., Maier, V., Znaleziona, J., Ševčík, J., and Douša, M. (2013). Enantiomeric separation of R,S-tolterodine and R,S-methoxytolterodine with negatively charged cyclodextrins by capillary electrophoresis. *J. Sep. Sci.*, **36**, 1561–1567.

248. Vshivkov, S., Pshenichnov, E., Golubenko, Z., Akhunov, A., Namazov, S., and Stipanovic, R.D. (2012). Capillary electrophoresis to quantitate gossypol enantiomers in cotton flower petals and seed. *J. Chromatogr. B Analyt. Technol. Biomed. Life Sci.*, **908**, 94–97.

249. Zhou, J., Li, Y., Liu, Q., Fu, G., and Zhang, Z. (2013). Capillary electrophoresis of clenbuterol enantiomers and NMR investigation of the clenbuterol/carboxymethyl-β-cyclodextrin complex. *J. Chromatogr. Sci.*, **51**, 237–241.

250. Kodama, S., Taga, A., Aizawa, S.I., Kemmei, T., Honda, Y., Suzuki, K., and Yamamoto, A. (2012). Direct enantioseparation of lipoic acid in dietary supplements by capillary electrophoresis using trimethyl-β-cyclodextrin as a chiral selector. *Electrophoresis*, **33**, 2441–2445.

251. Lomsadze, K., Vega, E.D., Salgado, A., Crego, A.L., Scriba, G.K., Marina, M.L., and Chankvetadze, B. (2012). Separation of enantiomers of norephedrine by capillary electrophoresis using cyclodextrins as chiral selectors: Comparative CE and NMR studies. *Electrophoresis*, **33**, 1637–1647.

252. Mantim, T., Nacapricha, D., Wilairat, P., and Hauser, P.C. (2012). Enantiomeric separation of some common controlled stimulants by capillary electrophoresis with contactless conductivity detection. *Electrophoresis*, **33**, 388–394.

253. Lee, S.C., Wang, C.C., Yang, P.C., and Wu, S.M. (2012). Enantioseparation of (±)-threo-methylphenidate in human plasma by cyclodextrin-

modified sample stackingcapillary electrophoresis. *J. Chromatogr. A,* **1232**, 302–305.

254. Kwan, H.Y., and Thormann, W. (2011). Enantioselective capillary electrophoresis for the assessment of CYP3A4-mediated ketamine demethylation and inhibition in vitro. *Electrophoresis,* **32**, 2738–2745.

255. Lammers, I., Buijs, J., Ariese, F., and Gooijer, C. (2010). Sensitized enantioselective laser-induced phosphorescence detection in chiral capillary electrophoresis. *Anal. Chem.,* **82**, 9410–9417.

256. Wang, M., Ding, X., Chen, H., and Chen, X. (2009). Enantioseparation of palonosetron hydrochloride by capillary zone electrophoresis with high-concentration β-CD as chiral selector. *Anal. Sci.,* **25**, 1217–1220.

257. Mikus, P., Maráková, K., Valásková, I., and Havránek, E. (2009). Enantiomeric purity control of levocetirizine in pharmaceuticals using anionic cyclodextrin mediated capillary electrophoresis separation and fiber-based diode array detection. *Pharmazie,* **64**, 423–427.

258. Choi, K., Kim, J., Jang, Y.O., and Chung, D.S. (2009). Direct chiral analysis of primary amine drugs in human urine by single drop microextraction in-line coupled to CE. *Electrophoresis,* **30**, 2905–2911.

259. Mikus, P., Kubacák, P., Valásková, I., and Havránek, E. (2006). Analysis of enantiomers in biological matrices by charged cyclodextrin-mediated capillary zone electrophoresis in column-coupling arrangement with capillary isotachophoresis. *Talanta,* **70**, 840–846.

Chapter 2

Recent Applications of Chiral Capillary Electrophoresis in Pharmaceutical Analysis

José María Saz and María Luisa Marina

Department of Analytical Chemistry, Physical Chemistry and Chemical Engineering,
Faculty of Biology, Environmental Sciences and Chemistry, University of Alcalá,
Ctra. Madrid-Barcelona, Km. 33.600, 28871 Alcalá de Henares (Madrid), Spain
josem.saz@uah.es

2.1   Introduction

Chirality has a big impact on drug analysis due to the different biological activities that enantiomers may have and the tendency to commercialize drugs as pure enantiomers. This fact makes necessary the development of sensitive and selective analytical methodologies enabling the individual determination of the enantiomers in raw materials and pharmaceutical formulations as well as in biological matrices. Chiral capillary electrophoresis (CCE) is a very powerful tool to develop enantiomeric analytical methodologies. Among the advantages of capillary electrophoresis (CE) for chiral analysis, its

Capillary Electrophoresis: Trends and Developments in Pharmaceutical Research
Edited by Suvardhan Kanchi, Salvador Sagrado, Myalowenkosi Sabela, and Krishna Bisetty
Copyright © 2017 Pan Stanford Publishing Pte. Ltd.
ISBN 978-981-4774-12-3 (Hardcover), 978-1-315-22538-8 (eBook)
www.panstanford.com

high efficiency, versatility, possibility for rapid screening of chiral selectors added in the background electrolyte (BGE), and the low consumption of reagents and samples can be highlighted [1].

Enantiomeric separations in CCE involve both electrophoretic and chromatographic principles owing to the interactions originated between the chiral selector and the analyte, which can give rise to different electrophoretic mobilities for the enantiomer–chiral selector complexes formed in the BGE in the CE mode named electrokinetic chromatography (EKC). A chiral stationary phase immobilized in the capillary is also employed to carry out enantiomeric separations by using capillary electrochromatography (CEC) [1].

This chapter describes the most recent progress and applications developed in the separation of chiral compounds of pharmaceutical interest by CCE during the last few years (from 2010 to 2015) using chiral selectors added to the BGE. With this aim, the effect of the type of chiral selector and concentration on the enantiomeric separation of drugs, the synergistic effects obtained with mixtures of chiral selectors, and the effect of buffer type, concentration, and pH will be discussed. Moreover, the strategies developed in order to improve the detection sensitivity of the developed methodologies and the most recent applications of CCE in the enantiomeric analysis of drugs in pharmaceutical formulations, in studies on drug metabolism, in stability studies of drugs under stress conditions, and in the analysis of biological samples and in forensic investigations will also be presented.

2.2 Chiral Selectors Added to BGE for Enantioseparation of Chiral Drugs by CCE

A great variety of chiral selectors have been employed for the enantiomeric separation of drugs by CCE. Among them, cyclodextrins (CDs), macrocyclic antibiotics, polysaccharides, crown ethers, calixarenes, proteins, surfactants, ligand-exchange complexes, and ionic liquids can be cited. Depending on the type of chiral selectors used, several separation mechanisms can take place, such as inclusion–complexation, micellar solubilization, affinity, ligand-exchange, ion-pairing interactions, etc. CDs are the most employed chiral selectors although antibiotics and polysaccharides have also been frequently used in the last few years to carry out the enantiomeric separation of drugs.

2.2.1 Cyclodextrins

Native CDs are macrocycle composed of α-D-glucopyranoside units linked between positions 1 and 4. The most widely used CDs in CCE separations are α-, β-, and γ-CDs, which contain six, seven, and eight glucose units, respectively, and form a ring with a cone shape. The size of the cone cavity of α-CD is smaller than that of the β-CD, and this, in turn, is smaller than that of the γ-CDs. CDs may form inclusion complexes inside their cone cavity with both enantiomers. The stability of these complexes depends on the size and other molecular properties of the analyte and the CD. Less frequently, external complexation is also possible. There are also a great variety of CD derivatives, and a broad spectrum of uncharged, positively and negatively charged CD derivatives is available. The substituent may interact with the analyte molecule contributing to the enantiorecognition mechanism. CDs also possess high solubility in aqueous solutions and low UV absorbance, which are important characteristics to be used into the BGE in CE. Natural and derivatized CDs were used during the last few years to separate enantiomers of a huge number of chiral drugs by using CCE [2].

The most important parameter to achieve an enantiomeric separation by CCE is the type of the CD used. The selection of an appropriate CD for a chiral separation is performed considering the molecular characteristics of the analyte (size, shape, functional groups, electric charge, etc.) and the CD. In general, the size of the hydrophobic cavity of α-CDs can accommodate, for example, a single phenyl ring, while β-CDs and γ-CDs can include substituted single- and multiple phenyl ring systems. Additionally, the interactions between substituents on the asymmetric center of the analyte and hydroxyl groups of the CD-rim are also responsible for chiral recognition. The presence of substituents at the secondary hydroxyl rim on the surface of the CD provides additional interaction points with the analyte, which can affect the resolution. Many investigations have been developed during the last few years that show the effect of the substituent type and the size of the cavity of the CDs on the enantioseparation of chiral drugs.

The effect of the polarity of neutral substituents of CD derivatives on the enantioseparation of drugs has been studied. In this regard, the enantioseparation capabilities of different neutral CD derivatives with substituents of various polarities (methyl, dimethyl, trimethyl, acetyl, triacetyl, hydroxymethyl, hydroxyethyl, hydroxypropyl, hydroxybutenyl) were compared toward different drug enantiomers such as iodiconazole and structurally related triadimenol analogues [3], duloxetine [4], and valsartan [5]. For example, hydroxyethyl-β-CD, hydroxypropyl-β-CD, dimethyl-β-CD, trimethyl-β-CD, γ-CD, and hydroxypropyl-γ-CD were investigated for the enantioseparation of iodiconazole and structurally related analogues, and it was observed that hydroxypropyl-γ-CD was the best chiral selector. This result was attributed to the cavity size and the hydrogen bonding between the analytes and the CD. Analyte structure-enantioseparation relationship studies showed the importance of the methyl substituent of a nitrogen atom present in the molecular structure of the analyte to the nitrogen can form hydrogen bonds with the CD.

An important factor to take into account to achieve a good enantioseparation in CCE is the CD and the analyte charges. In this regard, the enantioseparation power of CD derivatives with different types of ionizable substituents (carboxymethyl, sulfodiacetyl, sulfopropyl, sulfobutylether, sulfate, phosphate, etc.) was compared in the enantioseparation of different types of chiral drugs such as sibutramine [6], ofloxacin and ornidazole [7], ephedrine [8], bupivacaine [9], clenbuterol, bambuterol, tulobuterol, procaterol and salbutamol [10], hydroxyzine and cetirizine [11], 2-amino-1-phenyl-ethanol, 1-(4-methoxyphenyl)-2-(methylamine) ethanol, salbutamol sulfate and sotalol hydrochloride [12], amphetamine [13], isradipine [14], antimalarial drugs (primaquine, tafenoquine, mefloquine, chloroquine, and quinacrine) [15], cetirizine [16], cathinone derivatives [17], ornidazole [18], tolterodine and methoxytolterodine [19], meptazinol and its synthetic intermediates [20], zopiclone, N-desmethylzopiclone and zopiclone-N-oxide [21]. In all of these investigations, negatively charged CDs were used, and the analytes were neutral or positively charged. So that, the CD migrates to the anode and the analyte to the cathode in opposite directions (due to the EOF and its positive charge if it has). Therefore, the migration time is longer and there are more chances for the interaction between the analyte and the CD, and the enantioresolution improves. Furthermore, the CD and analyte charges may also affect the enantiorecognition mechanism. In general terms, it can be observed in the literature that a larger amount of chiral drugs have been separated with negative CDs than with neutral CDs. For those substituents such as carboxymethyl, which has a weak acid character, the pH of the BGE plays an important role in their ionization. As an example of the enantioseparation power of the negatively charged CDs, Fig. 2.1 shows the simultaneous separation within 12 min of all the enantiomers of zopiclone (ZO) and its metabolites *N*-desmethylzopiclone (*N*-Des-ZO) and zopiclone-*N*-oxide (*N*-Ox-ZO) with carboxymethyl-β-CD [21].

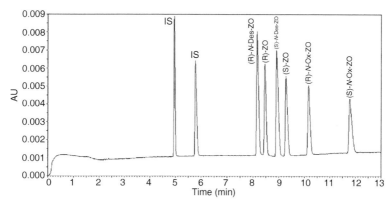

Figure 2.1 Electropherogram for the analysis of ZO, *N*-Des-ZO, and *N*-Ox-ZO. Electrophoretic conditions: background electrolyte: sodium phosphate buffer (pH 2.5; 50 mmol L^{-1}) plus 0.5% (w/v) of carboxymethyl-β-CD; applied voltage, +25 kV; capillary temperature, 20°C; injection pressure of 1 psi for 8 s; uncoated fused silica capillary, 50 cm (42 cm effective length) × 75 μm id; detection at 310 nm. Reprinted from Ref. [21], Copyright 2015, with permission from Elsevier.

Less frequently, neutral CDs give better enantiomeric separation than anionic CDs, depending on the molecular properties of the analyte and the CD. In this regard, for example, stereoselective separation of sotalol enantiomers was observed when hydroxypropyl-β-CD and randomly methylated-β-CD were used, but sulfobutylether-β-CD and carboxymethyl-β-CD conducted to low and no chiral resolution, respectively [22]. The optimum or good separation of carvedilol enantiomers was achieved by using natural β-CD, randomly methylated-β-CD or hydroxypropyl-β-CD. However, the use of sulfobutylether-β-CD led only to partial enantioseparation [23].

In several of the investigations mentioned above, the effect of the cavity size of the CD on the enantioseparation was also demonstrated. There are additional studies in the literature in the last few years about the importance of this parameter for the enantioseparation of chiral drugs. Thus, it was observed that highly sulfated-γ-CD provided the complete separation of all ketamine and norketamine enantiomers, while highly sulfated-β-CD could not separate ketamine enantiomers [24]. Various types of CDs with different cavity size (α-CD, hydroxypropyl-α-CD, β-CD, hydroxypropyl-β-CD, and γ-CD) were investigated as chiral selectors for the separation of trimipramine enantiomers. The good resolution obtained with α-CD was attributed to its small cavity size, which apparently provided better interaction with the trimipramine enantiomers [25]. Also, the effect of the cavity size of various CD derivatives (trimethyl-α-CD, trimethyl-β-CD, and trimethyl-γ-CD) on the separation of ketoprofen enantiomers was studied. Trimethyl-α-CD was required in higher concentrations compared to trimethyl-β-CD in order to provide adequate enantioseparation, and trimethyl-γ-CD appeared to be the best of the three chiral selectors investigated [26]. Four types of neutral CDs (α-CD, hydroxypropyl-α-CD, β-CD, and hydroxypropyl-β-CD) were investigated for the separation of propranolol enantiomers. α-CD and hydroxypropyl-α-CD did not provide a suitable resolution, and this result was attributed to the small size of the α-CD cavity [27]. Various natural β-CD and derivatives (randomly methylated-β-CD, hydroxypropyl-β-CD, and sulfobutylether-β-CD) allowed the chiral resolution of carvedilol enantiomers; however, no chiral separation was observed when using α- or γ-CDs. This result was attributed to the size of the cavities of α- and γ-CDs, which are too small and too large, respectively, to guest the carvedilol molecule [23].

When optimizing a chiral separation, not only the resolution but also the migration time and the signal-to-noise ratio should be considered. In this regard, taking into account both resolution and migration times, hydroxypropyl-β-CD or randomly methylated-β-CD gave better results than α-CD, β-CD, carboxymethyl-β-CD, or sulfobutylether-β-CD in the separation of amlodipine enantiomers [28]. In the case of the enantiomeric separation of valsartan, the best resolution was obtained with methyl-β-CD and acetyl-β-CD, but shorter analysis time and better signal-to-noise ratio were obtained with acetyl-β-CD [5].

In general terms, it can be concluded from the aforementioned investigations that the most employed CDs for the chiral separation of drugs in the last few years were anionic β-CD derivatives, probably because of the size of their cavity fits better with the size of the great majority of drugs, and because of the anionic β-CD derivatives migrate to the anode against the EOF improving the enantioseparation [2]. Furthermore, many drugs are basic compounds positively charged at acidic pHs.

There are cases where a non-satisfactory enantioseparation was observed when using a single CD system, and dual chiral selector systems were employed in order to improve the resolution. A combination of two different CDs can lead to an enhancement of enantioseparation owing to the differences in the complexation mechanisms of the two CDs with both enantiomers [2]. The use of dual CD systems improved the enantioseparation of tetrahydronaphthalenic derivatives [29], meptazinol intermediate III and intermediate IV [30].

The combined use of CDs and ionic liquids (IL) is also of considerable interest to enhance enantiomeric separations. Various types of CDs and ILs were evaluated for the enantiomeric separation of chiral drugs such as pindolol, oxprenolol, propranolol, ofloxacin, dioxopromethazine, isoprenaline, chlorpheniramine, liarozole, tropicamide, amlodipine, brompheniramine, homatropine, zopiclone, repaglinide, promethazine, carvedilol, venlafaxine, sibutramine, naproxen, pranoprofen, warfarin, miconazole, econazole, ketoconazole, and itraconazole [31–35].

The combination of CDs and antibiotics was also investigated. Antibiotics are a class of chiral selectors that are acquiring more attention for their high degree of selectivity. Clarithromycin was used in mixtures with various types of neutral CDs for the enantioseparation of nefopam, metoprolol, atenolol, propranolol, bisoprolol, esmolol, and ritodrine [36].

The use of CDs (β-CD and hydroxypropyl-β-CD) in combination with polysaccharides (glycogen) was also evaluated in the enantiomeric separation of several basic drugs (duloxetine, cetirizine, citalopram, sulconazole, laudanosine, amlodipine, propranolol, atenolol, and nefopam), but with a negative result [37].

An important parameter to optimize an enantiomeric separation is the concentration of the chiral selector. The study of the effect of the concentration of the CD on the enantiomeric separation of drugs showed that in many cases the resolution of enantiomers improved with the concentration of the CD, because when the CD concentration increased a greater number of interactions between the analytes and the CD molecules took place. Thus, an increase in the enantioresolution when increasing the CD concentration was observed for isradipine [14] and bupivacaine [9] with sulfobutyl-ether-β-CD, cetirizine [16] and cathinone derivatives [17] with sulfated-β-CD, ornidazole with sulfated-α-CD [18], and cathinone analogues with β-CD and highly sulfated-γ-CD [38].

Sometimes, increasing the CD concentration increases the resolution but the peak symmetry and the number of theoretical plates can worsen, as in the case of the separation of tolterodine and methoxytolterodine enantiomers with phosphated-γ-CD [19].

Some investigations revealed that when the CD concentration was continuously increased, a maximum value for the enantioresolution was reached, and the optimum CD concentration depended on the binding affinity of the analyte and the CD [2]. This was observed, for example, in the enantioseparation of trimipramine with α-CD [25], clenbuterol, bambuterol, tulobuterol, procaterol and salbutamol with carboxymethyl-β-CD [10], propranolol with hydroxypropyl-β-CD [27], and sotalol with randomly methylated-β-CD [22].

If the CD concentration is raised over the optimum value, a decrease in resolution is usually observed. In this case, all analyte molecules are complexed and the separation depends on the differences of electrophoretic molibities of both enantiomer-CD complexes, but generally the electrophoretic mobility for both complexes is similar since both of them possess the same mass and charge. Moreover, the increase in the BGE viscosity gives rise to longer migration times, leading to peak width broadening, which decreases further the resolution [2]. This behavior was observed, for example, in the chiral resolution of iodiconazole and triadimenol analogues with hydroxypropyl-γ-CD [3], meptazinol [20], vamlodipine [39] and carvedilol [40] with carboxymethyl-β-CD, and valsartan with acetyl-β-CD [5].

Sometimes, when increasing the concentration of the CD over a given value, a change in the migration order of enantiomers may be observed, as in the case of tolterodine and methoxytolterodine enantiomers separated using sulfated-α-CD and sulfated-β-CD [19]. This phenomenon is explained by the fact that at a high concentration of the chiral selector, both enantiomers of the analyte are bonded to the CD to a significant extent, and the migration order is determined by the mobilities of the transient diastereomeric complexes and not by the affinity of analyte enantiomers toward the CD. Therefore, a reversal of the migration order of the enantiomers can occur [2, 41].

Another important parameter to achieve a good enantiomeric separation is the buffer pH. There are various studies that were carried out in the last few years about the effect of the pH on the enantioseparation of drugs with CDs. For example, in the enantioseparation of ofloxacin, ornidazole, and cetirizine with sulfated-β-CD [7, 16], iodiconazole and triadimenol analogues with hydroxypropyl-γ-CD [3], bupivacaine, carvedilol, and amlodipine with sulfobutylether-β-CD [9, 23, 28], ornidazole with sulfated-α-CD [18], meptazinol, zopiclone, N-desmethylzopiclone, zopiclone-N-oxide, and amlodipine with carboxymethyl-β-CD [20, 21, 28, 29], cathinone analogues with β-CD and highly sulfated-γ-CD [38], amlodipine with α-CD, β-CD, randomly methylated-β-CD, and hydroxypropyl-β-CD [28], and valsartan with acetyl-β-CD [5]. The pH affects not only the ionization of the silanol groups of the capillary wall, but also determines the ionization of analytes and CDs that possess ionizable groups, so it may affect the enantiorecognition mechanism and the electrophoretic mobilities depending on the acid–base properties of both the analyte and the CD.

The effect of the type of buffer used in the enantioseparation of chiral drugs was also studied [6, 9, 10, 20, 23]. In general terms, it was shown that in several cases the type and concentration of the buffer used had a positive effect upon resolution, peak shape, and S/N ratio. Moreover, various studies showed the influence of the buffer co-ion on the enantioseparation of chiral drugs with CDs [18, 19]. The co-ion not only has an influence on the current, but also may affect the peak shape, since the electric field in the sample zone and in the pure BGE zone may be different, that is, the electrophoretic velocity of the analyte in both zones may be different, and this phenomenon can produce fronting or tailing peaks.

The addition of organic solvents such as acetonitrile (ACN) or methanol to the BGE may have an important effect on the chiral separations performed by CE. The polarity and viscosity of the BGE, the stability constants of the inclusion complexes and the EOF may be modified. Sometimes, the addition of organic solvents gave rise to a worse enantioseparation as in the case of cathinone derivatives [17] and amphetamine-like drugs [42] with sulfated-β-CD and cathinone analogues with β-CD [38]. In other cases, the enantioresolution did not change appreciably with the addition of organic modifiers, for example, in the enantioseparation of sotalol with randomly methylated-β-CD [22] and carvedilol with β-CD, methyl-β-CD, hydroxypropyl-β-CD, and sulfobutylether-β-CD [23], while in some examples, the enantioresolution did not change notably, but there were other separation parameters that improved as in the case of the separation of enantiomers of meptazinol with carboxymethyl-β-CD [20]. Finally, the enantioresolution may also be improved with the addition of organic modifiers, as it is the case of the enantioseparation of ketoconazole with trimethyl-β-CD [43], naproxen, pranoprofen, and warfarin with a dual chiral selector system of methyl-β-CD and a chiral ionic liquid [34]. Other additives such as organic cations have been used to avoid the adsorption of basic drugs (fluoxetine) onto the negatively charged capillary wall, improving the repeatability of migration time and peak area [44]. Surfactants such as SDS have been used as additives to improve the enantioseparation of ketoconazole. The hydrophobic tail of SDS monomers may be included into the CD cavity, and this phenomenon can change the nature of the solute–CD interaction [43]. Finally, the addition of PEG 4000 into the BGE improved the enantioseparation of various chiral drugs and intermediates [45].

New types of chemically modified capillaries (FunCap-CE/Type D (possessing diol groups), Type A (amino groups), Type C (carboxyl groups), and Type S (sulphate groups)) were developed in the last years and applied to the analysis of methamphetamine drugs by CCE with CDs [46–48]. The aim was to overcome the shift that can be observed in the migration times when using uncoated fused silica capillaries due to the difficulty in controlling the EOF.

Although aqueous buffers are commonly used in CE, the feasibility and potential of non-aqueous CE (NACE) has been demonstrated when using CDs as chiral selectors. Aqueous and non-aqueous media were compared for the analysis of chiral drugs such as propranolol [49], talinolol [50, 51], acebutolol, carazolol, carteolol, carvedilol and sotalol [51]. Also, the use of chiral NACE in the enantiomeric analysis of drugs was reviewed [52]. The main advantages of non-aqueous buffers for CE in pharmaceutical analysis are the lower conductivity, the ability to dissolve hydrophobic analytes, the feasibility of ion-pair formation and the compatibility with MS detection systems. Organic solvents possess lower dielectric constants than water, and intermolecular ionic interactions are promoted. This means that organic solvents constitute, in principle, a favorable environment for a chiral discrimination [2].

CDs have also been applied to chiral drug analysis using microfluidic chip-CE devices. Carboxymethyl-β-CD was used for the enantioseparation of anisodamine, atenolol, and metoprolol in human urine by using a microfluidic chip-CE device, which incorporated on-chip dilution and sensitive electrochemiluminescence detection [53].

2.2.2 Macrocyclic Antibiotics

Macrocyclic antibiotics (ansamycins, glycopeptides, aminoglycosides, polypeptides, and macrolides) are powerful chiral selectors. They exhibit a variety of interaction types with analytes, such as electrostatic, dipole–dipole, hydrophobic interactions and hydrogen bonding and inclusion, due to the diversity in chemical groups and their macrocyclic structure. The most useful structural types for CCE separations are the Ansa family (rifamycin B and SV) and the glycopeptide group (vancomycin, ristocetin, and teicoplanin). Applications of antibiotics as chiral selectors in CCE for the analysis of drugs have been reviewed in some articles [54, 55].

Erythromycin lactobionate is a macrolide-type antibiotic used for the enantioseparation of the basic drugs N,N-dimethyl-3-(2-methoxyphenoxy)-3-propylamine, propranolol, duloxetine, chloroquine, and nefopam. This chiral selector has a 14-membered ring structure with a shape like a basket, which contains several hydroxyl groups, one amino group, and one carboxyl group [56]. Clarithromycin lactobionate is a macrolide antibiotic, which has been investigated for the enantiomeric separation of several basic drugs (metoprolol, atenolol, propranolol, bisoprolol, esmolol, ritodrine, amlodipine, labetalol, and nefopam). It contains in its structure a macrocyclic lactone ring with 14 atoms and numerous functional groups that may contribute to the enantiorecognition through hydrogen bonding, charge–charge, hydrophobic, and steric interactions [57].

Macrocyclic antibiotics have been used in combination with other chiral selectors in dual systems to improve the enantiomeric separation of drugs. Synergistic systems based on the use of mixtures of antibiotics with chiral ionic liquids (IL) have been recently reported. A synergistic effect was observed when vancomycin (a macrocyclic glycopeptide antibiotic) was employed in combination with two chiral ILs (L-alanine and L-valine tert butyl ester bis (trifluoromethane) sulfonamide) for the enantioseparation of naproxen, carprofen, ibuprofen, ketoprofen, and pranoprofen [58]. Moreover, macrocylic antibiotics have also shown to enhance the chiral discrimination power of CDs. As mentioned above, when clarithromycin lactobionate was employed in combination with four different neutral CDs (methyl-β-CD, glucose-β-CD, hydroxyethyl-β-CD, and hydroxypropyl-β-CD), a synergistic effect was observed in all the four dual chiral selector systems in the enantioseparation of nefopam, metoprolol, atenolol, propranolol, bisoprolol, esmolol, and ritodrine [36]. Figure 2.2 illustrates the improvement in the enantiomeric separation of nefopam when using the four dual chiral systems in comparison with the use of the single systems of chiral selectors.

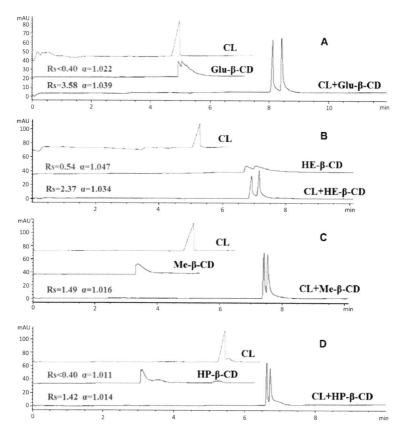

Figure 2.2 The electropherogram of nefopam in four dual chiral systems and single systems. Conditions: (A) clarithromycin lactobionate (CL) concentration, 50 mM; glucose-β-CD concentration, 40 mM; buffer pH, 5.5; (B) CL concentration, 50 mM; hydroxyethyl-β-CD concentration, 20 mM; buffer pH, 4.5; (C) CL concentration, 40 mM; methyl-β-CD concentration, 20 mM; buffer pH, 6.5; (D) CL concentration, 50 mM; hydroxypropyl-β-CD concentration, 15 mM; buffer pH, 7.0; fused silica capillary, 33 cm (24.5 cm effective length) ×50 μm id; BGE, 12.5 mM borax buffer with methanol (50% v/v); applied voltage, 20 kV; capillary temperature, 25°C. Reprinted from Ref. [36], Copyright 2015, with permission from John Wiley and Sons.

NACE was employed for the separation of chiral drugs using macrocyclic antibiotics. Thus, erythromycin lactobionate enabled the enantioseparation of propranolol and duloxetine by NACE. In this work, five organic solvents (methanol, ethanol, ACN, formamide, and N-methylformamide) were tested. Among them, methanol proved to be the most useful for the enantioseparation of propranolol and duloxetine. Since organic solvents differ in physical and chemical properties among themselves, changing the organic solvent or varying the proportions of two solvents in a mixture allows to improve a separation in NACE [59]. Azithromycin is a semi-synthetic macrolide antibiotic derived from erythromycin, which has a 15-membered ring structure having two sugar moieties, several hydroxyl groups, two tertiary amino groups, and one oxycarbonyl group [60]. It was useful in the enantioseparation of various chiral drugs (carvedilol, cetirizine, citalopram hydrobromide, darifenacin, and sertraline hydrochloride) by using NACE with ACN/methanol mixtures as non-aqueous media.

2.2.3 Polysaccharides

Polysaccharides are another important type of chiral selectors used for the separation of the enantiomers of chiral drugs. Among them, maltodextrins (MDs) have good structural characteristics, which make them suitable for chiral recognition. They are complex oligo- and polysaccharide mixtures obtained from starch by partial acid and/or enzymatic hydrolysis. MDs are neutral polysaccharides with hydroxyl groups that may give rise to multiple types of interactions (hydrogen bonding, ionic and hydrophobic) between chiral solutes and their helical structure, which constitute the basis of the enantio-recognition. The helical structure of MDs has a hydrophobic inside that mimics the cavity of CDs responsible for chiral recognition, but MDs represent a considerably more flexible entity than CDs with less steric restrictions toward a solute.

MDs were successfully employed as chiral selectors for the enantioseparation of cetirizine and hydroxyzine [61], tramadol [62], tolterodine [63], amlodipine [64], and citalopram [65]. In several of these investigations [63–65], MDs with various dextrose equivalent (DE) values (4–7, 13–17, and 16.5–19.5) were used, and it was observed that enantioresolution increased with decreasing the DE value. The higher the DE number, the higher the extent of starch hydrolysis and, consequently, the shorter the oligomeric chains present in a mixture. MDs with lower DE values and longer oligomeric chains have more sites to interact with the analyte molecules.

As mentioned above, charged chiral selectors such as sulfated-CDs display superior enantioselectivity in most cases, since they can offer not only an inclusion–complexation interaction, but also strong electrostatic interactions, and the apparent mobility difference between the two enantiomers will be greatest when the mobility of the analyte–chiral selector complex is in the opposite direction to that of the analyte itself. Furthermore, charged chiral selectors can be used for the enantioseparation of neutral chiral compounds. In this regard, sulfated-MD was evaluated for the enantioseparation of five basic drugs (amlodipine, hydroxyzine, fluoxetine, tolterodine, and tramadol), and a baseline separation was obtained for all of them. When neutral MD was used under the same conditions instead of sulfated-MD, no enantioseparation was observed [66].

Polysaccharides have been used in combination with chiral ionic liquids (IL) in dual chiral selector systems to improve enantioresolution. Improved enantioseparations of nefopam hydrochloride, citalopram hydrobromide, and duloxetine hydrochloride were obtained with dual chiral selector systems of glycogen and amino acid chiral ILs (tetramethylammonium-L-arginine and tetramethylammonium-L-aspartic acid). Tetramethylammonium-L-arginine exhibited better enantioselective properties than tetramethylammonium-L-aspartic acid toward these drugs [67]. As mentioned above, polysaccharides were also evaluated in dual chiral systems with CDs, but a negative effect was observed when adding glycogen to β-CD and hydroxypropyl-β-CD when separating enantiomerically several drugs [37].

2.2.4 Other Chiral Selectors

Sulfated cyclofructans (CF) showed exceptional selectivity toward many cationic analytes. Sulfated-CF6 was used for the enantiomeric separation of tamsulosin, tiropramide, bupivacaine, and norephedrine. The enantiomeric separation performance was compared with that obtained with sulfated α-, β-, and γ-CDs. Both of them, sulfated-CF6 and sulfated-CDs, were highly efficient and selective chiral selectors for these compounds. CF6 has a chiral

18-crown-6 structure and consists of six D-fructofuranose units [68].

The metal ion complex Cu(II)–clindamycin was used as a chiral selector for the enantioseparation of various chiral drugs (tropicamide, propranolol, sotalol, bisoprolol, epinephrine, esmolol, atenolol, and metoprolol). The chiral recognition is based on the formation of ternary mixed metal complexes between the chiral selector ligand and the analyte. A baseline enantioresolution was achieved for tropicamide, bisoprolol, propranolol, and sotalol, and a partial enantioseparation for epinephrine, esmolol, metoprolol, and atenolol [69].

The complex formed by the reaction between di-N-amyl L-tartrate and boric acid was used for the enantioseparation of chiral drugs (propranolol, sotalol, esmolol, atenolol, bisoprolol, metoprolol, terbutaline, clenbuterol, cycloclenbuterol, bambuterol, and tulobuterol) by using NACE with methanol as solvent. The enantioseparation mechanism is assumed to be based on ion-pair interactions between the negatively charged chiral counter-ion di-N-amyl L-tartrate–boric acid complex and the positively charged enantiomeric analytes, and the non-aqueous system is more favorable for the ion-pair formation [70].

Micellar electrokinetic chromatography (MEKC) was useful in the separation of enantiomers of drugs. Several polymeric surfactants poly(sodium N-undecanoyl-L-leucinate) (poly-L-SUL), poly(sodium N-undecanoyl-L,L-leucylleucinate) (poly-L,L-SULL), poly(sodium N-undecanoyl-L,Lleucyl-valinate) (poly-L,L-SULV), poly(sodium N-undecanoyl-L-valinate) (poly-L-SUV), poly(sodium N-undecanoyl-L-valylglycinate) (poly-L-SUVG), poly(sodium N-undecanoyl-L,Lalanyl-valinate) (poly-L,L-SUAV), poly(sodium N-undecanoyl-L,L-leucyl-alanate) (poly-L,L-SULA), and poly(sodium N-undecanoyl-L,L-valyl-valinate) (poly-L,L-SUVV) were studied as chiral selectors for the separation of Huperzine A enantiomers. A baseline enantiosepa-ration was achieved by using poly-L,L-SUAV [71].

Bacteria were also investigated as chiral selectors for the enantiomeric separation of drugs. *Escherichia coli*, *Pseudomonas aeruginosa*, and *Staphylococcus aureus* were used to separate ofloxacin (OFLX) enantiomers by partial-filling CCE. *E. coli* and *P. aeruginosa* showed better enantioseparation capabilities toward

OFLX than *S. aureus* (Fig. 2.3). This result is in good agreement with the antimicrobial activity of OFLX toward these bacteria. Partial-filling CE was used to avoid the high background signal of bacteria solution, since bacteria migrate toward the detector after OFLX [72].

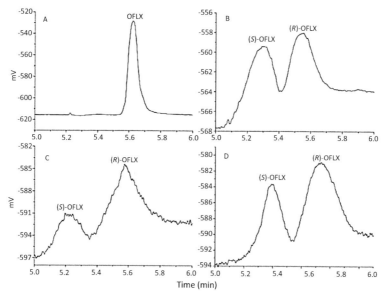

Figure 2.3 Electropherograms of the OFLX enantioseparation using bacteria as chiral selectors. Experimental conditions: running buffer: phosphate buffer composed of 10 mM containing disodium hydrogen phosphate and sodium dihydrogen phosphate at pH 7.4; applied voltage: 15 kV; (A) OFLX 5 s; (B) *E. coli* (6.0×10^8 cells/mL) were injected into the capillary by gravity with the injection height of 17.5 cm for 180 s; (C) *P. aeruginosa* (6.0×10^8 cells/mL) were injected into the capillary by gravity with the injection height of 17.5 cm for 300 s; (D) *S. aureus* (6.0×10^8 cells/mL) were injected into the capillary by gravity with the injection height of 17.5 cm for 300 s. Reprinted from Ref. [72], Copyright 2012, with permission from John Wiley and Sons.

2.3 Applications of CCE in Pharmaceutical Analysis

Table 2.1 groups the recent applications of CCE in pharmaceutical analysis. The drugs analyzed, the chiral selector employed, the detection system used, and the limit of detection (LOD) achieved are

Table 2.1 Recent applications of CCE in pharmaceutical analysis

Drug	Chiral Selector	Detection System	Precon. tech.	LOD (µg/mL)	Sample	Ref.
Amlodipine	Randomly methylated-β-CD	UV-Vis (238 nm)		2.31	Tablets	[28]
	Maltodextrin	UV-Vis (214 nm)		0.52	Tablets	[64]
	Hydroxypropyl-α-CD	UV-Vis (214 nm)	EME	3×10^{-3}	Human urine and plasma	[73]
Anisodamine atenolol metoprolol	Carboxymethyl-β-CD	ECL		9×10^{-2}	Human urine	[53]
Bupivacaine	Sulfobutylether-β-CD	Conductivity		5×10^{-2}	Rabbit serum and pharmaceutical injections	[9]
Bupropion	Sulfated-α-CD	Phosphorescence		5×10^{-2}	Tablets and human urine	[74]
Carnitine	Succinyl-γ-CD	IT-MS		10^{-2}	Ampoules, oral solutions, sachets, and tablets	[75]
Carvedilol	β-CD	UV-Vis (242 nm)		1.13	Tablets	[23]
	Carboxymethyl-β-CD	UV-Vis (241 nm)	DLLME + FASI	4×10^{-3}	Human plasma	[40]
Cathinone analogues	β-CD highly sulfated-γ-CD	UV-Vis (206 nm) TOF-MS		4×10^{-3} 10^{-3}	Seized drugs	[38]
	Sulfated-β-CD	UV-Vis (208 nm)			Synthetic solution	[17]

Drug	Chiral Selector	Detection System	Precon. tech.	LOD (µg/mL)	Sample	Ref.
Cetirizine hydroxyzine	Sulfated-β-CD	UV-Vis (214 nm)	DLLME	3.7×10^{-2} 7.5×10^{-2}	Liquid culture media	[11]
	Maltodextrin	UV-Vis (214 nm)	LLE	3×10^{-2} 10^{-2}	Human plasma	[61]
Chloroquine	Sulfobutylether-β-CD	UV-Vis (225 nm)		0.6	Standard solution	[76]
Citalopram	Maltodextrin	UV-Vis (214 nm)	HF-LPME	10^{-2}	Human urine	[65]
Duloxetine	Hydroxypropyl-β-CD	IT-MS		2×10^{-2}	Capsules	[4]
Eszopiclone	β-CD + [EMIm] [L-lactate]	UV-Vis (?)		0.3	Tablets	[33]
Iodiconazole	Hydroxypropyl-γ-CD	UV-Vis (200 nm)		4.6	Standard solution	[3]
Isradipine	Sulfobutylether-β-CD	UV-Vis (239 nm)		2.16	Capsules	[14]
Ketamine norketamine	Sulfated-β-CD	UV-Vis (200 nm)	LLE	2×10^{-2}	Equine liver microsomes	[77]
	Sulfated-β-CD	UV-Vis (195 nm)	LLE		Human	[78]
	Sulfated-β-CD highly sulfated-γ-CD		LLE		microsomal solution	[24]
	Highly sulfated-γ-CD					[79]
	Highly sulfated-γ-CD			0.18		[80]
	Sulfated-β-CD		LLE	8×10^{-3}	Brain and cerebrospinal fluid	[81]
Levamlodipine	Carboxymethyl-β-CD	UV-Vis (237 nm)		1	Bulk drugs	[39]

(*Continued*)

Table 2.1 (*Continued*)

Drug	Chiral Selector	Detection System	Precon. tech.	LOD (μg/mL)	Sample	Ref.
Levocetirizine	Sulfated-β-CD	UV-Vis (195 nm)		7.5×10^{-2}	Bulk drugs and tablets	[16]
Levornidazole	Sulfated-α-CD	UV-Vis (277 nm)		0.3	Starting material and injection solution	[18]
Methamphetamine heroin MDMA ketamine	β-CD	UV-Vis (200 nm)	DLLME	5×10^{-5}	Forensic samples (banknotes, kraft paper, plastic bag and silver paper)	[82]
Methamphetamine drugs	Dimethyl-β-CD	UV-Vis (195 nm)			Human urine	[46]
	Highly sulfated-γ-CD	UV-Vis (195 nm)			Synthetic solution	[47]
	Highly sulfated-γ-CD	IT-MS		2	Seized drugs	[48]
	Sulfated-β-CD	UV-Vis (210 nm)			Synthetic solution	[42]
Meptazinol	Carboxymethyl-β-CD	UV-Vis (271 and 237 nm)		2.5	Tablets	[20]
Naproxen	Vancomycin + L-ValC$_4$NTf$_2$	UV-Vis (235 nm)		4.5	Bulk drugs	[58]

Drug	Chiral Selector	Detection System	Precon. tech.	LOD (μg/mL)	Sample	Ref.
Ofloxacin	Hydroxypropyl-β-CD + [EMIm][L-lactate]	UV-Vis (?)		0.53	Bulk drugs	[32]
Ofloxacin ornidazole	Sulfated-β-CD	UV-Vis (230 nm)		0.46 0.54	Tablets	[7]
Pindolol oxprenolol propranolol	Dimethyl-β-CD + trimethyl-β-CD	UV-Vis (220 nm)	FASI	3×10^{-2}	Human urine	[31]
Propranolol	Hydroxypropyl-β-CD	UV-Vis (214 nm)	EME	7×10^{-3}	Human urine and plasma	[27]
Sibutramine	Methyl-β-CD	UV-Vis (223 nm)			Capsules	[6]
	Methyl-β-CD	UV-Vis (225 nm)		0.25	Capsules	[83]
Sotalol	Methyl-β-CD	UV-Vis (232 nm)		1.13	Tablets	[22]
Tolterodine	Maltodextrin	UV-Vis (214 nm)	EME	3×10^{-3}	Human urine and plasma	[63]

(*Continued*)

Table 2.1 (Continued)

Drug	Chiral Selector	Detection System	Precon. tech.	LOD (μg/mL)	Sample	Ref.
Tolterodine methoxytolterodine	Phosphated-γ-CD	UV-Vis (200 nm)		0.33 0.41	Tablets	[19]
Tramadol	Maltodextrin	UV-Vis (214 nm)		1.5	Tablets	[62]
Trimipramine	α-CD	UV-Vis (214 nm)	EME	7×10^{-3}	Human urine and plasma	[25]
Valsartan	Acetyl-β-CD	Acetyl-β-CD			Tablets	[5]
Verteporfin	Sodium cholate	UV-Vis (428 nm)	PAEKI	10^{-2}	Artificial urine	[84]
Zopiclone	Carboxymethyl-β-CD	UV-Vis (305 and 200 nm)		2×10^{-2}	Tablets	[85]
	Carboxymethyl-β-CD	UV-Vis (310 nm)	DLLME	1.5×10^{-2}	Liquid culture media	[21]

ECL: electrochemiluminescence; LLE: liquid–liquid extraction; DLLME: dispersive liquid–liquid microextraction; PAEKI: pressure-assisted electrokinetic injection stacking; EME: electromembrane extraction; MDMA: 3,4-methylenedioxymethamphetamine; FASI: field-amplified sample injection; HF-LPME: hollow fiber supported liquid-phase microextraction; [EMIm][L-lactate]: 1-ethyl-3-methylimidazolium-L-lactate; L-ValC$_4$NTf$_2$: L-valine tert butyl ester bis (trifluoromethane) sulfonamide included in this table together with the type of the sample analyzed. It is observed that, as stated before, the most frequently employed chiral selectors were cyclodextrins followed by macrocyclic antibiotics, polysaccharides, and other less applied chiral selectors such as cyclofructans, metal ion complexes, polymeric surfactants, and bacteria. A variety of drugs were analyzed; UV absorption was the most employed detection system, although the use of other detection systems such as electrochemiluminescence, phosphorescence, conductivity, or mass spectrometry was also reported. The highest sensitivity in terms of LODs corresponded to the use of time of flight-mass spectrometry (at the ng/mL level) and to the use of UV-Vis detection together with offline sample treatment techniques (DLLME, EME) or with the combination of them with in-capillary preconcentration techniques based on electrophoretic principles such as FASI (DLLME + FASI). UV-Vis detection combined with DLLME originated the lowest LODs at sub-ng/mL level.

As shown in Table 2.1, different analytical methodologies were developed by CCE and applied to the analysis of pharmaceutical formulations, illicit drugs, and biological samples for metabolic studies or forensic investigations.

2.3.1 Analysis of Pharmaceutical Formulations

Chiral drugs can be marketed as racemic mixtures or as pure enantiomers although in the last few years, there was a tendency to commercialize chiral drugs as pure enantiomers.

In the case of racemic drugs, applications of chiral CE include the determination of the active principles in different pharmaceutical formulations. When the drug is commercialized as pure enantiomer, application of chiral CE can be aimed to the quantitation of the majority enantiomer or to the control of the enantiomeric impurity, which cannot exceed 0.1% as established by the ICH regulations. In that case, the determination of a low amount of the enantiomeric impurity in the presence of the majority enantiomer can be difficult being paramount in some cases the migration order of the enantiomers. In fact, if the enantiomeric impurity is the first-migrating enantiomer, this fact usually precludes the overlapping of its peak with the big peak of the majority enantiomer, particularly when the enantiomeric resolution is not high. Sensitive analytical methodologies are needed when the control of the enantiomeric purity of pharmaceutical formulations is required.

Different analytical methodologies were developed to determine the enantiomers of chiral drugs in pharmaceutical formulations by CCE. A good agreement was shown between the determined values and the labeled amounts of different drugs in pharmaceutical formulations based on sibutramine [6], ofloxacin and ornidazole [7], meptazinol [20], sotalol [22], amlodipine [28, 64], carvedilol [23], isradipine [14], tramadol [62], bupropion [74], bupivacaine [9], and duloxetine [4].

Applications of CCE methodologies to degradation and racemization studies were also reported. Thus, the enantiomers of isradipine were separated from their degradation products and formulation excipients after degradation by oxidation, hydrolysis, and photolysis [14]. Also, the degradation and racemization of zopiclone under stress conditions (acid and alkaline hydrolysis studies, oxidative studies, and photo-degradation studies) was carried out by a CCE methodology enabling the simultaneous quantitation of zopiclone enantiomers and its impurities in tablets [85]. Moreover, a stability-indicating CE method with maltodextrin as chiral selector was developed and validated for the assay and stability evaluation of tramadol enantiomers in commercial tablets [62]. The stability evaluation of tramadol indicated to be more stable under thermolytic and photolytic stress condition in solid state in both tablets and API powder than in solution.

Regarding the determination of enantiomeric impurities in pharmaceutical formulations marketed as pure enantiomers, numerous analytical methodologies were reported enabling the control of these formulations by CCE, which were applied to the determination of a 0.24% of the (S)-enantiomer of chloroquine in standard solution [76], a 0.1% of the levocetirizine in tablets [16], a 0.05% of (S)-sibutramine in capsules [83], a 0.05% of the (R)-enantiomer of ornidazole as enantiomeric impurity in injection solutions (Fig. 2.4) [18], a 0.015% of (S)-tolterodine in tablets [19], a 0.015% of (R)-amlodipine in bulk drugs [39], or a 0.05% of the (R)-enantiomer impurity of valsartan in tablets [5]. Also, the amount of the enantiomeric impurity of (R)-ofloxacin in bulk drugs of (S)-ofloxacin was demonstrated to be below 0.2%. In this case, an interesting dual chiral system was employed in CE using as chiral selector a mixture of hydroxypropyl-β-CD and the ionic liquid 1-ethyl-3-methylimidazolium-L-lactate [32]. Other dual mixture of chiral selectors composed of β-CD and 1-ethyl-3-methylimidazolium-L-lactate showed its potential in the determination of the chiral purity of eszopiclone in commercial tablets. It was demonstrated that the concentration of the enantiomeric impurity of (R)-zopiclone was lesser than 0.1% [33]. Dual mixtures of chiral selectors constituted by macrocyclic antibiotics (vancomycine) and chiral ionic liquids (L-alanine and L-valine tert butyl ester bis (trifluoromethane) sulfonamide) were also employed as mentioned above [58].

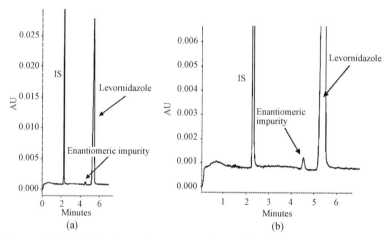

Figure 2.4 Typical electropherogram of 5 mg/mL levornidazole injection solution spiked with 2.5 μg/mL enantiomeric impurity (0.05%) (a) full size, (b) enlarged view. Experimental conditions: BGE: 20 mM Tris-phosphate buffer containing 2.0% (w/v) sulfated-α-CD at pH 2.1; voltage: 30 kV; temperature: 25°C; short end injection: 5 s at 0.5 psi. (From Ref. [18]).

Thus, a quantitative CE method for determining the chiral impurity (R)-naproxen in samples of (S)-naproxen was developed. The method allowed to determine the (R)-impurity at the level of 0.3%. Although UV was generally the most employed detection system, tandem mass spectrometry was also employed to carry out the sensitive determination of D-carnitine as enantiomeric impurity of L-carnitine in pharmaceutical formulations (ampoules, oral solutions, sachets, and tablets) [75].

2.3.2 Analysis of Illicit Drugs

For supporting drug crime investigations and forensic toxicology studies, CCE with CDs was used to characterize illicit drugs. Chemically modified capillaries of various types (FunCap-CE/ Type D (possessing diol groups), Type A (amino groups), Type C (carboxyl groups), and Type S (sulfate groups) in combination with sulfated-γ-CD were applied for the analysis of amphetamine-type stimulants in synthetic solutions with UV detection [47] and seized drugs with CE/MS/MS [48]. Figure 2.5 shows the extracted ion electropherograms and MS/MS spectra obtained for the CE/ MS/MS analysis of a group of amphetamine-type stimulants (amphetamine (AM), methamphetamine (MA), norephedrine (NEP), norpseudoephedrine (NpEP), ephedrine (EP), pseudoephedrine (pEP), dimethylamphetamine (DMA), and methylephedrine (MeEP)). The MS/MS spectra give important information for identification purposes [48]. Different types of native and derivatized CDs (β-CD, hydroxypropyl-β-CD, carboxymethyl-β-CD, sulfated-β-CD, and γ-CD) were used for the enantioseparation of 19 cathinone derivatives (or beta-keto amphetamines). Cathinone derivatives are substances of abuse, which are available for cheap money [17]. A CCE method using highly sulfated γ-CD was proposed for methamphetamine profiling (methamphetamine, amphetamine, ephedrine, pseudoephedrine, norephedrine, and norpseudoephedrine) of different synthetic solutions. Information about the chirality of a seized drug sample is essential for identifying its precursor, synthetic pathway, etc., and to determine the origin of the seized drug. In most cases, it is sufficient to separate both enantiomers and calculate the ratio [47]. A method for the enantioseparation of 13 new amphetamine-like designer drugs using sulfated-β-CD was also developed. Sulfated-β-CD was able to separate all amphetamines analyzed [42].

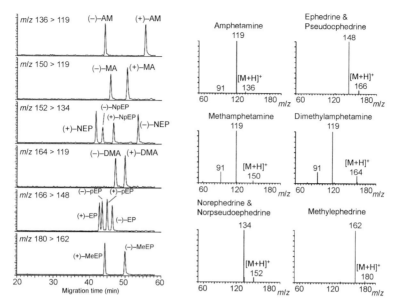

Figure 2.5 Extracted ion electropherograms and MS/MS spectra of a mixture of amphetamine-type stimulants (50 mg/mL for individual enantiomers). Reprinted from Ref. [48], Copyright 2015, with permission from Elsevier.

UV and TOF-MS detection were compared in the separation of 12 cathinone analogues by CCE using CDs in the BGE. UV detection values for LODs and LOQs were 4.2–7.0 ng/mL and 13–21 ng/mL, respectively. For MS detection, LODs and LOQs were 1.0–11 ng/mL and 3–33 ng/mL, respectively. Comparing to the figures of merit for CE-UV, the MS detector provides lower LODs for nine out of 12 analytes. The non-volatile phosphate buffer containing CDs was aspirated into a fraction of the capillary instead of fully flushing the capillary, so that the non-volatile portion of the separation buffer does not enter the mass spectrometer [38]. The method was applied to analyze seizes of drugs. A DLLME method, followed by a CE separation using UV detection with β-CD as chiral selector, was developed for the enantiomeric separation and determination of multiple illicit drugs on prepared samples, simulating that being associated with the trafficking of drugs. Figure 2.6 shows a typical electropherogram for a banknote sample before (a) and after (b) DLLME extraction. The LODs were between 50 and 200 pg/mL [82].

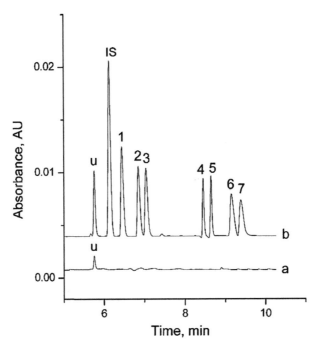

Figure 2.6 Typical electropherogram obtained for banknote sample before (a) and after (b) DLLME extraction. CE conditions: electrolyte, 100 mmol/L of potassium phosphate containing 20 mmol/L of β-CD (pH 3.23); injection, electrokinetic 5 kV for 7 s; voltage, 15 kV; capillary, 30 cm (effective length) 50 μm i.d.; detection, UV at 200 nm. Peak identification: u: unknown, IS: lidocaine, 1: heroin, 2 & 3: D,L-methamphetamine, 4 & 5: D,L-3,4-methylenedioxymethamphetamine, and 6 & 7: D,L-ketamine. Reprinted from Ref. [82], Copyright 2011, with permission from Elsevier.

2.3.3 Analysis of Biological Samples

CCE methodologies were applied to the analysis of drugs in different biological samples such as biological fluids (urine, serum, plasma, and cerebrospinal fluid) and tissue samples. Due to the complexity of biological samples and the sensitivity usually required in their analysis, different preconcentration strategies based on electrophoretic principles, offline sample preparation techniques, and sensitive detection systems were employed in different works.

The determination of anisodamine (AN), atenolol (AT), and metoprolol (ME) enantiomers in spiked human urine by using a microfluidic chip-CE device with electrochemiluminescence detection and carboxymethyl-β-CD as chiral selector was carried out (Fig. 2.7) [53]. Moreover, the enantiomers of bupropion were determined in spiked urine samples using a sensitized time-resolved phosphorescence detection mode in a home-built detection system, which allowed an LOD for each enantiomer of 2×10^{-7} M (0.048 µg/mL) [74]. CCE with UV detection with β-CD and dimethyl-β-CD and chemically modified capillaries having diol groups was applied to the chiral analysis of methamphetamine and related compounds in actual urine samples of methamphetamine addicts [46].

Conductivity detection is a universal and low cost technique for ionic compounds allowing non-UV-absorbing compounds to be analyzed with good sensitivity without chemical derivatization. A conductivity detection system was employed in the chiral separation and detection of the optical isomers of bupivacaine. A quantitative method was developed and validated. The method was applied to the analysis of rabbit serum with a LOD of 0.26 µg/mL [9].

Preconcentration strategies based on electrophoretic principles can be applied in CE. Field-amplified sample injection (FASI) was applied to improve detection limits of pindolol, oxprenolol, and propranolol using a dual chiral selector system formed by 2,6-dimethyl-β-CD and 2,3,6-trimethyl-β-CD with an achiral ionic liquid (glycidyltrimethylammonium chloride) as modifier. The method was applied to the analysis of a spiked urine sample [31]. A pressure-assisted electrokinetic injection stacking (PAEKI) for verteporfin was applied to achieve highly sensitive enantioseparation and detection of this drug in artificial urine samples [84].

Offline sample preparation techniques such as liquid–liquid extraction, liquid–electromembrane extraction (EME), dispersive liquid–liquid microextraction (DLLME), and hollow fiber supported liquid-phase microextraction (HF-LPME) were implemented to improve LODs and LOQs in the analysis of chiral drugs by CE. Thus, cetirizine (CTZ) and hydroxyzine (HZ) enantiomers were determined in plasma samples after liquid–liquid extraction by using CE with UV detection and maltodextrin as chiral selector (Fig. 2.8).

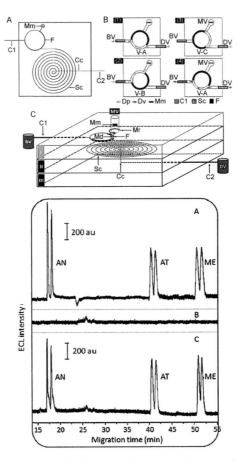

Figure 2.7 Left: Schematic diagrams showing the layout of the fabricated microfluidic chip-CE device. (A) Top view of the general layout. (B) The sequential four-step operation of urine sample loading for the microchip-CE device. The position of the ferrofluid plug in the circular valve is shown and labeled as V-A, valve A; V-B, valve B; V-C, valve C; Dp, pressure driving; Dv, high-voltage driving. (C) The perspective view of the microchip-CE device in operation. BV: buffer vial; MV: mixing vial; DV: detection vial; Mm: micromagnet; F: ferrofluid; Sc: spiral channel; Cc: connection channel; C1: capillary 1; C2: capillary 2; Mr: rotating magnet; Md: driving magnet. Right: Electropherograms of (A) 60 μM AN, 50 μM AT, and 30 μM ME standard solutions; (B) 30-fold diluted blank urine sample; (C) 30-fold diluted urine spiked with AN, AT, and ME to the same conditions as (A); microchip-CE-ECL conditions: separation channel, around 55 cm effective length; sample injection, 10 kV × 10 s; separation voltage, 17.5 kV; running buffer, 57.6 mM HAc-NaAc concentration (pH 5.3), 14.7 mM carboxymethyl-β-CD. Reprinted from Ref. [53], Copyright 2013, with permission from John Wiley and Sons.

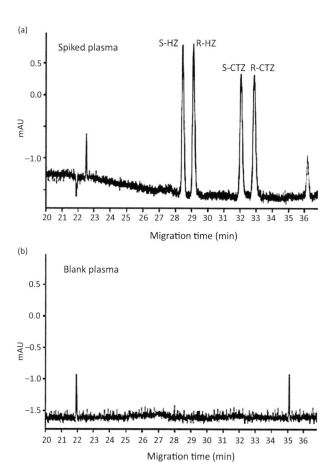

Figure 2.8 Electropherograms obtained from spiked and drug-free human plasma (spiked sample solution: 100 ng/mL of each enantiomer). Experimental conditions: capillary, 60 cm (50 cm effective length) µ75 mm i.d.; detection, 214 nm; applied voltage, 20 kV; temperature, 25°C; injection, 60 mbar × 10 s; separation solution, 75 mM phosphate solution containing 5% w/v MD. Reprinted from Ref. [61], Copyright 2011, with permission from John Wiley and Sons.

The LOQ was estimated to be 30 ng/mL, while the LOD was 10 ng/mL [61]. The organic solvent used as supported liquid membrane (SLM) must possess electrical conductivity. Clean extracts are obtained since the organic SLM and the electrical field prevent the migration and extraction of hydrophilic and neutral substances into the acceptor solution, that is, only substances that are ionized in the sample solution and able to migrate through the organic SLM are extracted into the acceptor phase. EME methods also have high enrichment factors. An EME method was developed for the extraction of amlodipine enantiomers from plasma and urine samples. A supported liquid membrane consisting of 2-nitrophenyl octyl ether impregnated on the wall of a hollow fiber was used. The enrichment factor achieved was about 124. The extract was analyzed by CCE using hydroxypropyl-α-CD into the BGE, and the LOD and LOQ for both enantiomers were 3–5 and 10–20 ng/mL, respectively [73]. An EME procedure was developed for the analysis of trimipramine enantiomers in biological samples (plasma and urine) using CCE with α-CD. The LOD and LOQ for both enantiomers were 7 and 20 ng/mL, respectively [25]. Propranolol enantiomers were analyzed in urine and plasma samples using an EME method with 2-nitro phenyl octyl ether as supported liquid membrane and CCE with hydroxypropyl-β-CD. Enrichment factors obtained ranged from 108 to 134 allowing LODs of 7 and 10 ng/mL, and LOQs of 20 and 30 ng/mL, for both enantiomers [27]. An EME method combined with maltodextrin-modified CE was used for the determination of tolterodine enantiomers in plasma and urine samples. 2-Nitro phenyl octyl ether, 1-ethyl-2-nitrobenzene, and nitrobenzene were tested as supported liquid membrane. Enrichment factors between 86 and 102 with plasma and urine samples were obtained. The LOQs and LODs for both enantiomers were 10 and 3 ng/mL, respectively [63].

The enantioselective analysis of zopiclone and its metabolite N-desmethylzopiclone was carried out using DLLME preconcentration and CCE with carboxymethyl-β-CD and UV detection after fungal biotransformation. A mixture of chloroform (extraction solvent) and methanol (dispersing solvent) was injected into the sample originating a cloud point. The enrichment factors for zopiclone and N-desmethylzopiclone were 20.5 and 12.7, respectively [21]. Moreover, the enantiomeric analysis of carvedilol in human plasma was possible by CCE with UV detection and carboxymethyl-β-CD as chiral selector using DLLME combined with FASI enabling an LOD of 4 ng/mL [40]. A DLLME method was also developed to extract hydroxyzine and cetirizine from liquid culture media previous to their analysis by CCE with sulfated-β-CD and UV detection, obtaining LODs of 75 and 37 ng/mL for hydroxyzine and cetirizine, respectively [11].

In an HF-LPME method, a porous hollow fiber made of polypropylene is used to protect the extraction solvent from the sample solution. Therefore, the extraction solvent is not in direct contact with the sample solution, and samples may be stirred or vibrated vigorously. Micro-pores of the hollow fiber prevent large molecules and other impurities from entering the extraction solvent in the lumen. The consumption of organic solvents in this technique is minimum. Various organic solvents (cyclohexane, toluene, hexadecane, 1-octanol, and 1-octane) were evaluated for their extraction efficiencies to analyze citalopram enantiomers in spiked urine samples by CCE with MDs and UV detection. The LOD and the LOQ were estimated to be 10 ng/mL and 30 ng/mL, respectively, for both enantiomers [65].

Since single drug enantiomers may significantly differ not only in activity and toxicity, but also in distribution, rate and products of metabolism, excretion, selectivity for transporters and enzymes, etc., the study of metabolism represents a key part of LADME (liberation, absorption, distribution, metabolism, excretion) tests of a potential drug. Chiral CE was used in metabolic studies of chiral drugs to investigate the stereoselectivity of metabolic steps.

CCE with sulfated-β-CD was employed to analyze the enantiomers of ketamine and its metabolites to identify cytochrome P450 enzymes involved in hepatic ketamine and norketamine biotransformation in vitro. The results suggested a stereoselective in vitro biotransformation of ketamine and norketamine [78]. Moreover, the enantioselective CE analysis of hepatic ketamine metabolism in different species was investigated in vitro using sulfated-β-CD as chiral selector [77]. Also, CCE with sulfated-β-CD and highly sulfated-γ-CD was applied to determine the N-demethylation kinetics of ketamine to norketamine catalyzed by CYP3A4, enzyme responsible for the metabolism of a half of the most commonly prescribed drugs. Its inhibition by ketoconazole in vitro was also studied. CYP3A4-catalyzed *N*-demethylation was shown to be stereoselective but not the inhibition by ketoconazole [24]. When investigating the stereoselectivity of the metabolism of ketamine to norketamine via CYP3A4 by CCE with highly sulphated-*γ*-CD using the separation capillary as a microreactor, an increase in the production of (*S*)-norketamine over the (*R*)-enantiomer was shown (Fig. 2.9) [79, 80]. Applications of these investigations included the study of metabolism and transport of ketamine and norketamine enantiomers in equine brain and cerebrospinal fluid by CE with sulfated-*β*-CD as chiral selector [81].

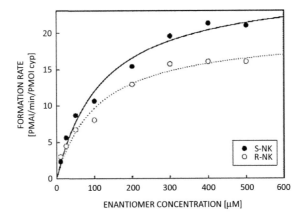

Figure 2.9 Norketamine enantiomer formation rate as a function of ketamine enantiomer concentration by CYP3A4 after 8 min incubation with racemic ketamine at 37°C. Symbols denote the mean of duplicates. Solid and dotted lines are predicted values based on nonlinear regression analysis using the Michaelis–Menten equation and assuming a twofold dilution of the enzyme. Key: • (*S*)-norketamine (S-NK) and ○ (*R*)-norketamine (R-NK). Reprinted from Ref. [79], Copyright 2012, with permission from John Wiley and Sons.

The use of microbial models offers several advantages, such as simplicity and low cost. CE with carboxymethyl-*β*-CD as chiral selector was applied to study the enantioselective fungal biotransformation of zopiclone into its active metabolite *N*-desmethylzopiclone. The results obtained demonstrated that *Cunninghamella* species of fungi perform enantioselective demethylation of zopiclone [21]. Microbial transformation of organic compounds also constitutes an economical technology to obtain drugs, since biological transformation can be enantioselective, producing pure drug enantiomers from racemic mixtures. The fungal biotransformation of hydroxyzine to cetirizine was investigated by CCE using sulfated-β-CD. Three of the six fungi species studied were able to convert hydroxyzine to cetirizine enantioselectively. The pharmacological activity of cetirizine is primarily due to the (R)-enantiomer, also known as levocetirizine [11].

From all the information presented in this chapter, it can be concluded that CCE is a powerful tool to achieve the enantiomeric analysis of drugs due to the interesting characteristics of this technique such as high efficiency, versatility, and low consumption of reagents and samples. In addition to the low cost of CCE compared with other analytical techniques, the low consumption of reagents also reduces the environmental impact related to the use of this technique, allowing considering it as a green analytical technique.

Optimization of the enantioseparation of a drug requires in most cases the selection of an appropriate chiral selector and other experimental conditions that have relevant influence on the chiral separation such as the concentration of the chiral selector, the nature, concentration, and pH of the buffer or the presence of additives. A variety of chiral selectors can be employed in CCE for the separation of chiral drugs. However, the most employed are CDs followed in the last years by macrocyclic antibiotics and polysaccharides. Other chiral selectors were less frequently employed.

UV-Vis was the detection system chosen in most cases to carry out the enantiomeric analysis of drugs. However, to obtain high sensitivity in terms of LODs, the use of offline sample preparation techniques or in-capillary preconcentration strategies based on electrophoretic principles was required. The combination of both strategies can also be an alternative to increase the sensitivity. Other detection systems such as chemiluminescence, phosphorescence, or mass spectrometry (time of flight and ion-trap) were employed; the lowest LODs were obtained with time of flight-mass spectrometry detection and with UV-Vis detection combined with the use of DLLME or EME, or with a combination DLLME+FASI enabling to reach LODs at ng/mL level and sub-ng/mL level.

Applications of CCE to the analysis of the enantiomers of drugs in pharmaceutical formulations, illicit drugs, and biological samples such as urine, plasma, cerebrospinal fluid, or brain tissue were demonstrated and some trends such as miniaturization by incorporating CCE to microchip devices were also implemented.

Acknowledgments

The authors thank the Ministry of Economy and Competitiveness (Spain) for research project CTQ2013-48740-P.

References

1. Marina, M. L., Ríos, A., and Valcárcel, M. (eds.) (2005). *Analysis and Detection in Capillary Electrophoresis* (Comprehensive Analytical Chemistry (CAC) Series, Elsevier, The Netherlands).

2. Saz, J. M. and Marina, M. L. (2016). Recent advances on the use of cyclodextrins in the chiral analysis of drugs by capillary electrophoresis, *J. Chromatogr. A*, **1467**, pp. 79–94.

3. Li, W. H., Zhao, L. A., Tan, G. G., Sheng, C. Q., Zhang, X. R., Zhu, Z. Y., Zhang, G. Q., and Chai, Y. F. (2011). Enantioseparation of the new antifungal drug iodiconazole and structurally related triadimenol analogues by CE with neutral cyclodextrin additives, *Chromatographia*, **73**, pp. 1009–1014.

4. Sánchez-López, E., Montealegre, C., Marina, M. L., and Crego, A. L. (2014). Development of chiral methodologies by capillary electrophoresis with ultraviolet and mass spectrometry detection for duloxetine analysis in pharmaceutical formulations, *J. Chromatogr. A*, **1363**, pp. 356–362.

5. Lee, K. R., Nguyen, N. T., Lee, Y. J., Choi, S., Kang, J. S., Mar, W., and Kim, K. H. (2015). Determination of the R-enantiomer of valsartan in pharmaceutical formulation by capillary electrophoresis, *Arch. Pharm. Res.*, **38**, pp. 826–833.

6. Zhu, H., Wu, E., Chen, J., Men, C., Jang, Y. S., Kang, W., Choi, J. K., Lee, W., and Kang, J. S. (2010). Enantioseparation and determination of sibutramine in pharmaceutical formulations by capillary electrophoresis, *Bull. Korean Chem. Soc.*, **31**, pp. 1496–1500.

7. Al Azzam, K. M., Saad, B., Adnan, R., and Aboul-Enein, H. Y. (2010). Enantioselective analysis of ofloxacin and ornidazole in pharmaceutical

formulations by capillary electrophoresis using single chiral selector and computational calculation of their inclusion complexes, *Anal. Chim. Acta*, **674**, pp. 249–255.

8. Vega, E. D., Lomsadze, K., Chankvetadze, L., Salgado, A., Scriba, G. K. E., Calvo, E., Lopez, J. A., Crego, A. L., Marina, M. L., and Chankvetadze, B. (2011). Separation of enantiomers of ephedrine by capillary electrophoresis using cyclodextrins as chiral selectors: Comparative CE, NMR and high resolution MS studies, *Electrophoresis*, **32**, pp. 2640–2647.

9. Li, Y., Yu, Y. J., Zhu, P. Y., Duan, G. L., Li, Y., and Song, F. Y. (2012). Chiral separation of bupivacaine hydrochloride by capillary electrophoresis with high frequency conductivity detection and its application to rabbit serum and pharmaceutical injection, *Pharmazie*, **67**, pp. 25–30.

10. Zhou, J., Yao, H. C., Shao, H., Li, Y. H., and Zhang, Z. Z. (2012). Enantioseparation of beta-agonists with carboxymethyl-beta-cyclodextrin by CE, *J. Liq. Chromatogr. R. T.*, **35**, pp. 50–58.

11. Fortes, S. S., Barth, T., Furtado, N. A. J. C., Pupo, M. T., de Gaitani, C. M., and de Oliveira, A. R. M. (2013). Evaluation of dispersive liquid–liquid microextraction in the stereoselective determination of cetirizine following the fungal biotransformation of hydroxyzine and analysis by capillary electrophoresis, *Talanta*, **116**, pp. 743–752.

12. Li, L. B., Li, X., Luo, Q., and You, T. Y. (2015). A comprehensive study of the enantioseparation of chiral drugs by cyclodextrin using capillary electrophoresis combined with theoretical approaches, *Talanta*, **142**, pp. 28–34.

13. Wongwan, S., Sungthong, B., and Scriba, G. K. E. (2010). CE assay for simultaneous determination of charged and neutral impurities in dexamphetamine sulfate using a dual CD system, *Electrophoresis*, **31**, pp. 1475–1481.

14. Aguiar, F. A., de Gaitani, C. M., and Borges, K. B. (2011). Capillary electrophoresis method for the determination of isradipine enantiomers: Stability studies and pharmaceutical formulation analysis, *Electrophoresis*, **32**, pp. 2673–2682.

15. Nemeth, K., Tarkanyi, G., Varga, E., Imre, T., Mizsei, R., Ivanyi, R., Visy, J., Szeman, J., Jicsinszky, L., Szente, L., and Simonyi, M. (2011). Enantiomeric separation of antimalarial drugs by capillary electrophoresis using neutral and negatively charged cyclodextrins, *J. Pharmaceut. Biomed.*, **54**, pp. 475–481.

16. Deng, X. L., De Cock, B., Vervoort, R., Pamperin, D., Adams, E., and van Schepdael, A. (2012). Development of a validated capillary electrophoresis method for enantiomeric purity control and quality control of levocetirizine in a pharmaceutical formulation, *Chirality*, **24**, pp. 276–282.

17. Mohr, S., Pilaj, S., and Schmid, M. G. (2012). Chiral separation of cathinone derivatives used as recreational drugs by cyclodextrin-modified capillary electrophoresis, *Electrophoresis*, **33**, pp. 1624–1630.

18. Deng, X. L., Yuan, Y. Z., Adams, E., and van Schepdael, A. (2013). Development and validation of a sensitive enantiomeric separation method for new single enantiomer drug levornidazole by CD-capillary electrophoresis, *Talanta*, **106**, pp. 186–191.

19. Lehnert, P., Pribylka, A., Maier, V., Znaleziona, J., Sevcik, J., and Dousa, M. (2013). Enantiomeric separation of R,S-tolterodine and R,S-methoxytolterodine with negatively charged cyclodextrins by capillary electrophoresis, *J. Sep. Sci.*, **36**, pp. 1561–1567.

20. Yu, J., Jiang, Z., Sun, T. M., Ji, F. F., Xu, S. Y., Wei, L., and Guo, X. J. (2014). Enantiomeric separation of meptazinol and its three intermediate enantiomers by capillary electrophoresis: Quantitative analysis of meptazinol in pharmaceutical formulations, *Biomed. Chromatogr.*, **28**, pp. 135–141.

21. de Albuquerque, N. C. P., de Gaitani, C. M., and de Oliveira, A. R. M. (2015). A new and fast DLLME-CE method for the enantioselective analysis of zopiclone and its active metabolite after fungal biotransformation, *J. Pharmaceut. Biomed.*, **109**, pp. 192–201.

22. Hancu, G., Samarghitan, C., Rusu, A., and Mircia, E. (2014). Sotalol chiral separation by capillary electrophoresis, *J. Chil. Chem. Soc.*, **59**, pp. 2559–2563.

23. Hancu, G., Carje, A., Iuga, I., Fulop, I., and Szabo, Z. I. (2015). Cyclodextrine screening for the chiral separation of carvedilol by capillary electrophoresis, *Iranian J. Pharm. Res.*, **14**, pp. 425–433.

24. Kwan, H. Y. and Thormann, W. (2011). Enantioselective capillary electrophoresis for the assessment of CYP3A4-mediated ketamine demethylation and inhibition *in vitro*, *Electrophoresis*, **32**, pp. 2738–2745.

25. Fakhari, A. R., Tabani, H., Nojavan, S., and Abedi, H. (2012). Electromembrane extraction combined with cyclodextrin-modified

capillary electrophoresis for the quantification of trimipramine enantiomers, *Electrophoresis*, **33**, pp. 506–515.

26. Samakashvili, S., Salgado, A., Scriba, G. K. E., and Chankvetadze, B. (2013). Comparative enantioseparation of ketoprofen with trimethylated α-, β-, and γ-cyclodextrins in capillary electrophoresis and study of related selector–selectand interactions using nuclear magnetic resonance spectroscopy, *Chirality*, **25**, pp. 79–88.

27. Tabani, H., Fakhari, A. R., Shahsavani, A., and Alibabaou, H. G. (2014). Electrically assisted liquid-phase microextraction combined with capillary electrophoresis for quantification of propranolol enantiomers in human body fluids, *Chirality*, **26**, pp. 260–267.

28. Hancu, G., Budau, M., Kantor, L. K., and Carje, A. (2015). Cyclodextrine screening for the chiral separation of amlodipine enantiomers by capillary electrophoresis, *Adv. Pharm. Bull.*, **5**, pp. 35–40.

29. Lipka, E., Danel, C., Yous, S., Bonte, J. P., and Vaccher, C. (2010). Dual CD system in capillary electrophoresis for direct separation of the four stereoisomers of agonist and antagonist melatoninergic ligands, *Electrophoresis*, **31**, pp. 1529–1532.

30. Yu, J., Zhao, Y. F., Song, J. X., and Guo, X. J. (2014). Enantioseparation of meptazinol and its three intermediate enantiomers by capillary electrophoresis using a new cationic β-cyclodextrin derivative in single and dual cyclodextrin systems, *Biomed. Chromatogr.*, **28**, pp. 868–874.

31. Jin, Y., Chen, C., Meng, L. C., Chen, J. T., Li, M. X., and Zhu, Z. W. (2012). Simultaneous and sensitive capillary electrophoretic enantioseparation of three β-blockers with the combination of achiral ionic liquid and dual CD derivatives, *Talanta*, **89**, pp. 149–154.

32. Cui, Y., Ma, X. W., Zhao, M., Jiang, Z., Xu, S. Y., and Guo, X. J. (2013). Combined use of ionic liquid and hydroxypropyl-β-cyclodextrin for the enantioseparation of ten drugs by capillary electrophoresis, *Chirality*, **25**, pp. 409–414.

33. Zuo, L. H., Meng, H., Wu, J. J., Jiang, Z., Xu, S. Y., and Guo, X. J. (2013). Combined use of ionic liquid and β-CD for enantioseparation of 12 pharmaceuticals using CE, *J. Sep. Sci.*, **36**, pp. 517–523.

34. Zhang, J. J., Du, Y. X., Zhang, Q., Chen, J. Q., Xu, G. F., Yu, T., and Hu, X. Y. (2013). Investigation of the synergistic effect with amino acid-derived chiral ionic liquids as additives for enantiomeric separation in capillary electrophoresis, *J. Chromatogr. A*, **1316**, pp. 119–126.

35. Zhao, M., Cui, Y., Yu, J., Xu, S. Y., and Guo, X. J. (2014). Combined use of hydroxypropyl-β-cyclodextrin and ionic liquids for the simultaneous

enantioseparation of four azole antifungals by CE and a study of the synergistic effect, *J. Sep. Sci.*, **37**, pp. 151–157.

36. Yu, T., Du, Y. X., Chen, J. Q., Xu, G. F., Yang, K., Zhang, Q., Zhang, J. J., Du, S. J., Feng, Z. J., and Zhang, Y. J. (2015). Study on clarithromycin lactobionate based dual selector systems for the enantioseparation of basic drugs in capillary electrophoresis, *J. Sep. Sci.*, **38**, pp. 2900–2906.

37. Chen, J. Q., Du, Y. X., Zhu, F. X., and Chen B. (2010). Evaluation of the enantioselectivity of glycogen-based dual chiral selector systems towards basic drugs in capillary electrophoresis, *J. Chromatogr. A*, **1217**, pp. 7158–7163.

38. Merola, G., Fu, H. Z., Tagliaro, F., Macchia, T., and McCord, B. R. (2014). Chiral separation of 12 cathinone analogs by cyclodextrin-assisted capillary electrophoresis with UV and mass spectrometry detection, *Electrophoresis*, **35**, pp. 3231–3241.

39. Qi, Y. D. and Zhang, X. J. (2014). Determination of enantiomeric impurity of levamlodipine besylate bulk drug by capillary electrophoresis using carboxymethyl-β-cyclodextrin, *Cell Biochem. Biophys.*, **70**, pp. 1633–1637.

40. Hamidi, S., Soltani, S., and Jouyban, A. (2015). A dispersive liquid–liquid microextraction and chiral separation of carvedilol in human plasma using capillary electrophoresis, *Bioanalysis*, **7**, pp. 1107–1117.

41. Chankvetadze, B. (1997). *Capillary Electrophoresis in Chiral Analysis*, (John Willey & Sons, Chichester).

42. Burrai, L., Nieddu, M., Pirisi, M. A., Carta, A., Briguglio, I., and Boatto, G. (2013). Enantiomeric separation of 13 new amphetamine-like designer drugs by capillary electrophoresis, using modified-β-cyclodextrins, *Chirality*, **25**, pp. 617–621.

43. Ibrahim, W. A. W., Arsad, S. R., Maarof, H., Sanagi, M. M., and Aboul-Enein, H. Y. (2015). Chiral separation of four stereoisomers of ketoconazole drugs using capillary electrophoresis, *Chirality*, **27**, pp. 223–227.

44. Sattary, F., Shafaati, A., and Zarghi, A. (2013). Improvement of capillary electrophoretic enantioseparation of fluoxetine by a cationic additive, *Iranian J. Pharm. Res.*, **12**, pp. 71–76.

45. Xue, H. B., Jiao, Y. N., and Li, H. (2013). Capillary electrophoresis separation of the six pairs of chiral pharmaceuticals enantiomers, *Asian J. Chem.*, **25**, pp. 2742–2746.

46. Iwamuro, Y., Iio-Ishimaru, R., Chinaka, S., Takayama, N., Kodama, S., and Hayakawa, K. (2010). Reproducible chiral capillary electrophoresis

of methamphetamine and its related compounds using a chemically modified capillary having diol groups, *Forensic Toxicol.*, **28**, pp. 19–24.

47. Iwata, Y. T., Mikuma, T., Kuwayama, K., Tsujikawa, K., Miyaguchi, H., Kanamori, T., and Inoue, H. (2013). Applicability of chemically modified capillaries in chiral capillary electrophoresis for methamphetamine profiling, *Forensic Sci. Int.*, **226**, pp. 235–239.

48. Mikuma, T., Iwata, Y. T., Miyaguchi, H., Kuwayama, K., Tsujikawa, K., Kanamori, T., and Inoue, H. (2015). The use of a sulfonated capillary on chiral capillary electrophoresis/mass spectrometry of amphetamine-type stimulants for methamphetamine impurity profiling, *Forensic Sci. Int.*, **249**, pp. 59–65.

49. Servais, A. C., Rousseau, A., Fillet, M., Lomsadze, K., Salgado, A., Crommen, J., and Chankvetadze, B. (2010). Separation of propranolol enantiomers by CE using sulfated beta-CD derivatives in aqueous and non-aqueous electrolytes: Comparative CE and NMR study, *Electrophoresis*, **31**, pp. 1467–1474.

50. Chankvetadze, L., Servais, A. C., Fillet, M., Salgado, A., Crommen, J., and Chankvetadze, B. (2012). Comparative enantioseparation of talinolol in aqueous and non-aqueous capillary electrophoresis and study of related selector–selectand interactions by nuclear magnetic resonance spectroscopy, *J. Chromatogr. A*, **1267**, pp. 206–216.

51. Feng, Y., Wang, T. T., Jiang, Z. J., Chankvetadze, B., and Crommen, J. (2015). Comparative enantiomer affinity pattern of β-blockers in aqueous and nonaqueous CE using single-component anionic cyclodextrins, *Electrophoresis*, **36**, pp. 1358–1364.

52. Ali, I., Sanagi, M. M., and Aboul-Enein, H. Y. (2014). Advances in chiral separations by nonaqueous capillary electrophoresis in pharmaceutical and biomedical analysis, *Electrophoresis*, **35**, pp. 926–936.

53. Guo, W. P., Rong, Z. B., Li, Y. H., Fung, Y. S., Gao, G. Q., and Cai, Z. M. (2013). Microfluidic chip capillary electrophoresis coupled with electrochemiluminescence for enantioseparation of racemic drugs using central composite design optimization, *Electrophoresis*, **34**, pp. 2962–2969.

54. Dixit, S. and Park, J. H. (2014). Application of antibiotics as chiral selectors for capillary electrophoretic enantioseparation of pharmaceuticals: A review, *Biomed. Chromatogr.*, **28**, pp. 10–26.

55. Prokhorova, A. F., Shapovalova, E. N., and Shpigun, O. A. (2010). Chiral analysis of pharmaceuticals by capillary electrophoresis using

antibiotics as chiral selectors, *J. Pharmaceut. Biomed.*, **53**, pp. 1170–1179.

56. Xu, G. F., Du, Y. X., Chen, B., and Chen, J. Q. (2010). Investigation of the enantioseparation of basic drugs with erythromycin lactobionate as a chiral selector in CE, *Chromatographia*, **72**, pp. 289–295.

57. Yu, T., Du, Y. X., and Chen, B. (2011). Evaluation of clarithromycin lactobionate as a novel chiral selector for enantiomeric separation of basic drugs in capillary electrophoresis, *Electrophoresis*, **32**, pp. 1898–1905.

58. Zhang, J. J., Du, Y. X., Zhang, Q., and Lei, Y. T. (2014). Evaluation of vancomycin-based synergistic system with amino acid ester chiral ionic liquids as additives for enantioseparation of non-steroidal anti-inflamatory drugs by capillary electrophoresis, *Talanta*, **119**, pp. 193–201.

59. Chen, B., Du, Y. X., and Wang, H. (2010). Study on enantiomeric separation of basic drugs by NACE in methanol-based medium using erythromycin lactobionate as a chiral selector, *Electrophoresis*, **31**, pp. 371–377.

60. Kumar, A. P. and Park, J. H. (2011). Azithromycin as a new chiral selector in capillary electrophoresis, *J. Chromatogr. A*, **1218**, pp. 1314–1317.

61. Nojavan, S. and Fakhari, A. R. (2011). Chiral separation and quantitation of cetirizine and hydroxyzine by maltodextrin-mediated CE in human plasma: Effect of zwitterionic property of cetirizine on enantioseparation, *Electrophoresis*, **32**, pp. 764–771.

62. Mohammadi, A., Nojavan, S., Rouini, M., and Fakhari, A. R. (2011). Stability evaluation of tramadol enantiomers using a chiral stability-indicating capillary electrophoresis method and its application to pharmaceutical analysis, *J. Sep. Sci.*, **34**, pp. 1613–1620.

63. Fakhari, A. R., Tabani, H., Behdad, H., Nojavan, S., and Taghizadeh, M. (2013). Electrically-enhanced microextraction combined with maltodextrin-modified capillary electrophoresis for quantification of tolterodine enantiomers in biological samples, *Microchem. J.*, **106**, pp. 186–193.

64. Nojavan, S., Pourmoslemi, S., Behdad, H., Fakhari, A. R., and Mohammadi, A. (2014). Application of maltodextrin as chiral selector in capillary electrophoresis for quantification of amlodipine enantiomers in commercial tablets, *Chirality*, **26**, pp. 394–399.

65. Abolhasani, J., Jafariyan, H. R., Khataei, M. M., Hosseinzadeh-khanmiri, R., Ghorbani-kalhor, E., and Hassanpour, A. (2015). Hollow fiber

supported liquid-phase microextraction combined with maltodextrin-modified capillary electrophoresis for the determination of citalopram enantiomers in urine samples, *Anal. Methods*, **7**, pp. 2012–2019.

66. Tabani, H., Mahyari, M., Sahragard, A., Fakhari, A. R., and Shaabani, A. (2015). Evaluation of sulfated maltodextrin as a novel anionic chiral selector for the enantioseparation of basic chiral drugs by capillary electrophoresis, *Electrophoresis*, **36**, pp. 305–311.

67. Zhang, Q. and Du, Y. X. (2013). Evaluation of the enantioselectivity of glycogen-based synergistic system with amino acid chiral ionic liquids as additives in capillary electrophoresis, *J. Chromatogr. A*, **1306**, pp. 97–103.

68. Zhang, Y. J., Huang, M. X., Zhang, Y. P., Armstrong, D. W., Breitbach, Z. S., and Ryoo, J. J. (2013). Use of sulfated cyclofructan 6 and sulfated cyclodextrins for the chiral separation of four basic pharmaceuticals by capillary electrophoresis, *Chirality*, **25**, pp. 735–742.

69. Wu, J. F., Liu, P., Wang, Q. W., Chen, H., Gao, P., Wang, L., and Zhang, S. Y. (2011). Investigation of enantiomeric separation of chiral drugs by CE using Cu(II)–clindamycin complex as a novel chiral selector, *Chromatographia*, **74**, pp. 789–797.

70. Wang, L. J., Hu, S. Q., Guo, Q. L., Yang, G. L., and Chen, X. G. (2011). Di-n-amyl L-tartrate–boric acid complex chiral selector in situ synthesis and its application in chiral nonaqueous capillary electrophoresis, *J. Chromatogr. A*, **1218**, pp. 1300–1309.

71. Tsioupi, D. A., Nicolaou, I. N., Moore, L., and Kapnissi-Christodoulou, C. P. (2012). Chiral separation of huperzine A using CE–method validation and application in pharmaceutical formulations, *Electrophoresis*, **33**, pp. 516–522.

72. Li, L. X., Xia, Z. N., Yang, F. Q., Chen, H., and Zhang, Y. L. (2012). Enantioselective analysis of ofloxacin enantiomers by partial-filling capillary electrophoresis with bacteria as chiral selectors, *J. Sep. Sci.*, **35**, pp. 2101–2107.

73. Nojavan, S. and Fakhari, A. R. (2010). Electro membrane extraction combined with capillary electrophoresis for the determination of amlodipine enantiomers in biological samples, *J. Sep. Sci.*, **33**, pp. 3231–3238.

74. Castro-Puyana, M., Lammers, I., Buijs, J., Gooijer, C., and Ariese, F. (2010). Sensitized phosphorescence as detection method for the enantioseparation of bupropion by capillary electrophoresis, *Electrophoresis*, **31**, pp. 3928–3936.

75. Sánchez-Hernández, L., García-Ruiz, C., Crego, A. L., and Marina, M. L. (2010). Sensitive determination of D-carnitine as enantiomeric impurity of levo-carnitine in pharmaceutical formulations by capillary electrophoresis–tandem mass spectrometry, *J. Pharmaceut. Biomed.*, **53**, pp. 1217–1223.

76. Wongwan, S. and Scriba, G. K. E. (2011). Development and validation of a capillary electrophoresis assay for the determination of the stereoisomeric purity of chloroquine enantiomers, *Electrophoresis*, **32**, pp. 2669–2672.

77. Schmitz, A., Thormann, W., Moessner, L., Theurillat, R., Helmja, K., and Mevissen, M. (2010). Enantioselective CE analysis of hepatic ketamine metabolism in different species *in vitro*, *Electrophoresis*, **31**, pp. 1506–1516.

78. Portmann, S., Kwan, H. Y., Theurillat, R., Schmitz, A., Mevissen, M., and Thormann, W. (2010). Enantioselective capillary electrophoresis for identification and characterization of human cytochrome P450 enzymes which metabolize ketamine and norketamine *in vitro*, *J. Chromatogr. A*, **1217**, pp. 7942–7948.

79. Kwan, H. Y. and Thormann, W. (2012). Electrophoretically mediated microanalysis for characterization of the enantioselective CYP3A4 catalyzed N-demethylation of Ketamine, *Electrophoresis*, **33**, pp. 3299–3305.

80. Reminek, R., Glatz, Z., and Thormann, W. (2015). Optimized on-line enantioselective capillary electrophoretic method for kinetic and inhibition studies of drug metabolism mediated by cytochrome P450 enzymes, *Electrophoresis*, **36**, pp. 1349–1357.

81. Theurillat, R., Larenza, M. P., Feige, K., Wolfensberger, R., and Thormann, W. (2014). Development of a method for analysis of ketamine and norketamine enantiomers in equine brain and cerebrospinal fluid by capillary electrophoresis, *Electrophoresis*, **35**, pp. 2863–2869.

82. Meng, L., Wang, B., Luo, F., Shen, G. J., Wang, Z. Q., and Guo, M. (2011). Application of dispersive liquid–liquid microextraction and CE with UV detection for the chiral separation and determination of the multiple illicit drugs on forensic samples, *Forensic Sci. Int.*, **209**, pp. 42–47.

83. Lee, Y. J., Choi, S., Lee, J., Nguyen, N., Lee, K., Kang, J. S., Mar, W., and Kim, K. H. (2012). Chiral discrimination of sibutramine enantiomers by capillary electrophoresis and proton nuclear magnetic resonance spectroscopy, *Arch. Pharm. Res.*, **35**, pp. 671–681.

84. Xu, Z. Q., Li, A. M., Wang, Y. L., Chen, Z. L., and Hirokawa, T. (2014). Pressure-assisted electrokinetic injection stacking for verteporfin drug to achieve highly sensitive enantioseparation and detection in artificial urine by capillary electrophoresis, *J. Chromatogr. A*, **1355**, pp. 284–290.

85. Tonon, M. A. and Bonato, P. S. (2012). Capillary electrophoretic enantioselective determination of zopiclone and its impurities, *Electrophoresis*, **33**, pp. 1606–1612.

Chapter 3

A Mini-Review on Enantiomeric Separation of Ofloxacin using Capillary Electrophoresis: Pharmaceutical Applications

Suvardhan Kanchi, Myalowenkosi Sabela, Deepali Sharma, and Krishna Bisetty

Department of Chemistry, Durban University of Technology, P.O. Box 1334, Durban 4000, South Africa
ksuvardhan@gmail.com

3.1 Introduction

Among thousands of drugs in the world, more than half have chiral characteristics, ~20% are single enantiomers, and the rest of the drugs are racemic in nature. Even though enantiomers are similar in many aspects, they frequently exhibit various bioactivities. The main purpose of chiral separation has been complimentary in pharmaceuticals to avoid the harmful form of drug [1, 2]. Ofloxacin [(±)-9-fluro-2,3-dihydro-3-methyl-10-(4-methyl-1-piperazinyl)-7-oxo-7H-pyrido-[1,2,3-de]-1,4-benzoxazine-6-carboxylic] is fluo-

---

Capillary Electrophoresis: Trends and Developments in Pharmaceutical Research
Edited by Suvardhan Kanchi, Salvador Sagrado, Myalowenkosi Sabela, and Krishna Bisetty
Copyright © 2017 Pan Stanford Publishing Pte. Ltd.
ISBN 978-981-4774-12-3 (Hardcover), 978-1-315-22538-8 (eBook)
www.panstanford.com

roquinolone, a broad-spectrum antibacterial drug belonging to the third-generation synthetic class of quinolone. The structure of ofloxacin is shown in Fig. 3.1.

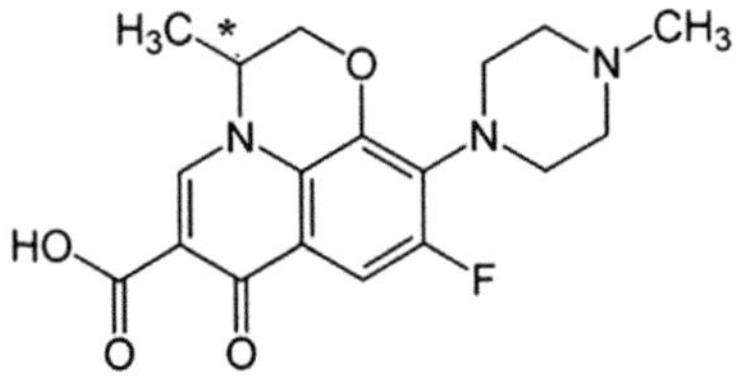

Figure 3.1 Chemical structure of Ofloxacin (*: Chiral center).

(S)-(–)-ofloxacin isomer (levofloxacin) is twofold active as ofloxacin and 8–128 times active as (R)-(+)-ofloxacin (dextrofloxacin) [3, 4]. Consequently, research activities on the chiral separation of ofloxacin in pharmaceuticals have attracted more attention in the recent years.

3.2 Role of CE in Pharmaceutical Analysis

Since the 1980s, capillary electrophoresis (CE) has been used in the pharmaceutical industry for the separation of large and small biological molecules as drugs. CE has conventionally been applied mainly to proteins and nucleic acids (DNA, RNA), but it is also applicable over a wide range of analytes covering small inorganic ions to cell organelles and even complete cells and viruses. The CE technique is mainly being used to replace traditional chromatographic techniques and also gel electrophoresis. Subsequently, it works quite well for protein analysis; therefore, it has recently gained an increased interest within the field of biotechnology as well.

Chiral CE is a treasured instrument for the separation of chiral drugs, as most of the pharmaceutical products are chiral and each enantiomer exhibits different toxicity and pharmacological behavior. For example, enantiomers of anti-hypertensive and non-steroidal anti-inflammatory drugs show different activities, side effects, or even toxicity. Therefore, chiral separation in pharmaceutical analysis results to get the non-toxic and appropriate enantiomer.

CE has proven to be a powerful separation tool for the stereoselective analysis of several classes of compounds using different separation mechanisms due to its simplicity, high resolution, and speed when compared with other analytical techniques such as high-performance liquid chromatography (HPLC), thin-layer chromatography (TLC), and gas chromatography (GC).

CE is important for quality control of pharmaceutical entities and products (e.g., qualitative and quantitative analyses, purity testing, chiral purity, and related substance and stoichiometric determination). Quantitative analysis is mainly determined by comparison of mobility or migration time of the compounds of interest with those of the standards. It is calculated from the peak height or peak area based on a calibration curve of a standard. Purity testing is usually required for drugs with low stability, whereas the determination of related substances is essential to ensure that precursors or reagents from synthesis processes do not exceed limits. Chiral purity is now routinely performed for racemic drugs, in which each enantiomer shows different pharmacological activity or toxicity. CE is currently an established method for pharmaceutical analysis and is recommended in several pharmacopeias.

3.3 Challenges Involved in the Enantiomeric Separation of Ofloxacin

Ofloxacin enantiomer analysis is very important in pharmaceuticals, especially in the department of quality control to investigate the pharmacokinetics, designing of in vitro studies and development of novel chiral products. Therefore, numerous chromatographic techniques, including CE [5, 6] and HPLC [7, 8], using chiral stationary phases or chiral mobile phases for the analysis of ofloxacin enantiomers [9, 10] have been developed. In most cases, HPLC analysis requires specially designed chiral column for chiral stationary phase, which involves the cost factor. Therefore, to overcome these drawbacks, efficient and cost-effective RP18 columns eluted with ligand mobile phase containing amino acid (L-leucine or L-phenylalamine) were reported in the literature [11–13] for the separation of ofloxacin enantiomers. Nevertheless, none of these reported methods were able to distinguish the separation and quantification of ofloxacin enantiomers individually in pharmaceutical formulations. Using HPLC, chiral separation of ofloxacin in biological samples has been reported [11–13] with the addition of chiral selector to the mobile phase or using chiral stationary phase. However, this reported method suffers from a few drawbacks such as elaborate extraction or derivatization, tedious sample preparation, and requirement of expensive chiral columns. For the last two centuries, CE has proven to be a versatile and alternative technique to HPLC for enantioselective analysis [14, 15]. When compared to HPLC, CE offers many advantages, such as high separation efficiency, less consumption of chiral selector, ease of operation, and quick analysis.

3.4 Developments in Enanatiomeric Separation of Ofloxacin

Several methods of CE have been reported on the chiral separation of ofloxacin enantiomers using various chiral selectors such as vancomycin [16], cyclodextrins, [17, 18], bovine serum albumin [19], or by combining β-cyclodextrin (β-CD) with D-phenylalanine and zinc sulfate [20]. However, the resolution factors remained poor, not reproducible, and required long migration time [21]. In later 1990s, microchip capillary electrophoresis (MCE) attracted much attention due to its good sensitivity, less consumption of reagents, shorter analysis time, and high integration in the field of chemistry and biochemistry [22–28]. A multidisciplinary field of microfluidic chip, micro-total analysis systems (μ-TAS), or lab-on-chip (LOC) was reported for the enantiomeric separation of ofloxacin [29]. A novel laser-induced fluorescence (LIF) detector based on CE was developed to improve quantitative assay of ofloxacin enantiomers in human urine [24]. Although LIF detectors are sensitive in terms of chiral separation, they are not widely reported in the literature because they are also expensive.

Electrochemical detectors such as conductometry, potentiometry, and amperometry are often used when the optical path of the optical detectors is short [22–27]. These detectors are cost effective and lead to miniaturization due to their intrinsic properties. The conductometry detector has either contact or contactless (oscillometric) patterns, which is facile and sensitive for the enantiomeric separation and their detection [28, 29]. The contactless pattern has several merits when compared to the contact pattern, due to:

- Its effective separation at high voltages,
- Prevention of electrode corrosion, and
- Compatibility with the microfluidic chip method.

3.4.1 Recent Developments in Enantiomeric Separation of Ofloxacin

Chen et al. [30] developed a novel method for the enantiomeric separation of ofloxacin with MCE using capacitively coupled contactless conductivity detection. The operation conditions were as follows: 1 mM 2-(*N*-morpholino)ethanesulfonic acid (MES) and 1 mM Tris (pH 8.0) as a running buffer, separation voltage of 1.5 kV, and injection time of 10 s. Under these experimental conditions, the enantiomers were achieved within 1 min, which is one of the advantages of the method. The LODs were found to be 18 and 21 µg mL^{-1} for levofloxacin and dextrofloxacin, respectively, with a relative standard deviations of 3.4% ($n = 6$) for levofloxacin and 4.0% ($n = 6$) for dextrofloxacin in terms of peak areas. Furthermore, the method has been applied to the ofloxacin eye drops for chiral separation. The typical electropherograms of ofloxacin (a) standard and (b) ofloxacin eye drops are illustrated in Fig. 3.2.

Another CE method for the separation of the enantiomers of ofloxacin using carboxymethyl-β-cyclodextrin (CM-β-CD) as chiral selector was described by Elbashir et al. [31]. The effect of the type of cyclodextrin, its concentration, and its pH, as well as instrumental parameters, such as applied voltage and temperature, were systematically studied. The highest resolution between the ofloxacin enantiomers was around 2.8. This was achieved using Tris-citrate buffer (pH 4.5) that contained 3 mg mL^{-1} CM-β-CD using UV detection (254 nm), applied voltage (12 kV), and capillary temperature of 25°C. The obtained results suggested that the recoveries were found to be ranging from 98.3% to 103.4% for the enantiomeric separation of ofloxacin from spiked samples. The method is fast, sensitive, and inexpensive, and its usefulness was demonstrated for the analysis of five pharmaceutical preparations, two of which just contained the (*S*)-ofloxacin, while the other three contained both isomers as racemic mixtures, as shown in Fig. 3.3.

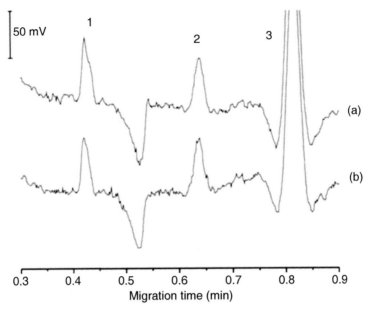

Figure 3.2 Electropherogram of ofloxacin (a) standard solution and (b) ofloxacin eye drops. Conditions: buffer, 1 mmol L^{-1} MES and 1 mmol L^{-1} Tris (pH 8.0); separation voltage, 1.5 kV; injection time, 10 s; sample concentration in standard solution, 0.15 mg mL^{-1}. Reprinted with permission from Ref. [30], Copyright 2014, John Wiley and Sons.

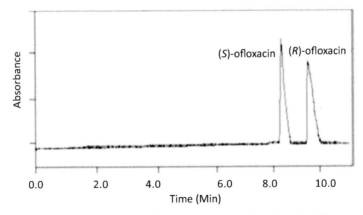

Figure 3.3 Electropherogram obtained from the injection of 100 μg mL^{-1} standard racemic ofloxacin sample. Reprinted with permission from Ref. [31], Copyright 2007, Taylor & Francis.

Xue et al. reported a fast, facile, and sensitive CE method coupled with chemiluminescence (CL) detector for the analysis of ofloxacin enantiomers [32]. In this study, sulfonated β-cyclodextrin (β-CD) was used as the chiral selector and added to the running buffer of luminol-diperiodatocuprate (DPC) (III)(K5[Cu(HIO6)2], DPC) chemiluminescence system. The basic experiment demonstrated that luminol could be oxidized by DPC to produce CL emission in alkaline solution, and this CL emission could be greatly enhanced in the reaction in the presence of ofloxacin. The kinetic characteristics of both DPC-luminol-ofloxacin and DPC-luminol CL reactions were tested by a static mode, and the typical response curves (CL intensity versus times) are shown in Fig. 3.4. In this method, β-CD, tosyl-β-CD, and sulfated-β-CD were also tested for the enantiomeric separation of OFs. However, they showed limited solubility and slight interaction with the analytes. The electropherograms obtained from the three chiral selectors are shown in Fig. 3.5. A satisfactory chiral recognition was achieved when sulfated-β-CD was chosen as the chiral selector. The concentration of sulfated-β-CD was a key factor influencing the chiral resolution of ofloxacins. The highest signal intensity and the best peak shapes were obtained when the concentration of sulfated-β-CD was 3.0 mg mL^{-1} [32]. DPC was used as an efficient oxidant in this CL reaction and showed a great effect on the generation of CL signal. The decreased CL intensity at higher DPC concentration could be attributed to the self-absorption of the DPC solution, since DPC showed a peak absorbance at 415 nm, which was very near to the maximum emission wavelength of luminol (425 nm) NaOH. The results indicated that the maximal relative CL intensity was reached when the NaOH concentration was 40 mM, providing required alkaline environment in the CL system. At higher concentrations of NaOH, the CL signal decreased. Thus, 40 mM was chosen as the optimal NaOH concentration. In order to avoid the electrical discrimination resulting from the electromigration injection, the sample was loaded into the capillary column by an altitude difference. The increased loading time and height difference led to a high CL signal owing to a large sample loading amount, which also resulted in a broadened peak and a decreased resolution [32]. This proposed method was successfully applied to the separation and analysis of ofloxacin enantiomers with the detection limit (S/N=3) of 8.0 and 7.0 nM for levofloxacin and dextrofloxacin, respectively.

The method was utilized for analyzing ofloxacin in urine where the results obtained were satisfactory and recoveries were ranging from 89.5% to 110.8%, which demonstrated the reliability of the method. This approach can also be further extended to analyze different commercial medicines for ofloxacin [32].

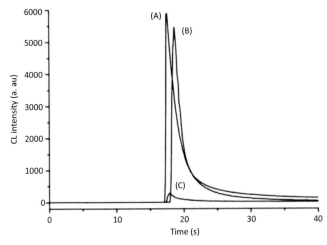

Figure 3.4 Kinetic curves for (A and B) luminol-DPC-ofloxacin and (C) luminol-DPC-CL systems. 100 mL DPC solution (5.0 µM) was injected into a mixture of 0.50 mL of 1.0 µM luminol (in 40 mM NaOH), 10 µL of 3.0 mg mL^{-1} sulfated-β-CD, and 1.0 mL of 5.0 µM (A) levofloxacin or (B) dextrofloxacin. (C) 100 mL DPC solution (5.0 µM) was injected into a mixture of 0.50 mL of 1.0 µM luminol (in 40 mM NaOH), 10 µL of 3.0 mg mL^{-1} sulfated-β-CD and 1.0 mL of water. Reprinted from open access article in Ref. [32]. Open Access funded by Xi'an Jiaotong University, under a Creative Commons license.

Ofloxacin enantiomers were separated and quantified using capillary zone electrophoresis in extremely diluted samples. Electrokinetic injection mode was adopted after liquid–liquid extraction of ofloxacin from physiological solution to improve the sensitivity. The method was optimized using a central composite design for four experimental factors, which include background electrolyte concentration, methyl-β-cyclodextrin concentration, pH, and the temperature. The use of experimental design strategies for optimization and robustness testing allowed for a development of an analytical method to quantify the ofloxacin enantiomers in the ppb range where methyl-β-CD is used as a chiral selector. The possibility of dealing with several responses at a time, obtaining the best compromise between different goals, was a very important aspect, since separation techniques usually require a simultaneous optimization of several responses. In this method, increasing the buffer concentration improves the resolution and increases the amount of the analyte migrating into the capillary; however, it produces unwanted effects on the migration time and current. While the pH has a positive effect on the resolution, it increases the migration time. The chiral selector concentration shows similar effects as pH on all responses. The resolution between the ofloxacin enantiomers, the migration time, and the generated current were evaluated as responses depend on the amount of sample migrated into the capillary, which will be determined by the peak area. The quantification limits were found to be 11.4 ng mL^{-1} for (S)-ofloxacin and 10.8 ng mL^{-1} for (R)-ofloxacin.

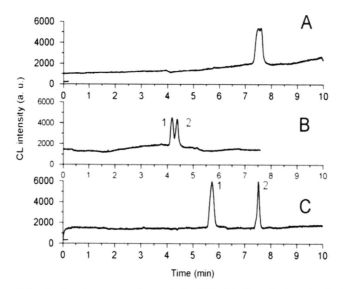

Figure 3.5 Electropherograms for analysis of 40 µM ofloxacin. Running buffer: 2.0 mM phosphate buffer (pH 4.0) containing 0.10 mM luminol and 3.0 mg mL^{-1} β-CD (A), or 3.0 mg mL^{-1} tosyl-β-CD (B), or 3.0 mg mL^{-1} sulfated-β-CD (C). The solution of 0.10 mM DPC in 40 mM NaOH was siphoned into the detection window with an altitude difference of 25 cm, 1=dextrofloxacin, 2=levofloxacin. Reprinted from open access article in Ref. [32]. Open Access funded by Xi'an Jiaotong University, under a Creative Commons license.

Simultaneous separation and enantioseparation of ofloxacin and its metabolities (desmethyl ofloxacin and ofloxacin N-oxide) with CE were developed and validated in human urine [33, 34]. Sulfobutyl-β-cyclodextrin was used as a running buffer for enantioseparation of ofloxacin. The analytes were detected by LIF detection, which uses an HeCd-laser with an excitation wavelength of 325 nm. In comparison with conventional UV detection, LIF detection provides higher sensitivity and selectivity. The separation was performed after direct injection of urine into the capillary without any sample preparation, and there was no interference effect with the assay. Additionally, the high sensitivity of this method allowed the quantification of the very low concentrations of enantiomers of both metabolites. The limit of quantification was 250 ng mL^{-1} for ofloxacin enantiomers and 100 ng mL^{-1} for corresponding metabolite's enantiomers. The method was applied to the analysis of human urine samples collected from a volunteer after oral administration of 200 mg of (6)-ofloxacin to elucidate stereoselective differences in the formation and excretion of the metabolites. With reference to Fig. 3.6, it could be demonstrated that the renal excretion of the (S)-configured metabolites, especially (S)-desmethyl ofloxacin, within the first 20 h after dosage was significantly lower than that of the (R)-enantiomers [33, 34].

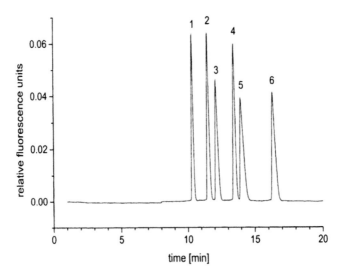

Figure 3.6 Electropherogram of an aqueous standard solution containing 5 mg mL^{-1} of S-(2)-OFLX (2), R-(1)-OFLX (5), S-DMOF (1), R-DMOF (3), S-OFNO (4), and R-OFNO (6); electrophoretic conditions: 30/37 cm capillary; applied voltage, 25 kV; BGE: 100 mM phosphate-TEA buffer, pH 2.0, containing 0.65 mg mL^{-1} sulfobutyl-β-cyclodextrin. Reprinted from Ref. [34], Copyright 2001, with permission from Elsevier.

3.5 Applications of Nanotechnology in Enantiomeric Separation of Ofloxacin

Separation sciences progress with size, and extensive efforts devoted to miniaturization of analytical instruments have significantly promoted the development of this field. The wide applications of novel materials with size at the sub-micron level as separation media have offered more opportunities for solving separation challenges. The potential of nanostructured materials, e.g., nanoparticles (NPs), in chromatographic and electrophoretic techniques has gradually been discovered in recent years. NPs are referred to particles in the size range of 1–100 nm with a large surface-to-volume ratio and other fascinating properties derived from the "quantum size effect." Existing NPs comprise the following [35–37]: fullerene NPs, silica NPs, noble metal NPs, metal oxide NPs, semiconductor quantum dots, and polymer-based NPs (i.e., polymer NPs, molecular micelles, molecularly imprinted polymers, and dendrimers). Surface modification of the NPs with functional groups and/or molecules is a key procedure to prevent them from aggregation and to control the particle size, thereby providing extra selectivity when the NPs are used as separation media. NPs are mostly stabilized by steric exclusion and electrostatic repulsion.

Various NPs have been employed in capillary and chip-based electrophoresis for improved separation with higher efficiency, selectivity, and better reproducibility due to their large surface-to-volume ratio and favorable surface chemistry. They serve either as inner surface coating in permanent or dynamic mode or as a pseudo-stationary phase added to the buffer and used in partial filling or continuous filling form. For example, AuNPs were taken to demonstrate the applications of NPs in CE. Citrate-stabilized AuNPs with average diameters of 18 nm and 10 nm have been used as anionic surface coating in normal and chip-based CE, respectively, with poly(diallyldimethyl ammonium chloride) as the first cationic layer [38, 39].

Novel chiral ligand-exchange capillary electrophoresis (LE-CE) was developed with polymeric nanoparticles as a chiral ligand and Cu^{2+} as a central ion [40]. It is based on the formation of complexes between ligands and central ions, typically a divalent metal ion. Enantioseparation by ligand exchange was restricted to analytes with two or three electron-donating groups such as amino acids, hydroxy acids, amino alcohols, or diols [41]. For example, according to the LE-CEC principle L-phenylalanine amide centered with Cu^{2+} was used as a chiral selector in the monolithic silica columns. It was shown that the chiral discrimination based on the principle of ligand exchange attributed to the exchange of one ligand in the Cu^{2+} complex on the stationary phase, an analyte ligand, forming ternary mixed copper complexes. Therefore, in this study, the enantioselectivity depended on the different interactions between NPs and the complex consisting of Cu^{2+} and ofloxacin enantiomers [42]. NPs were prepared by the polymerization of N-methacryloyl-L-histidine methyl ester (MAH) and ethylene dimethacrylate (EDMA) and were characterized by elemental analysis, Fourier transform infrared spectroscopy (FTIR), and atomic force microscopy (AFM). CE systems that contain NPs in running buffer can be thought of as pseudo-capillary electrochromatography. Using this approach, enantiomer separation of ofloxacin was carried out by using LE-CE. The results demonstrated that NPs with chiral-functionalized group interacted differently with structural enantiomers of ofloaxcin. Ofloxacin enantiomers were successfully enantioseparated by LE-CE using polymer NPs with chiral-functionalized group. Cu^{2+} was used as the central ion. Chromatographic conditions such as concentrations of NPs and Cu^{2+}, pH, and ACN content were optimized. Separation factor up to 2.9 was achieved. The enantioseparation efficiency of ofloxacin enantiomers in this study using NPs was far better than those obtained by both electromigration and liquid techniques [40]. The optimum running conditions for the enantioseparation of ofloxacin were found to be a background electrolyte (BGE) (pH 4.7) containing 70% ACN, 10 mM $CuSO_4$, 40 mM $(NH_4)_2SO_4$, and 30 mg mL^{-1} NPs. Under these conditions, the enantioseparation of ofloxacin was successfully achieved. With this system, (R)-ofloxacin and (S)-ofloxacin (levofloxacin) were used for the analysis capsules in the ofloxacin tablets [40].

Molecularly imprinted magnetic nanoparticles (MIP–MNPs) as stationary phase were developed for rapid enantioseparation by capillary electrochromatography. The nanoparticles were synthesized by the co-polymerization of methacrylic acid and ethylene glycol dimethacrylate on 3-(methacryloyloxy)propyltrimethoxysilane-functionalized magnetic nanoparticles (25 nm diameter) in the presence of template molecule, and characterized with infrared spectroscopy, thermal gravimetric analysis, and transmission electron microscope. The imprinted nanoparticles (200 nm diameter) could be localized as stationary phase in the microchannel of microfluidic device with the tunable packing length by the help of an external magnetic field [43]. Fabrication of MIP–MNPs–MD can be carried as follows: First, the slurry of MIP–MNPs was introduced into the capillary (80 mm length) using a syringe with microinjection pump, and packed in a region where the magnets were located, as shown in Fig. 3.7. The dense MIP–MNPs packing could be used as a stationary support for highly efficient separation.

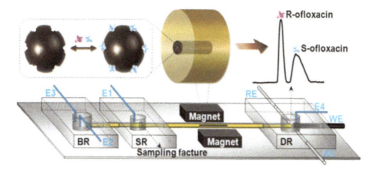

Figure 3.7 Schematic representation of enantioseparation on MIP–MNPs–MD. BR, buffer reservoir; SR, sample reservoir; DR, detection reservoir; WE, working electrode; RE, reference electrode; AE, auxiliary electrode; E1, E2, E3, and E4, electrodes for applying sampling and separation voltages. Reprinted from Ref. [43], Copyright 2010, with permission from Elsevier.

Briefly, a poly(dimethylsiloxane) matrix with an inner channel of 365 μm diameter was divided into three segments as the polymer retainers for the preparation of buffer reservoir (BR), sampling reservoir (SR), and detection reservoir (DR), respectively. The MIP–MNPs-packed capillary with two short magnets was then inserted into the SR retainer until a small scratch made previously appeared in the area of SR. After the BR and DR retainers were assembled to the two sides of the SR retainer by the capillary, the sampling fracture was obtained in the SR by sonicating the small scratch. Electrodes E1, E2, E3, and E4 were inserted into SR, BR, BR, and SR to achieve the sampling and separation, respectively. The sampling voltage was applied between E1 and E2 with E3 and

E4 in floating. The separation voltage was applied between E3 and E4 with E1 and E2 in floating. The voltage of 1000 V was applied to both ends of the capillary for 30 min to increase the packing density and remove the untrapped particles [45, 46]. Subsequently, the working electrode was mounted in the guide channel of the DR retainer, exactly opposite to the end of the separation channel at an optimum distance of 15 ± 5 μm [44, 47], and an Ag/AgCl reference electrode and a Pt wire as the auxiliary electrode were inserted into two sides of the DR to obtain an integrated three-electrode system for amperometric detection [43]. Using (S)-ofloxacin as the template molecule, the preparation of imprinted NPs, the composition and pH of mobile phase, and the separation voltage were optimized to obtain baseline separation of ofloxacin enantiomers within 3.25 min. The analytical performance could be conveniently improved by varying the packing length of NPs zone, showing an advantage over the conventional packed capillary electrochromatography. The linear ranges for amperometric detection of the enantiomers using carbon fiber microdisk electrode at +1.0 V (vs. Ag/AgCl) were from 1.0 to 500 μM and 5.0 to 500 μM with the LODs of 0.4 and 2.0 μM, respectively. The magnetically tunable microfluidic device could be expanded to localize more than one kind of template-imprinted magnetic NPs for realizing simultaneous analysis of different kinds of chiral compounds.

3.6   Role of Computational Techniques in Enantiomeric Separation of Ofloxacin

The resolution mechanism when CDs are used as chiral selectors is usually based on inclusion complexation where the analyte fits into the CD cavity, i.e., guest–host mechanism. Thus, the cavity of the CD and the shape and structure of the analyte are of paramount importance in the chiral recognition process. However, spatial compatibility alone is not sufficient for chiral recognition; interactions between substituents on the asymmetric center of the analyte and the hydroxyl groups on the CD-rim are also responsible for the chiral recognition. Detail mechanisms underlying the separation in the presence of CD as a chiral selector for many analytes have not been fully clarified.

Therefore, it is required to understand how the enantioseparation was achieved. With molecular simulation, determination of the differences in intermolecular forces and energies of the guest–host systems allows better understanding of the separation mechanism and can assist in predicting the suitable chiral selector for a given enantioseparation system.

Enantiomers of ofloxacin and ornidazole were described using CE. Various parameters affecting the separation were also studied, which included the type and concentration of chiral selector, pH, voltage, and temperature. Good chiral separation of the racemic mixtures was achieved in less than 16 min with resolution factors R_s = 5.45 and 6.28 for ofloxacin and ornidazole enantiomers, respectively. Separation was conducted using a bare fused-silica capillary and a background electrolyte (BGE) of 50 mM H_3PO_4– 1M Tris solution; pH 1.85; containing 30 mg mL^{-1} of sulfated-β-cyclodextrin (S-β-CD). The separation was carried out in reversed polarity mode at 25°C, +18 kV, detection wavelength at 230 nm and using hydrodynamic injection for 15 s. The acceptable validation criteria for selectivity, linearity, precision, and accuracy were studied. The LOD and LOQ of the enantiomers (ofloxacin enantiomer 1 (OF-E1), ofloxacin enantiomer 2 (OF-E2), ornidazole enantiomer-1 (OR-E1), and ornidazole enantiomer-2 (OR-E2)) were (0.52, 0.46, 0.54, 0.89) and (1.59, 1.40, 3.07, 2.70) $\mu g\ mL^{-1}$, respectively [48]. Additionally, molecular modeling using semiempirical calculations (Parameterized Model, PM3) was used to aid in the understanding of the mechanism of the separation.

The starting geometries of ornidazole and ofloxacin enantiomer structures were built based on their structures generated from the crystallographic parameters provided by the Cambridge Structural Database (CSD) [49, 50]. The initial structure was built based on the molecular structure of the β-CD, and three sulfate groups were added to the structure as proposed by Ma et al. [51]. The starting geometries of the inclusion complexes were constructed using HyperChem [52]. All the structures were fully optimized using the semiempirical method, PM3, implemented in the Gaussian 03 software package [53].

The coordinate system used to define the complexation process was based on constructing the β-CD with the seven identical glucose units positioned symmetrically around the z-axis, such that all the glycosidic oxygens were in the xy-plane and their center was defined as the center of the coordination system. Multiple starting points were generated by moving the guest molecules along the $-z$ and $+z$-axis. The inclusion complexes were emulated by moving the guest molecule from $+10$ Å to -10 Å, at 1 Å intervals, and by rotating the guest molecules from $0°$ to $360°$ at $30°$ intervals. The complexation energy ΔE_{comp} was calculated for the minimum energy structures by the following equation:

$$\Delta E_{comp} = E_{comp} - E_{fg} - E_{S-\beta-CD}$$

where E_{comp}, E_{fg}, and $E_{S\,\beta\,CD}$ represent the total energy of the host–guest complex, the free guest molecule, and the free host molecule, respectively. The magnitude of the energy change is an indication of the driving force toward complexation. The more negative the complexation energy change is, the more thermodynamically favorable the inclusion complex.

To further understand how enantiodifferentiation takes place at the atomic level, molecular modeling techniques were used to complement the experimental results. The geometries were optimized for the lowest energy conformation for the inclusion complexes of the ornidazole and ofloxacin enantiomers with S-β-CD. The inclusion complexes of S-β-CD with (S)- and (R)-ofloxacin enantiomers had larger binding energies when compared to the ornidazole counterparts. The large negative binding energies indicated that the S-β-CD can form more stable inclusion complexes with both the (R)- and (S)-enantiomers. The obtained results showed that the (R)-ornidazole/S-β-CD complex was significantly more favorable than the (S)-ornidazole/S-β-CD by an energy difference of -7.76 kJ mol^{-1}, and similar results were observed for the ofloxacin. However, the difference in the complexation energy between the (R)- and (S)-enantiomers is smaller (-4.28 kJ mol^{-1}). The negative values for $\Delta G°$ for all the complexes indicated the spontaneity of the binding of the guest molecule to the host, i.e., S-β-CD. The equilibrium constant K at 298.15 K was determined using the following equation:

$$\Delta G° = -RT \ln K$$

The results indicate that the equilibrium favors the formation of the complexes for both ofloxacin and ornidazole molecules. The lowest energy conformation for the ornidazole complexes shows the involvement of one sulfate group and the external part of the S-β-

CD, which provides a more favorable environment for recognition. A close-up view into the ornidazole/S-β-CD complexes shows that one of the sulfate groups forms hydrogen bonding with the H atom from the methyl as well as chloropropyl groups in ornidazole. Both the ornidazole enantiomers only slightly touch the secondary rim of the cyclodextrin, and encapsulation rather than inclusion of the guest molecule takes place through the sulfate group. A different complexation behavior was observed for ofloxacin enantiomers. The (R)-ofloxacin molecule penetrates through the piperazine ring, which fits tightly and deeply inside the center of the cavities of the CD. However, the piperazine ring of (S)-ofloxacin enantiomer is only partially included into the ring. Interestingly, the contribution from the sulfate group that forms hydrogen bonding through the hydrogen atom of the $-CH_2-$ in the oxazine ring and the oxygen atom in the sulfate group was also observed. In addition, coulombic bondings of CH–F types were also detected in both the (R)- and (S)-ofloxacin complexes. The completely symmetrical CD host structure adopted nonsymmetrical structure upon complexation with both ornidazole and ofloxacin molecules. As ornidazole is a smaller guest molecule compared to ofloxacin, the different complexation and recognition behaviors are expected. A view of the potential energy profile of ornidazole/S-β-CD complexes shows that the central part of the CD cavity represents a highly unfavorable environment for trapping the ornidazole enantiomers [48].

Hydroxypropyl-β-cyclodextrin (HP-β-CD) has been used successfully as a chiral additive to separate ofloxacin enantiomers in a CE system. Using electrospray-mass spectrometry (ESI–MS), it was revealed that ofloxacin forms an inclusion complex with HP-β-CD at 1:1 stoichiometry. The interaction of enantiomers with the host was also investigated by molecular modeling using molecular mechanics dockings, PM7 semiempirical calculations, and molecular dynamics simulations. Calculations using PM7 semiempirical methods indicated that the separation was brought about by a large difference in the binding energies (DDE) of 15 kcal mol^{-1} between (R)-ofloxacin-HP-β-CD and (S)-ofloxacin-HP-β-CD inclusion complexes. (S)-ofloxacin was predicted to be eluted first by the PM7 method, which corroborates the experimental results. Moreover, the molecular dynamic simulations show the formation of stable (R)-ofloxacin-HP-β-CD as it is more deeply inserted into the cavity of the host. The study also revealed the absence of the role of strong hydrogen bonding in the enantioseparations [54].

The starting structures of (R)- and (S)-ofloxacin as well as β-CD were extracted from the crystallographic parameters provided by the Structural Data Base System of the Cambridge Crystallographic Data Center. All molecules were fully optimized with the semiempirical method PM7 [55] using the MOPAC2012 package (www.openmopac.net). Further, the structure of 2-hydroxypropyl-β-cyclodextrin (HP-β-CD) was built on the β-CD structure by substitutions of 2-hydroxypropyl moieties randomly at O2 and O6 positions.

The most stable complex between (R)- and (S)-ofloxacin and the host molecule molecular docking studies were performed using the Autodock program (version 4.2) [56]. In this study, Lamarckian genetic algorithm (LGA) for the docking study was used to generate the inclusion complexes. Autodock usually defines a conformational space by implementation of grids over all the possible search space. Grid maps of 58 × 58 × 58 Å grid box with 0.375 Å spacing were obtained using the Autogrid 4 program. The center of mass of the HP-β-CD was set as the center of the box. The initial torsions and positions of (R)- or (S)-ofloxacin were generated randomly. With the help of Autodock tools, the partial charges were calculated using the Gasteiger–Marsili method [57, 58]. For the search, population of 100 LGA runs with a maximum number of energy evaluations of 2.5×10^7 and a maximum number of generations of 27,000. An elitism value of 1 was used and a probability of mutation and cross-over of 0.02 and 0.08 were used, respectively. At the end of each run, the solutions are separated into clusters based on their lowest root mean square deviation (RMSD) and the best score based on a free energy function. Cluster solutions with average scores that are over 1.0 kcal mol^{-1} with respect to the best energy obtained in the respective run were selected. The final structures obtained were refined and optimized by the semiemperical PM7 method using the MOPAC2012. Using the same method, thermodynamic parameters of the inclusion were estimated at a pressure of 1 atm and a temperature of 298 K.

In this study, each system consisting of a guest and host molecule was solvated in a sphere of TIP3P water molecules [59] using periodic boundary conditions. Langevin dynamics were implemented to maintain the temperature and the pressure close to 300 K and 1 atm, respectively. The solvated complexes were equilibrated and energy was minimized prior to MD simulations. Solvated complexes were heated to 300 K followed by an equilibration step of 500 ps at 300 K and 1 atm. MD production runs were performed at constant temperature and pressure for an additional 5 ns. The dynamics were performed using NAMD package [60], using CHARMM22 parametrization [61]. The time step for MD simulation was 2 fs. The initial geomoetries, topologies, and force–field parameter files were generated by VEGA ZZ software [62], which was also used for the analysis of the MD trajectories. Also VMD software [63] was used to perform analysis of the trajectories.

3.6.1 Molecular Modeling

It is known that a precise description of the structure of chemically modified CD such as HP-β-CD is not realistic because of the wide range of random substitution patterns occurring during derivatization. HP-β-CD is customarily synthesized via condensation reaction between propylene oxide and β-CD. Depending on the alkalinity, different substituents at O2 and O6 positions of β-CD were obtained [64]. In the work reported by Rusu et al. [65] and Babic et al. [66], HP-β-CD species were used, in which two substituents at second position and four substituents at sixth position are available. The obtained structure was fully optimized by the newly introduced semiempirical method PM7. The acid–base properties of ofloxacin and other fluoroquinolones have been described in the literature [65, 66]. The presence of a carboxylic acid group (pK_a 6.2) at C11 and a piperazinyl group (pK_a 8.2) at C6 indicates that ofloxacin can exist in aqueous solutions, as cation, zwitterion, or anion depending on the pH of the solution. The monoprotonated cation, where the piperazinyl group was protonated, predominates in the pH range 2–4 and was used in this study.

Initially, molecular docking of the inclusion of ofloxacin into HP-β-CD nanocavity was performed using Autodock 4.2 [56]. By cluster analysis, all complexes differing by 1.0 Å in a positional root mean square deviation (RMSD) were clustered together. The results showed that the lowest energy structure (average DG = −6.50 kcal mol^{-1}) for (R)-ofloxacin-HP-β-CD corresponds to a frequency of 75%, whereas that for (S)-ofloxacin–HP-β-CD, the lowest energy structure (average DG = −6.55 kcal mol^{-1}) corresponds to 90% of ofloxacin conformation, indicating convergence of the docking procedure for both systems. The predominant conformations of the inclusion complexes obtained for both enantiomers were shown in Fig. 3.8. It is evident from this figure that both enantiomers are introduced into the HP-β-CD through the secondary hydroxyl group side (wider side). This is expected as the side of cyclodextrin bearing the primary hydroxyl is affected by the steric blockage by the 2-hydroxypropyl substituents. It is clear that the large size of the fused ring system of ofloxacin can easily fit into the extended and distorted CD cavity of HP-β-CD. These findings were also in line with the reported NMR results [67] and in all cases, the guest molecule is tilted toward the sides of the CD in order to maximize host–guest interactions such as hydrogen bonding.

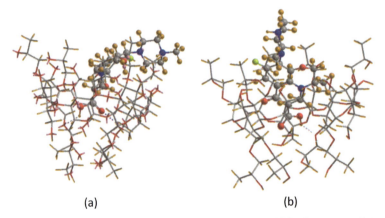

Figure 3.8 Geometries of the inclusion complexes of (a) (S)-ofloxacin-HP-β-CD and (b) (R)-ofloxacin-HP-β-CD. Reprinted from Ref. [67], Copyright 2010, with permission of Springer.

The differences in binding energies obtained by Autodock were insignificantly compared to the standard errors reported for the docking simulations; therefore, further calculations using the semiempirical PM7 were also conducted. Quantum mechanical calculations were then performed on the structures of the complexes obtained from the docking simulation using the PM7 method, a modified version of PM6 where errors associated with large molecules were reduced [55]. To obtain the binding energies ($\Delta E_{binding}$) of the 1:1 inclusion complexes, the following equation was used:

$$\Delta E_{\text{binding}} = E_{\text{complex}} - E_{\text{ofloxacin}} - E_{\text{HP-}\beta\text{-CD}}$$

where E_{complex}, $E_{\text{ofloxacin}}$, and $E_{\text{HP-}\beta\text{-CD}}$ are the energies of the inclusion complex, the free ofloxacin, and the free HP-β-CD molecules, respectively. These results clearly show that interaction of guest with the host is associated with an energy that was always lower than the sum of the energies of the free molecules, indicative of the formation of stable inclusion complexes in both cases. The obtained results indicate that the binding energy obtained by quantum mechanics calculations of HP-β-CD/(R)-ofloxacin complex is higher than that of HP-β-CD/(S)-ofloxacin by 15.0 kcal mol^{-1}. These results were in line with the CE results, indicating that a less stable complex is obtained with the (S)-ofloxacin, which travels faster through the column and, therefore, HP-β-CD efficiently separates the two enantiomers.

Using the parametric model PM7, statistical thermodynamic calculations were performed at 1 atm and 298 K. The complex formation is clearly associated with a relatively large negative ΔH values, indicating that it is an enthalpy-driven process. The replacement of high enthalpy water molecules from the cavity of the CD by suitable, less polar molecules accompanied by the formation of strong hydrophobic interaction remarkably lower the energy of the system. Deep insertion of guest molecules inside the host results in strong van der Waals effects, a process characterized by negative enthalpy changes and negative entropy changes. Similar results were experimentally obtained for the inclusion complex formation of ofloxacin with β-CD [68]. It was reported that the complexation was associated with an enthalpy change, ΔH of -30.5 kcal mol^{-1} and an entropy change ΔS of -87.2 J mol^{-1} K^{-1}. These results are in line with those obtained by the theoretical procedure used in this report. Evidently, there are no simple models that could describe the mechanism of enantioseparation of racemic mixture by supramolecular assembly as it involves several forces working together, such as hydrogen bonding, van der Waals interactions, release of ring strain, and hydrophobic interactions among others.

3.6.2 Molecular Dynamic Simulations

In order to further rationalize recognition of (R)- and (S)-ofloxacin enantiomers by HP-β-CD and to have more insight into the mechanism of separation, Suliman et al. [54] performed MD simulations on their complexes with HP-β-CD. These calculations were aimed at investigating the intermolecular interactions between the guest and the host molecules. In fact, the analysis was focused on the hydrogen bond type of interactions. Nevertheless, in macrocycle molecules such as cyclodextrin, guest molecules were customarily held strongly in the hydrophobic cavity by various interactions such as van der Waal forces, hydrophobic interactions and electrostatic forces, with hydrogen bonds playing a dominant role in enantiomer recognitions.

The authors Suliman et al. [54] conducted 5 ns MD simulation in order to investigate the various structural features that contribute to the stability of the host–guest complexes of the drug with HP-β-CD. Analysis of these trajectories along the MD runs at constant temperature and pressure was performed, showing that the stable complexes were obtained.

The root mean square deviations (RMSD) of the complex conformations from a given reference frame were found to be stable after a simulation times of 1 ns for both systems. The time evolutions of the distances from the center of mass of the macrocycle host and that of the drug (dCn–Cn) clearly suggest that the relative placements of (R)- and (S)-ofloxacin guest molecule with respect to the host cavity are well preserved throughout the simulation time with values of dCn–Cn of 1.7 ± 0.2 and 3.3 ± 0.5 Å for (R)- and (S)-ofloxacin, respectively. Moreover, the radius of gyration (rgyr) was found to be 6.26 ± 0.07 and 6.48 ± 0.11 Å for (R)- and (S)-ofloxacin-HP-β-CD complexes, respectively. On the other hand, rgyr values for HP-β-CD was found to be 6.50 ± 0.07 Å and that of ofloxacin was 4.0 ± 0.04 Å. Inspection of these results reveals the fact that the values of rgyr of the complexes were always lower than the sum of the two components and close to that of HP-β-CD, indicating the formation of stable complexes during the simulation time. These results suggested that (R)-ofloxacin forms a more stable complex with HP-β-CD compared to (S)-isomer [54].

3.7 Conclusion

Chiral CE is a treasured instrument for the separation of chiral drugs, as most of the pharmaceutical products are chiral and each enantiomer exhibits different toxicity and pharmacological behavior. CE is important for quality control of pharmaceutical entities and products. For the last two centuries, it has proven to be a versatile and alternative technique to HPLC for enantioselective analysis. When compared to HPLC, CE offers many advantages such as (a) high separation efficiency, (b) easy to operate, (c) less consumption of chiral selector, and (d) quick analysis. In this chapter, several methods were reviewed for the enantioseparation of ofloxacin, but interestingly β-CD is proved as an excellent candidate for the chiral selector not only for ofloxacin, but also for several organic molecules of pharmaceutical importance. Separation sciences progress with size, and extensive efforts devoted to miniaturization of analytical instruments have significantly promoted the development of this field. The wide applications of novel materials with size at the sub-micron level as separation media have offered more opportunities for solving separation challenges. However, to overcome the challenges in using NPs to enhance the enantioseparation, the following requirements are to be implemented:

- High surface area
- Selective interaction with the analyte
- Different mobility from that of the electroosmotic flow
- Equal mobility to prevent peak broading

Additionally, molecular modeling and molecular dynamic simulations can be used by adopting semiempirical calculations, which are powerful tools to predict the mechanism of separation. However, molecular docking assists in predicting the suitable chiral selector for a given enantioseparation system.

Acknowledgment

The authors gratefully acknowledge the financial assistance from the Durban University of Technology, South Africa.

References

1. Elbashir, A. A., Saad, B., and Aboul-Enein, H. Y. (2010). Recent developments of enantioseparations for fluoroquinolones drugs using

liquid chromatography and capillary electrophoresis, *Current Pharm. Anal.*, **6**, pp. 246–255.

2. Tian, M., Row, H. S., and Row, K. H. (2010). Chiral separation of ofloxacin enantiomers by ligand exchange chromatography, *Monatsh Chem.*, **141**, pp. 285–290.

3. The Pharmacopoeia Commission of the People's Republic of China. *Pharmacopoeia of the People's Republic of China*, Chemical Industry Press: Beijing, 2010; pp. 821–825.

4. Arai, T. and Kuroda, H. (1991). Distribution behavior of some drug enantiomers in an aqueous two-phase system using counter-current extraction with protein, *Chromatographia*, **32**, pp. 56–60.

5. Al Azzam, K. M., Saad, B., Adnan, R., and Aboul-Enein, H. Y. (2010). Enantioselective analysis of ofloxacin and ornidazole in pharmaceutical formulations by capillary electrophoresis using single chiral selector and computational calculation of their inclusion complexes, *Anal. Chim. Acta*, **674**, pp. 249–255.

6. de Boer, T., Mol, R., de Zeeuw, R. A., de Jong, G. J., and Ensing, K. (2001). Enantioseparation of ofloxacin in urine by capillary electrokinetic chromatography using charged cyclodextrins as chiral selectors and assessment of enantioconversion, *Electrophoresis*, **22** (68), pp. 1413–1418.

7. Sun, X., Wu, D., Shao, B., and Zhang, J. (2009). High-performance liquid-chromatographic separation of ofloxacin using a chiral stationary phase, *Anal. Sci.*, **25**, pp. 931–938.

8. Zeng, S., Zhong, J., Pan, L., and Li, Y. (1999). High-performance liquid chromatography separation and quantitation of ofloxacin enantiomers in rat microsomes, *J. Chromatogr. B*, **728**, pp. 151–155.

9. Yan, H. and Row, K. H. (2007). Rapid chiral separation and impurity determination of levofloxacin by ligand-exchange chromatography, *Anal. Chim. Acta*, **584**, pp. 160–165.

10. Bi, W., Tian, M., and Row, K. H. (2011). Chiral separation and determination of ofloxacin enantiomers by ionic liquid-assisted ligand-exchange chromatography, *Analyst*, **136**, pp. 379–387.

11. Sherrington, L. A., Abba, M., Hussain, B., and Donnelly, J. (2005). The simultaneous separation and determination of five quinoline antibiotics using isocratic reversed phase HPLC: Application to stability studies on an ofloxacin tablet formulation, *J. Pharm. Biomed. Anal.*, **39**, pp. 769–775.

12. Lin, W. F., Kang, X. H., Chen, Z. G., Yang, L., Cai, P. X., and Mo, J. Y. (2005). Content determination of ofloxacin and levofloxacin in their preparations by capillary electrophoresis, *Chin. J. Antibiot.*, **30**, pp. 756–758.

13. Fierens, C., Hillaert, S., and Bossche, W. V. D. (2002). The qualitative and quantitative determination of quinolones of first and second generation by capillary electrophoresis, *J. Pharm. Biomed. Anal.*, **22**, pp. 763–772.

14. Blanco, M. and Valverde, I. (2003). Choice of chiral selector for enantioseparation by capillary electrophoresis, *Trends Anal. Chem.*, **22**, pp. 428–439.

15. Chankevetadze, B. (2001). Enantioseparation of chiral drugs and current status of electromigration techniques in this field, *J. Sep. Sci.*, **24**, pp. 691–705.

16. Arai, T., Nimura, N., and Kinoshita, T. (1996). Investigation of enantioselective separation of quinolonecarboxylic acids by capillary zone electrophoresis using vancomycin as a chiral selector, *J. Chromatogr. A*, **736**, pp. 303–311.

17. Zhou, S., Ouyang, J., Baeyyens, W. R. G., Zhao, H., and Yang, Y. (2006). Chiral separation of four fluoroquinlones compounds using capillary electrophoresis with hydroxypropyl-b-cyclodextrin as chiral selector, *J. Chromatogr. A*, **1130**, pp. 296–301.

18. Boer, T. D., Mol, R., Zeeuw, R. A. D., Jong, G. J. D., and Ensing, K. (2001). Enantioseparation of ofloxacin in urine by capillary electrokinetic chromatography using charged cyclodextrins as chiral selectors and assessment of enantioconversion, *Electrophoresis*, **22**, pp. 1413–1418.

19. Zhu, X. F., Ding, Y. S., Lin, B. C., Jakob, A., and Koppenhoefer, B. (1999). Study of enantioselective interactions between chiral drugs and serum by capillary electrophoresis, *Electrophoresis*, **20**, pp. 1869–1877.

20. Horimai, T., Ohara, M., and Ichinose, M. (1997). Optical resolution of new quinoline drugs by capillary electrophoresis with ligand-exchange and host-guest interactions, *J. Chromatogr. A*, **760**, pp. 235–244.

21. Horstkotter, C. and Blaschke, G. (2001). Stereoselective determination of ofloxacin and its metabolites in human urine by capillary electrophoresis using laser induced fluorescence detection, *J. Chromatogr. B*, **754**, pp. 169–178.

22. Wang, J., Escarpa, A., Pumera, M., and Feldman, J. (2002). Capillary electrophoresis-electrochemistry microfluidic system for the determination of organic peroxides, *J. Chromatogr. A*, **952**, pp. 249–254.

23. Nan, H. and Lee, S. W. (2012). Fast screening of rice knockout mutants by multi-channel microchip electrophoresis, *Talanta*, **97**, pp. 249–255.

24. Xu, L., Guo, Q. H., Yu, H., Huang, J. S., and You, T. Y. (2012). Simultaneous determination of three β-blockers at a carbon nanofiber paste electrode by capillary electrophoresis coupled with amperometric detection, *Talanta*, **97**, pp. 462–467.

25. Huang, B. R., Huang, C. G., Liu, P. P., Wang, F. F., Na, N., and Jin, O. Y. (2011). Fast haptoglobin phenotyping based on microchip electrophoresis, *Talanta*, **85**, pp. 333–338.

26. Yang, F., Li, X. C., Zhang, W., Pan, J. B., and Chen, Z. G. (2011). A facile light-emitting-diode induced fluorescence detector coupled to an integrated microfluidic device for microchip electrophoresis, *Talanta*, **84**, pp. 1099–1106.

27. Krizkova, S., Hrdinova, V., Adam, V., Burgess, E. P. J., Kramer, K. J., Masarik, M., and Kizek, R. (2008). Chip-based CE for Avidin determination in transgenic tobacco and its comparison with square-wave voltammetry and standard gel electrophoresis, *Chromatographia*, **67**, pp. S75–S81.

28. Horváth, G., Kovács, K., Kocsis, B., and Kustos, I. (2009). Effect of thyme (*Thymus vulgaris* L.) essential oil and its main constituents on the outer membrane protein composition of Erwinia strains studied with microfluid chip technology, *Chromatographia*, **70**, pp. 1645–1650.

29. Manz, A., Graber, N., and Widmer, H. M. (1990). Miniaturized total chemical analysis systems: A novel concept for chemical sensing, *Sens. Actuators B*, **1**, pp. 244–248.

30. Chen, B., Zhang, Y., Xie, H.-L., Chen, Q.-M., and Mai, Q.-H. (2014). Chiral separation of ofloxacin enantiomers by microchip capillary electrophoresis with capacitively coupled contactless conductivity detection, *J. Chin. Chem. Soc.*, **61**, pp. 432–436.

31. Elbashir, A. A., Saad, B., Ali, A. S. M., Saleh, M. I., and Aboul-Enein, H. Y. (2008). Determination of ofloxacin enantiomers in pharmaceutical formulations by capillary electrophoresis, *J. Liq. Chromatogr. Relat. Technol.*, **31**, pp. 348–360.

32. Xie, H.-Y., Wang, Z.-R., and Fu, Z.-F. (2014). Highly sensitive trivalent copper chelate–luminol chemiluminescence system for capillary electrophoresis chiral separation and determination of ofloxacin enantiomers in urine samples, *J. Pharm. Anal.*, **4**, pp. 412–416.

33. Awadallah, B., Schmidt, P. C., and Wahl, M. A. (2003). Quantitation of the enantiomers of ofloxacin by capillary electrophoresis in the parts

per billion concentration range for in vitro drug absorption studies, *J. Chromatogr. A*, **988**, pp. 135–143.

34. Horstkötter, C. and Blaschke, G. (2001). Stereoselective determination of ofloxacin and its metabolites in human urine by capillary electrophoresis using laser-induced fluorescence detection, *J. Chromatogr. B*, **754**, pp. 169–178.

35. Guihen, E. and Glennon, J. D. (2003). More recent progress in the preparation of Au nanostructures, properties, and applications, *Anal Lett.*, **36**, pp. 3309–3336.

36. Nilsson, C. and Nilsson, S. (2006). Nanoparticle-based pseudostationary phases in capillary electrochromatography, *Electrophoresis*, **27**, pp. 76–83.

37. Nilsson, C., Birnbaum, S., and Nilsson, S. (2007). Use of nanoparticles in capillary and microchip electrochromatography, *J. Chromatogr. A*, **1168**, pp. 212–224.

38. Neiman, B. Grushka, E., and Lev, O. (2001). Use of gold nanoparticles to enhance capillary electrophoresis, *Anal. Chem.*, **73**, pp. 5220–5227.

39. Pumera, M., Wang, J., Grushka, E., and Polsky, R. (2001). Gold nanoparticle-enhanced microchip capillary electrophoresis, *Anal. Chem.*, **73**, pp. 5625–5628.

40. Aydoğan, C., Karakoç, V., Yılmaz, F., Shaikh, H., and Denizli, A. (2013). Enantioseparation of ofloxacin by ligand exchange capillary electrophoresis using L-histidine modified nanoparticles as chiral ligand, *Hacettepe, J. Biol. Chem.*, **41**(1), pp. 29–36.

41. Scriba, G. K. E. (2012). Chiral recognition mechanisms in analytical separation sciences, *Chromatographia*, **75**, pp. 815–838.

42. Bi, W., Tian, M., and Row, K. H. (2011). Chiral separation and determination of ofloxacin enantiomers by ionic liquid-assisted ligand-exchange chromatography, *Analyst*, **136**, pp. 379–387.

43. Qu, P., Lei, J., Zhang, L., Ouyang, R., and Ju, H. (2010). Molecularly imprinted magnetic nanoparticles as tunable stationary phase located in microfluidic channel for enantioseparation, *J. Chromatogr. A*, **1217**, pp. 6115–6121.

44. Zhai, C., Qiang, W., Lei, J. P., and Ju, H. X. (2009). Highly resolved separation and sensitive amperometric detection of amino acids with an assembled microfluidic device, *Electrophoresis*, **30**, pp. 1490–1496.

45. Wang, Y. C., Zhang, Z. C., Zhang, L., Li, F., Chen, L., and Wan, Q. H. (2007). Magnetically immobilized beds for capillary electrochromatography, *Anal. Chem.*, **79**, pp. 5082–5086.

46. Okamoto, Y., Ikawa, Y., Kitagawa, F., and Otsuka, K. (2007). Preparation of fritless capillary using avidin immobilized magnetic particles for electrochromatographic chiral separation, *J. Chromatogr. A*, **1143**, pp. 264–269.

47. Zhai, C., Li, C., Qiang, W., Lei, J. P., Yu, X. D., and Ju, H. X. (2007). Amperometric detection of carbohydrates with a portable silicone/quartz capillary microchip by designed fracture sampling, *Anal. Chem.*, **79**, pp. 9427–9432.

48. Al Azzam, K. M., Saad, B., Adnan, R., and Aboul-Enein, H. Y. (2010). Enantioselective analysis of ofloxacin and ornidazole in pharmaceutical formulations by capillary electrophoresis using single chiral selector and computational calculation of their inclusion complexes, *Anal. Chim. Acta*, **674**, pp. 249–255.

49. Deng, L., Wang, W., and Lv, J. (2007). Ornidazole hemihydrate, *Acta Cryst. E*, **63**, O4204, doi:10.1107/S1600536807045680.

50. Yoshida, A. and Moroi, R. (1991). Crystal structure of ofloxacin perchlorate, *Anal. Sci.*, **7**, pp. 351–352.

51. Ma, S., Shen, S., Haddad, N., Tang, W., Wang, J., Lee, H., Yee, N., Senanayake, C., and Grinberg, N. (2009). Chromatographic and spectroscopic studies on the chiral recognition of sulfated β-cyclodextrin as chiral mobile phase additive enantiomeric separation of a chiral amine, *J. Chromatogr. A*, **1216**, pp. 1232–1240.

52. Hyper Chem Release 7.5 for Windows, HyperCube, Inc. USA. 2003.

53. Frisch, M. J., Trucks, G. W., Schlegel, H. B., Scuseria, G. E., Robb, M. A., Cheeseman, J. R., Montgomery Jr., J. A., Vreven, T., Kudin, K. N., Burant, J. C., Millam, J. M., Iyengar, S. S., Tomasi, J., Barone, V., Mennucci, B., Cossi, M., Scalmani, G., Rega, N., Petersson, G. A., Nakatsuji, H., Hada, M., Ehara, M., Toyota, K., Fukuda, R., Hasegawa, J., Ishida, M., Nakajima, T., Honda, Y., Kitao, O., Nakai, H., Klene, M., Li, X., Knox, J. E., Hratchian, H. P., Cross, J. B., Adamo, C., Jaramillo, J., Gomperts, R., Stratmann, R. E., Yazyev, O., Austin, A. J., Cammi, R., Pomelli, C., Ochterski, J. W., Ayala, P. Y., Morokuma, K., Voth, G. A., Salvador, P., Dannenberg, J. J., Zakrzewski, V. G., Dapprich, S., Daniels, A. D., Strain, M. C., Farkas, O., Malick, D. K., Rabuck, A. D., Raghavachari, K., Foresman, J. B., Ortiz, J. V., Cui, Q., Baboul, A. G., Clifford, S., Cioslowski, J., Stefanov, B. B., Liu, G., Liashenko, A., Piskorz, P., Komaromi, I., Martin, R. L., Fox, D. J., Keith, T., Al-Laham, M. A., Peng, C. Y., Nanayakkara, A., Challacombe, M., Gill, P. M. W., Johnson, B., Chen, W., Wong, M. W., Gonzalez, C., and Pople, J. A. (2003). Gaussian 03, Revision B.03, Gaussian, Inc., Pittsburgh, PA.

54. Suliman, F. O., Elbashir, A. A., and Schmitz, O. J. (2015). Study on the separation of ofloxacin enantiomers by hydroxylpropyl-β-cyclodextrin as a chiral selector in capillary electrophoresis: A computational approach, *J. Incl. Phenom. Macrocycl. Chem.*, **83**, pp. 119–129.

55. Stewart, J. J. P. (2013). Optimization of parameters for semiempirical methods VI: More modifications to the NDDO approximations and re-optimization of parameters, *J. Mol. Model.*, **19**, pp. 1–32.

56. Morris, G. M., Goodsell, D. S., Halliday, R. S., Huey, R., Hart, W. E., Belew, R. K., and Olson, A. J. In. The Scripps Research Institute, La Jolla (2009).

57. Sanner, M. F., Huey, R., Dallakyan, S., Lindstorm, W., Morris, G. M., Norledge, A., Omelchenko, A., Stoffler, D., and Vareille, G. In. The Scripps Institute, La Jolla (2007).

58. Gasteiger, J. and Marsili, M. (1980). Iterative partial equalization of orbital electronegativity: A rapid access to atomic charges, *Tetrahedron*, **36**, pp. 3219–3228.

59. Jorgensen, W. L., Chandrasekhar, J., Madura, J. D., Impey, R. W., and Klein, M. L. (1983). Comparison of simple potential functions for simulating liquid water, *J. Chem. Phys.*, **79**, pp. 926–935.

60. Phillips, J. C., Braun, R., Wang, W., Gumbart, J., Tajkhorshid, E., Villa, E., Chipot, C., Skeel, R. D., Kale, L., and Schulten, K. (2005). Scalable molecular dynamics with NAMD, *J. Comput. Chem.*, **26**, pp. 1781–1802.

61. MacKerell Jr., A. D., Bashford, D., Bellott, M., Dunbrack Jr., R. L., Evanseck, J. D., Field, M. J., Fischer, S., Gao, J., Guo, H., Ha, S., Joseph-McCarthy, D., Kuchnir, L., Kuczera, K., Lau, F. T. K., Mattos, C., Michnick, S., Ngo, T., Nguyen, D. T., Prodhom, B., Reiher, W. E., Roux, B., Schlenkrich, M., Smith, J. C., Stote, R., Straub, J., Watanabe, M., Wiorkiewicz-Kuczera, J., Yin, D., and Karplus, M. (1998). All-atom empirical potential for molecular modeling and dynamics studies of proteins, *J. Phys. Chem. B*, **102**, pp. 3586–3616.

62. Pedretti, A., Villa, L., and Vistoli, G. (2002). VEGA: A versatile program to convert, handle and visulize molecular structure on Windows based PCs, *J. Mol. Graph. Model.*, **21**, pp. 47–49.

63. Humphrey, W., Dalke, A., and Schulten, K. (1996). VMD-visual molecular dynamics, *J. Mol. Graph. Model.*, **14**, pp. 33–38.

64. Rao, C. T., Pitha, J., Lindberg, B., and Lindberg, J. (1992). Distribution of substituents in O-(2-hydroxypropyl) derivatives of cyclomaltooligosaccharides (cyclodextrins): Influence of increasing substitution, of the base used in the preparation, and of macrocyclic size, *Carbohydr. Res.*, **223**, pp. 99–107.

65. Rusu, A., Toth, G., Szocs, L., Kokosi, J., Kraszni, M., Gyresi, A., and Noszal, B. (2012). Triprotic site-specific acid-base equilibria and related properties of fluoroquinolone antibacterials original research article, *J. Pharm. Biomed. Anal.*, **66**, pp. 50–57.

66. Babic, S., Horvat, A. J. M., Pavlovic, D. M., and Kastelan-Macan, M. (2007). Determination of pK_a values of active pharmaceutical ingredients, *Trends Anal. Chem.*, **26**, pp. 1043–1138.

67. Li, J. and Zhang, X. (2011). Preparation and characterization of the inclusion complex of ofloxacin with β-CD and HP-β-CD, *J. Incl. Phenom. Macrocycl. Chem.*, **69**, pp. 173–179.

68. Ghosh, B. C., Deb, N., and Mukherjee, A. K. (2010). Determination of individual proton affinities of ofloxacin from its UV-Vis absorption, fluorescence and charge-transfer spectra: Effect of inclusion in β-cyclodextrin on the proton affinities, *J. Phys. Chem. B*, **114**, pp. 9862–9871.

Chapter 4

Nano-Stationary Phases for Capillary Electrophoresis Techniques

Chaudhery Mustansar Hussain and Sagar Roy
Department of Chemistry and Environmental Sciences,
New Jersey Institute of Technology, University Heights, Newark, NJ 07102
chaudhery.m.hussain@njit.edu

Capillary electrophoresis (CE) is a rapidly growing technique in separation sciences due to its high-resolution power, short analysis times, and low consumption of reagents and samples. Recently, nanomaterials have made great contributions to the innovation of new stationary phases for various chromatographic technologies in separation sciences. The introduction of nanomaterials has remarkably enhanced the performance of these stationary phases with regard to retentive capability and selectivity. In this chapter, we present an up-to-date overview of the developments of stationary phases based on nanomaterials for CE. We devote a large part of this review to the accomplishments in the past 5 years with regard to retention and selectivity offered by new nano-CE stationary phases, predominantly including carbon nanotubes, fullerenes, nano-

---

Capillary Electrophoresis: Trends and Developments in Pharmaceutical Research
Edited by Suvardhan Kanchi, Salvador Sagrado, Myalowenkosi Sabela, and Krishna Bisetty
Copyright © 2017 Pan Stanford Publishing Pte. Ltd.
ISBN 978-981-4774-12-3 (Hardcover), 978-1-315-22538-8 (eBook)
www.panstanford.com

graphene, and nano-diamonds. We have also detailed the different synthetic strategies for nano-stationary phase immobilization procedures. In the end, future prospects of new nano-CE stationary phases are also commented.

4.1 Introduction

CE is considered a special case of electrokinetic separation technique using an electrical field through submillimeter diameter capillaries and in micro- and nanofluidic channels to separate the components of a mixture. CE also often refers to capillary zone electrophoresis (CZE), and other electrophoretic techniques such as capillary gel electrophoresis (CGE), capillary isoelectric focusing (CIEF), capillary isotachophoresis, and micellar electrokinetic chromatography (MEKC) are considered in this class [1].

The research on CE was instigated with the use of glass U tubes and trials of both gel and free solutions as early as the late 1800s. Arnes Tiselius, in 1930, first revealed the potentiality of electrophoresis in separating proteins in free solution. Later, in the 1960s, Hjerten introduced the use of capillaries, followed by Jorgenson and Lukacs, who exhibited some extraordinary separation performances that seemed unachievable with other conventional techniques. It was found that with narrow capillaries, diameter reduced to 80 μm and the voltages as high as 30 kV could be applied without experiencing overheating problems and further, the separation time was also shortened to less than 30 min, which is comparable to modern chromatographic methods. Introduction of narrow capillary in electrophoresis had introduced several advantages over common problems in conventional electrophoresis; for instance, the thin dimensions of the capillaries significantly increased the surface-to-volume ratio and also eliminated overheating by high voltages, enhanced the separation efficiency, reduced time, etc. The incredible separation abilities of CE, the fabrication of low-cost, narrow-bore capillaries and the development of highly sensitive online detection methods collectively urged increasing interest among the scientific community to achieve further developments in this technique [2, 3].

4.2 Different Types of Nanoparticles/ Nanomaterials

Recently, researchers have demonstrated extensive applications of nanoparticles (NPs), which have drawn significant attention in various fields of chemistry, biology, and physics [2–5]. The unique characteristics of NPs that arise from the gradual transition of molecular level to solid state properties have fascinated scientists. Nanomaterials (NMs), in principle, are extremely small, having at least one dimension 100 nm or less. For the past 20 years, hundreds of different types of novel NPs have been developed, including quantum dots, nanocrystals, dendrimers, clusters, metal and metal oxide NPs, fullerenes, nanofibers, and nanotubes. Depending on their size, NMs can be classified into zero-dimensional, one-dimensional, two-dimensional, and three-dimensional nanostructures. Zero-dimensional (0D) nanostructures are the simplest building block used for NM design and are often denoted by NPs, nanoclusters, or nanocrystals. The dimensions of one-dimensional (1D) nanostructures are equivalent in all but one direction. The diameters of these materials are in the 1–100 nm range, whereas their lengths are, typically, few microns and are often termed nanotubes, nanofibers, nanowires, and nanorods. Similarly, two-dimensional (2D) and three-dimensional (3D) nanostructured species include nanoplates, nanosfilms, nanowalls, nanodisks, nanoprism, etc., and nanoballs, nanocoils, nanopillers, nanoflowers, etc., respectively. The high aspect ratio of these NMs in combination with unique physical, optical, sorption, and diffusion properties intrigues scientists and engineers to explore various fundamental and commercial applications [6–9].

Until now, very limited research articles have been published for the application of NPs for chemical separation via CE. However, research works reveal the utility and versatility of NPs in CE. The introduction of NMs into the pseudostationary or stationary phases further improves the selectivity mainly via three mechanisms [9]. NMs hold pleasant properties for incorporation into CE as pseudostationary and stationary phases, as these materials are extremely "small" and can be contained in very low concentration compared to traditional pseudostationary or stationary phases. The overall volume of the NMs possesses less than 1% of the total capillary volume [8]. Second, the large surface area of NPs in CE serves as the immobilization platform for the organo-functional groups via covalent–non-covalent bonding that interact with the capillary surface and/or the analytes that leads to enhanced selectivity via alteration of the apparent mobilities of target analytes, as well as the electro-osmotic flow. The main purpose of the NPs is to offer additional interaction sites with which the solutes can interact.

NMs perfectly play the role of pseudostationary and stationary phases in CE because of their small sizes, large surface-area-to-volume ratios versus bulk materials, and customizable surface chemistries. The stability of the NMs as pseudostationary and stationary phases is highly dependent on their core composition and surface chemistry. The classification of NMs' pseudostationary and stationary phases in various capillary-based separation techniques is shown in Fig. 4.1.

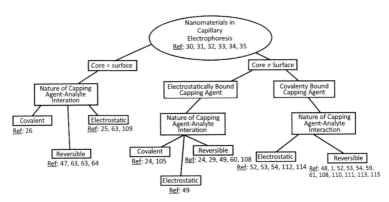

Figure 4.1 Classification of nanomaterials used in capillary electrophoresis. Nanomaterials are divided into two general categories: Core = Surface and Core ≠ Surface. Reprinted from Ref. [9], Copyright 2011, with permission from the Royal Society of Chemistry.

The NMs can be divided into two main categories: (1) where the core of the NMs has been modified by the capping agent; i.e., the surface composition is different from the core and (2) NMs are modified by inherent surface functionalization. The surface chemistry of NMs performs two important functions; first, it can regulate the separation mechanisms by controlling both electro-osmotic flow and capillary surface and control the elution order of targeted analytes. It is observed that effective capillary surface charge increases as the concentration of silica nanoparticles (diameter, d = 60 nm, pH~3) increases [10]. With increase in concentration (~50–180 nM), the agglomeration behavior of the NMs changes, which affects the interaction between the capillary wall and the analytes, resulting in a change in migration times of the analytes. These may also affect the detection sensitivities, unstable baselines, and irreproducible separations. Second, modification at the surface of the NMs could further enhance the stability by changing the NM–NM interactions, interactions between the NMs and the separation environment, and interactions between the NMs and analytes (Fig. 4.2).

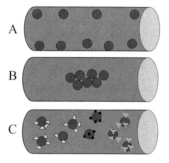

Figure 4.2 Nanomaterials impact separations via three mechanisms: (A) nanoparticle–capillary, (B) nanoparticle–nanoparticle, and (C) nanoparticle–analyte interactions. In all cases, the spheres represent nanoparticles. In part C, the squares, triangles, and rods represent various analytes. Drawings are not to scale. Adapted from Ref. [13].

It has been found that the use of citrate not only stabilizes gold NPs (d = 13 nm) but also enhances the detection limits by a factor of ~4000 during the extraction of indoleamines from solution [11–13].

Some examples of NPs that already exhibit their potentiality in CE analysis include gold, Fe_3O_4, graphene oxide (GO), carbon nanotubes (CNTs), etc. [14]. Nanostructures developed by the carbon moiety exhibit astoundingly versatile ability to interact with other functional groups in various ways to generate novel materials possessing myriad diverse properties. A wide variety of nanocarbons (NCs), including graphite, diamond, spherical fullerenes, CNTs, cup-stacked nanotubes, nanohorns, nanotori, nanobuds, and nano-onions [15, 16] have emerged recently in the field of separation and purifications.

The adsorption efficiency of NPs in matrix-assisted laser desorption/ionization mass spectroscopy (MALDI-MS) and high performance liquid chromatography–mass spectroscopy (HPLC-MS) analyses has been proven highly efficient during the extraction/enrichment of analytes from the sample matrices. Gold NPs exhibited stronger catalytic activity than gold's ionic forms when added to the CE running buffer. The chemiluminescence (CL) signal of the luminol–H_2O_2–gold colloids system has been inhibited by certain organic compounds containing OH, NH_2, and SH groups, which shows the potentiality of the sensitive determination of CL of these compounds. The gold NPs-catalyzed luminol CL also showed in situ detection of DNA and immunoassay of immunoglobulin G (IgG). Recently, studies on the AuNP-enhanced CE-CL technique and its analytical potentials have been reported. Metal NPs and polymer-based NPs have been used to coat fused silica capillaries and silica gel NPs as run buffer additives for use in CE [17–19]. GO is considered an outstanding adsorbent due to its high surface area and π–π electrostatic stacking property. GO supported on Fe_3O_4 showed excellent adsorption performance for solid phase extraction for various analytes ranging from small molecules of pollutants to biomolecules such as proteins and peptides. Moreover, the introduction of magnetic properties into graphene opens new windows for separation by combining the high adsorption capacity with the separation characteristics of the magnetic materials. These composites have been applied to arsenic removal of water, drug delivery and photothermal therapy, magnetic resonance imaging, extraction and determination of sulfonamide antibiotics, and carbamate pesticides in water samples [20–23].

CNTs have demonstrated their effectiveness and versatility in different fields of science and engineering, including CE. Depending on the method of synthesis, CNTs may vary in length, diameter, and structure. Many of their fascinating properties depend on these distinctive structural phenomena. These structural variations impart blending of physicochemical properties that include high aspect ratio, structural strength with high flexibility, and large modifiable surface area. Although CNTs are chemically inert, surface functionalization with oxygen-containing groups such as carboxyl, carbonyl, and hydroxyl moieties can be possible via oxidation of nanotubes using, e.g., a mixture of sulfuric and nitric acid that provides a foundation for covalent sidewall connection of other molecules. On the other hand, interactions of CNTs with various compounds, including gases, small organic molecules, surfactants, polymers, carbohydrates, nucleic acids, peptides, and proteins, via non-covalent bonding are also very simple and effective. Functionalization with diazonium salts can be very effective in controlling the conductivity of CNTs. The unique tubular structure of CNTs can form a network structure in the run buffer as a pseudostationary phase at a definite concentration yielding different separation mechanism from other conventional CE systems. The incorporation of functional groups onto CNT surfaces could affect the electrophoretic behavior by altering the interaction with the analytes. The aggregations of CNTs play the role of anti-convective media that minimize the solute diffusion, resulting in zone broadening. Again, the dispersibility and conductivity of CNTs are also affected by surface modifications. In general, CNTs are non-polar and insoluble in water. However, oxidization by concentrated acids yields hydroxyl, carbonyl, or carboxyl head groups, which increase their solubility in water. This strategy has been widely accepted to incorporate CNTs as a pseudostationary phase in electrokinetic chromatography since 2003 after successful application [24, 25].

Another variety of NPs has generated significant attention in CE: latex NPs that improve the separation of several pharmaceutical compounds. These latex NPs inherently contain separate surface chemistry that makes them water soluble, which are nonconductive. The incorporation of latex NPs as pseudostationary stage enhanced the separation efficiency dramatically with slightly lowering the average plate numbers. Silica NPs are also used widely in CE. The negatively charged surface silanol groups of silica NPs, above pH~2.3, force the analytes to migrate toward the anode, thus improving separation efficiency. The selectively induced hydrogen bonding between the analytes and the NP surface improves the separation of polar organic compounds, such as aromatic acids and antibacterial quinolones [25–27].

The role of nanoparticles as run buffer additives is quite similar to that of the use of micelle additives in MEKC. However, in the MEKC, the solutes can penetrate to the core of the micelles, whereas in NP-enhanced CE, the interactions with the solutes always occur at the outer surface of the NPs. These interactions could be either with the NP surface or with the organo-functional moieties attached to the surface of the NPs. The presence of NPs in the run buffer acts as pseudostationary phase, which helps to avoid the need to pack the capillary with a conventional stationary phase, thus eliminating the frits and other retaining techniques. During the movement of NPs through the capillary in the run buffer, the electro-osmotic flow and their own native electrophoretic mobility affect the apparent mobility, which yields continuous turnover in the interacting media. On the other hand, the movements of the solutes are influenced by the applied electrical field and are separated due to the differences in their effective charges and also by differential partitioning between the aqueous buffer and the NP phase. The migration times of the solutes also vary significantly with the change in apparent and electro-osmotic mobilities and the selectivities between them, which reflect their distinguished peaks' shape. This partitioning effect of NPs allows significant separation performances with neutral species and offers to control the nature of the interactions between the NPs and the solutes and customize the CE technique for a particular analyte system. Additionally, the application of NPs increased the operating electrical field range. The thermal effect induced by the applied electrical field limits the CE to a large extent. Thus, it is greatly advantageous to operate the CE at low ionic strength [28–30].

4.3 Synthetic Strategies for Immobilization for Nano-Stationary Phases

NM-enhanced CE is considered an effective technique for the liquid-phase separation of various species ranging from metal ions to organic molecules to biomolecules. The advancement in NM-immobilized pseudostationary and stationary phases (NISP) has drawn enormous attention in electrically driven separations. The high surface area of NPs along with diverse surface chemistry and multiple core compositions offers unique characteristics, which are very much appropriate for the separation techniques by CE. NPs also influence the viscosity, pH, and conductivity of the buffer solution significantly. However, one of the major disadvantages associated with the incorporation of NPs is its extremely high surface energy that induces agglomeration and may lead to incompetent, irreproducible, and unpredictable separation and detection performances. It is thus highly important to incorporate NPs into capillary system systematically and efficiently, which requires accurate control over the synthesis and characterization of NPs and capillary surface.

Initial use of NMs in electrokinetic chromatography employed continuous full filling methods of NMs into the background electrolyte solution, where the analytes interact throughout the capillary. In another partial filling technique, discontinuous pseudostationary phases of NMs have been generated by injecting the NM plugs into a capillary prior to the sample matrix. The NPs in pseudostationary phase enhance the analyte selectivity by influencing the mobility of the target molecules. Irrespective of the method, the injection can influence the NM concentration that drastically impacts the NM separation performances. Later, scientists developed a new technique of utilization of NMs by coating the capillary via covalent and/or electrostatic interactions [32], demonstrated as a stationary phase.

Surface modification of NPs can be utilized to prevent disorganized or produce organized aggregation of solution-phase NPs by creating an electrostatically induced steric barrier [33]. The modification of NM surfaces via self-assembled monolayers (SAMs) revolutionized the application of NMs in a wide variety of technologies. Various functional groups, including carboxylic acid, amine and amides, thiols, hydroxyl, isocyanide, phosphine, etc., have been used to fabricate SAMs on NM surface. The enhanced stability of dispersed NMs increased their usefulness in CE. The modified surface chemistry of modified NPs allowed to interact with target molecules differently during CE.

The dispersion of metallic NPs can be prepared in various organic and aqueous solutions using diverse procedures. The solvent itself sometimes can stabilize the dispersed nanophase, or it can be stabilized by the addition of surfactants, specific ligands. The dispersion of NMs into aqueous phase is more complicated and requires special treatment, such as capping of NPs via stabilization with various ligands, including citrate, 3-mercaptopropionate, coenzyme A, tiopronin, glutathione, 4-hydroxytiophenol, etc. These NPs can also be stabilized by steric stabilization through using polyelectrolytes, such as, poly(vinylpyrrolidine), poly-(ethylenoxide), or by formation of stable silicate shell around the particles. Gold NPs have been widely used as pseudostationary phase in CE. Modification of Au NPs with thioctic acid (Au-TA), carboxylic acid (Au-COOH), and citrate (Au-CT) influenced the zeta potential of the NPs, thus affecting the overall separation performances [34].

In a very well-known method, aqueous gold colloidal dispersions were prepared by the reduction of $AuCl^{4-}$ by the citrate ions, often used with sodium 3-mercaptopropionate, as a stabilizer, and all of the solutes were dissolved at 2×10^{-3} M concentration. The run buffer for this was prepared by adjusting the pH to 6.4 by using 0.005 M sodium phosphate dibasic heptahydrate solution with concentrated (15 M) phosphoric acid. In some other cases, the inner surface of the capillary was also modified with poly(diallyldimethylammonium) chloride (PDADMAC, having very high molecular weight). In this process, fused silica capillary was first washed with 1M NaOH solution to activate silanol, followed by water and buffer solution. The run buffer solution was then pumped through the capillary, and the high voltage (10 kV) was turned on for about 1 h. Then a run buffer solution containing PDAADMAC was pumped through the capillary for 30 min without voltage operation. The capillary was then flushed with the acetate run buffer solution containing PDADMAC at 1:5000 dilutions. The modified capillary was then ready to use [35].

The fabrication of GO-incorporated monolithic column via one-step room temperature polymerization technique for capillary electrochromatography (CEC) exhibits great separation performances. A GO-based novel stationary phase for CEC is expected considering the superior separation/adsorption properties of GO. Nevertheless, formation of GO nano film into the cartridge/column allows homogenous separation. The GO was grafted onto the 3-aminopropyl-trimethoxysilane (APTMS) treated capillary through the reaction between amino groups of capillary and epoxy, hydroxyl, carboxyl groups of GO. This GO-capillary exhibited successful baseline separation of several benzene compounds. Incorporation of GO significantly enhanced their CEC separation by improving the interactions between the tested neutral analytes (alkyl benzenes and polycyclic aromatics) and the stationary phase. Often fabrication of graphene-based magnetic NPs by the help of an external magnetic field is very useful to fabricate the stationary phase of the OT-CEC of tunable packing length [36, 37].

4.4 Evaluation of Nano-Stationary Phases

The applications of nano-stationary phases are rapidly rising and are changing the science of separation techniques. The large surface-area-to-volume ratio enhances the ability of the NPs to selectively interact with the capillary surface and/or the analyte yielding high mass transfer and radically changes the separation performances. In addition, the interaction of the NPs with the capillary surface could lead to a change in the electro-osmotic mobility of the run buffer, the selectivities between solute and capillary surfaces, and even alter its direction of movement. These changes significantly modify the migration times of the solutes and thus their peaks' shape [38].

Although in some cases traditional CE has been observed to be fast and highly efficient, but the small capillary used in CE possesses some inherent drawbacks that affect the performances of the traditional CE significantly. The shorter path length available with small-diameter capillaries reduces the sensitivity of the absorbance detector. On the other hand, the sensitivity may be enhanced by using larger-diameter capillaries and longer path length. However, the increase in inner diameter and path length is associated with poor dissipation of heat and the increment of Joule heating effect that results broadening of bands and poor resolution [39].

Polymer NPs have been used in CE separations for the last two decades. Wallingford, in 1989, demonstrated a pseudostationary phase containing sulfonated polymer nanoparticles (diameter 20 nm) that exhibited the improved separation performances of five catecholamines [39]. Although the results did show a poor resolution, the usefulness of NMs in CE was clearly demonstrated.

Gold ions (Au^{+3}) are widely used as catalyst in luminol CL reaction. However, it has been found that gold nanoparticles possess stronger catalytic activity on luminol CL than Au(III) did when added to the CE running buffer. The pseudostationary NP phase also improves the separation efficiency and reproducibility for acidic and basic proteins separation. Also the highly selective and sensitive AuNP-enhanced CE-CL analytical technique has been employed in clinical and biomedical applications for the detection of uric acid in physiological fluids, which are related to certain medical conditions such as diabetes. Researchers observed that AuNP-catalyzed luminol-H_2O_2 CL emission is inhibited by the presence of trace amount of uric acid, which led to the advancement of CE-CL assay of uric acid in human serum. The sensitivity of the detection, as well as the detection selectivity, was improved in the presence of Au NPs into running buffer. Medical test reports showed significant variation in uric acid levels when human serum samples were taken from normal and diabetic patients.

Another application of nano-CE is the detection and quantitation of Alzheimer's disease (AD) biomarker. Medical research reports estimate that several million people in the United States are suffering from AD and the total annual care cost associated with AD was more than \$226 billion in 2015, which is continuously increasing. This neurodegenerative disease is associated primarily with the cognitive loss of memory due to the formation of senile plaques containing amyloid-beta (Aβ) peptides, degeneration of the cholinergic neurons, and the development of neurofibrillary tangles in the brains of the affected people. Till date, there is no specific clinical test to diagnose AD. However, scientific study reveals that Aβ peptides and tau proteins could be reliable biomarkers that can be used for AD diagnosis and tracking disease progression. Recently, scientists have tried CE as a fast and efficient analytical technique for the detection and quantification of AD biomarker. They have successfully separated Aβ1-41 and Aβ1-42 fibrils using CE and a laser-induced fluorescence detector (CE-LIF) in 5 min, although the real-time application of this technique with patient samples is much complex and requires more sample cleanup procedures.

The benefit of CNTs in CE was further improved by modifying with surfactants such as sodium dodecyl sulfate (SDS), which improves the separation resolution of three mixtures containing chlorophenols, non-steroidal anti-inflammatory drugs, or penicillin derivatives. The CNT-SDS exhibits molecule-dependent separation selectivity, by decreasing the mobility of the penicillin derivative penicillin G when it is added in the separation buffer, indicating a strong affinity of penicillin G to the surface of the NMs in comparison with the other penicillin derivatives.

In the case of homologues of caffeine and theobromine, distinct changes in the electrophoretic parameters occur at a critical concentration of c-SWNT in the run buffer. The presence of charged c-SWNT suppressed the electro-diffusion and decreased the adsorption between capillary wall and solutes, which led to better peak shapes of isomers. The performances of CNTs for the separation of peptides, aniline derivatives, and water-soluble vitamins showed highly promising. However, the main drawbacks of CNTs are primarily the interference with detectable signals from both ultraviolet (UV) and laser-induced fluorescence (LIF) detectors, and secondly, raw CNTs are highly conductive, which can lead to irreproducible currents and separations. Moreover, the surface of CNTs is highly hydrophobic, difficult to dissolve in water, and commonly used in separation.

4.5 Conclusion

This chapter highlighted the recent advancements in nano-stationary phases for CE techniques and further applications. A wide range of NMs have been applied to the stationary phases of CEC, including CNTs, graphene, mesoporous silica, and metal NPs, to improve column efficiency and selectivity. The immobilization strategies for each type of NP on various CEC modes mainly consist of physical adsorption, covalent bonding, and other methods. Covalent bonding is the most used method, due to its good stability, favorable reproducibility, and moderate lifetime; however, its disadvantage is evident (i.e., tedious preparation procedures for some NPs). In contrast, physical adsorption requires only mild conditions and a simple process, but it obtains a short lifetime. So development of a simple, stable, and effective method is still necessary to fabricate NP-modified stationary phases. Furthermore, because the capillary monolithic column is prepared in situ via synthesis with organic or silica polymers, the NPs could be directly incorporated in the monolith matrix to achieve functionalization of the NPs. This approach is currently suitable for a large range of NPs, as long as they can be dissolved in the prepolymerization mixture.

However, some issues have not been resolved, such as the specific mechanism for NPs to improve the separation performance. Most previous reports hold a vague view that the large surface area and specific chemical and physical properties of NPs contribute to the enhanced separation ability in NP-modified CEC for complex samples. Furthermore, as NPs are constantly evolving, it is important to develop NPs with more targeted and specific utilities and to satisfy the requirements for more efficient analytical separations.

Websites of Interest

http://pubs.acs.org/doi/abs/10.1021/ac101151k
http://www.ncbi.nlm.nih.gov/pubmed/12458933
http://www.shsu.edu/chm_tgc/primers/pdf/CE.pdf
http://www.wiley.com/WileyCDA/WileyTitle/productCd-0470178515.html
http://link.springer.com/article/10.1007%2Fs10311-015-0547-x
http://www.sciencedirect.com/science/article/pii/S0021967307012083
http://store.elsevier.com/Gold-Nanoparticles-in-Analytical-Chemistry/isbn-9780444632852/
http://pubs.rsc.org/en/Content/ArticlePDF/2011/AN/c0an00458h
http://www.uv.es/clecem/
https://www.chem.tamu.edu/rgroup/russell/research/capillaryCE-MS.html

References

1. Graham, K. (1998). Capillary electrophoresis: A versatile family of analytical techniques, *Biotechnol. Appl. Biochem.*, **27**, pp. 9–17.

2. Bradley, J. S. (1994). *Clusters and Colloids*; Schmid, G., Ed.; VCH: Weinheim, pp. 459–554.

3. Schmid, G. (1992). Large clusters and colloids: Metals in the embryonic state, *Chem. Rev.*, **92**, pp. 1709–1712.

4. Margel, S., Burdygin, I., Reznikov, V., Nitzan, B., Melamed, O., Kedem, M., Gura, S., Mandel, G., Zuberi, M., and Boguslavsky, L. (1997). *Recent Res. Dev. Polym. Sci.*, **1**, pp. 51–55.

5. Godovsky, D. Y. (2000). Device applications of polymer-nanocomposites, *Adv. Polym. Sci.*, **153**, pp. 163–166.

6. Yu, C. J., Su, C. L., and Tseng, W. L. (2006). Separation of acidic and basic proteins by nanoparticle-filled capillary electrophoresis, *Anal Chem.*, **78**, pp. 8004–8010.

7. Hsieh, Y. L., Chen, T. H., Liu, C. P., and Liu, C. Y. (2005). Titanium dioxide nanoparticles-coated column for capillary electrochromatographic separation of oligopeptides, *Electrophoresis,* **26,** pp. 4089–4097.

8. Freitag, R. (2004). Comparison of the chromatographic behavior of monolithic capillary columns in capillary electrochromatography and nano-high-performance liquid chromatography, *J. Chromatogr. A,* **1033,** pp. 267–273.

9. Ivanov, M. R. and Haes, A. J. (2011). Nanomaterial surface chemistry design for advancements in capillary electrophoresis modes, *Analyst,* **136,** pp. 54–63.

10. Kuo, I., Huang, Y., and Chang, H. (2005). Silica nanoparticles for separation of biologically active amines by capillary electrophoresis with laser-induced native fluorescence detection, *Electrophoresis,* **26,** pp. 2643–2651.

11. Wang, W., Zhao, L., Zhou, F., Zhu J. J., and Zhang, J. R. (2007). Electroosmotic flow-switchable poly(dimethylsiloxane) microfluidic channel modified with cysteine based on gold nanoparticles, *Talanta,* **73,** pp. 534–539.

12. Huang, Y. F., Chiang, C. K., Lin, Y. W., Liu, K., Hu, C. C., Bair, M. J., and Chang, H. T. (2008). Capillary electrophoretic separation of biologically active amines and acids using nanoparticle-coated capillaries. *Electrophoresis,* **29,** pp. 1942–1951.

13. Ivanov, M. R. (2011). Doctor of Philosophy Thesis, University of Iowa.

14. Liu, F. K., Hsu, Y. T., and Wu, C. H. (2005). Open tubular capillary electrochromatography using capillaries coated with films of alkanethiol-self-assembled gold nanoparticle layers, *J. Chromatogr. A,* **1083,** pp. 205–210.

15. Delgado, J. L., Herranz, M. A., and Martín, N. (2008). The nano-forms of carbon, *J. Mater. Chem.,* **18,** pp. 1417–1420.

16. Vanderpuije, B. N. Y., Han, G., Rotello, V. M., and Vachet, R. W. (2006). Mixed monolayer-protected gold nanoclusters as selective peptide extraction agents for MALDI-MS analysis, *Anal. Chem.,* **78,** pp. 5491–5496.

17. Song, Y., Zhao, S., Tchounwou, P., and Liu, Y. M. (2007). A nanoparticle-based solid-phase extraction method for liquid chromatography-electrospray ionization-tandem mass spectrometric analysis, *J. Chromatogr. A,* **1166,** pp. 79–84.

18. Wang, Z., Hu, J., Jin, Y., Yao, X., and Li, J. (2006). In situ amplified chemiluminescent detection of DNA and immunoassay of IgG using

special-shaped gold nanoparticles as label, *J. Clin. Chem.*, **52**, pp. 1958–1961.

19. Kleindienst, G., Huber, C. G., Gjerde, D. T., Yengoyan, L., and Bonn, G. K. (1998). Capillary electrophoresis of peptides and proteins in fused-silica capillaries coated with derivatized polystyrene nanoparticles, *Electrophoresis*, **19**, pp. 262–265.

20. Tang, L. A. L., Wang, J., and Loh, K. P. (2010). Graphene-based SELDI probe with ultrahigh extraction and sensitivity for DNA oligomer, *J. Am. Chem. Soc.*, **132**, pp. 10976–10979.

21. Liu, Q., Shi, J. B., Sun, J. T., Wang, T., Zeng, L. X., and Jiang, G. B. (2011). Graphene and graphene oxide sheets supported on silica as versatile and high-performance adsorbents for solid-phase extraction, *Angew. Chem. Int. Ed.*, **50**, pp. 5913–5917.

22. Chandra, V., Park, J., Chun, Y., Lee, J. W., Hwang, I. C., and Kim, K. S. (2010). Water-dispersible magnetite-reduced graphene oxide composites for arsenic removal, *ACS Nano*, **4**, pp. 3979–3986.

23. Guo, Z., Sadler, P. J., and Tsang, S. C. (1998). Thermal reduction occurs as a result of reduction by decomposing organic matter, *Adv. Mater.*, **10**, pp. 701–705.

24. Wang, Z., Luo, G., Chen, J., Xiao, S., and Wang, Y. (2003). Carbon nanotubes as separation carrier in capillary electrophoresis, *Electrophoresis*, **24**, pp. 4181–4188.

25. Palmer, C., Hilder, E. F., and Quirino, J. P. (2010). Electrokinetic chromatography and mass spectrometric detection using latex nanoparticles as a pseudostationary phase, *Anal. Chem.*, **82**, pp. 4046–4054.

26. Fujimoto, C. (1997). Electrokinetic chromatography using nanometer-sized silica particles as the dynamic stationary phase, *J. High Reso. Chromatogr.*, **20**, pp. 400–402.

27. Liu, D. N., Wang, J., Guo, Y. G., Yuan, R. J., Wang, H. F., and Bao, J. J. (2008). Separation of aromatic acids by wide-bore electrophoresis with nanoparticles prepared by electrospray as pseudostationary phase, *Electrophoresis*, **29**, pp. 863–870.

28. Terabe, S. (1993). *Capillary Electrophoresis Technology*; Guzman, N. A., Ed.; Marcel Dekker: New York.

29. Shamsi, S. A., Palmer, C. P., and Warner, I. M. (2001). Molecular micelles: Novel pseudostationary phases for CE, *Anal. Chem.*, **73**, pp. 140A–149A.

30. Verwey, E. J. and Overbeek, J. T. (1948). *Theory of the Stability of Lyophilic Colloids*, 1st Ed., Elsevier: Amsterdam, pp. 1–20.

31. Wallingford, R. A. and Ewing, A. G. (1989). Separation of serotonin from catechols by capillary zone electrophoresis with electrochemical detection, *Anal Chem.*, **61**, pp. 98–100.

32. Yu, C. J., Su, C. L., and Tseng, W. L. (2006). Separation of acidic and basic proteins by nanoparticle-filled capillary electrophoresis, *Anal. Chem.*, **78**, pp. 8004–8010.

33. Hsieh, Y.-L., Chen, T.-H., Liu, C.-P., and Liu, C.-Y. (2005). Titanium dioxide nanoparticles-coated column for capillary electrochromatographic separation of oligopeptides, *Electrophoresis*, **26**, pp. 4089–4097.

34. Hunter, R. J. (1993). *Introduction to Modern Colloid Science*; Oxford University Press.

35. Neiman, B., Grushka, E., and Lev, O. (2001). Use of gold nanoparticles to enhance capillary electrophoresis, *Anal. Chem.*, **73**, pp. 5220–5227.

36. Xu, Y. Y., Niu, X. Y., Dong, Y. L., Zhang, H. G., Li, X., Chen, H. L., and Chen, X. G. (2013). Preparation and characterization of open-tubular capillary column modified with graphene oxide nanosheets for the separation of small organic molecules, *J. Chromatogr. A*, **1284**, pp. 180–187.

37. Nilsson, C., Birnbaum, S., and Nilsson, S. (2007). Use of nanoparticles in capillary and microchip electrochromatography, *J. Chromatogr. A*, **1168**, pp. 212–224.

38. Heiger, D. N. (1992). *High Performance Capillary Electrophoresis: An Introduction*, France: Hewlett Packard Company.

39. Wallingford, R. A. and Ewing, A. G. (1989). Capillary electrophoresis, *Adv. Chromatogr.*, **29**, pp. 1–76.

Chapter 5

Capillary Electrophoresis Coupled to Mass Spectrometry for Enantiomeric Drugs Analysis

Vítězslav Maier, Martin Švidrnoch, and Jan Petr
Department of Analytical Chemistry, Regional Centre of Advanced Technologies and Materials, Faculty of Science, Palacký University, 17. listopadu 12, CZ-77146, Czech Republic
vitezslav.maier@upol.cz

Capillary electrophoresis–mass spectrometry (CE-MS) is a powerful analytical technique that combines high efficiency, short analysis time, high selectivity, and separation versatility of the CE with the selectivity, sensitivity, and possibility for identification and structure elucidation of unknown analytes from the MS detection point of view. These benefits predetermine CE-MS as a method having high potential as an analytical tool for the enantioseparation of drugs and their metabolites in various matrices and samples (e.g., bulk pharmaceutical materials, final products, urine, blood, and serum).

This chapter covers the current state of the art in CE-MS focused on the enantioseparation of chiral drugs. A short discussion of

---

Capillary Electrophoresis: Trends and Developments in Pharmaceutical Research
Edited by Suvardhan Kanchi, Salvador Sagrado, Myalowenkosi Sabela, and Krishna Bisetty
Copyright © 2017 Pan Stanford Publishing Pte. Ltd.
ISBN 978-981-4774-12-3 (Hardcover), 978-1-315-22538-8 (eBook)
www.panstanford.com

CE Coupled to MS for Enantiomeric Drugs Analysis

commercially available CE-MS instruments and interface design is followed by a detailed review of various modes of chiral CE-MS. The different strategies for enantioseparation (namely partial filling technique, counter-current method, utilization of chiral stationary phases, and chiral or achiral micellar phases) are described and documented by the appropriate works published in the recent years.

5.1 Enantioseparation of Pharmaceutical Products

National authorities such as the U.S. Pharmacopeial Convention, Food and Drug Administration, and International Council for Harmonisation, which regulate the development and introduction of chiral pharmaceuticals, emphasize the analysis of enantiomers in order to:

- Determine the relative pharmacological contribution of each enantiomer of the optically active drug comparing to the racemic mixture, in both humans and animals.
- Compare the toxicological profile of the racemic mixture with the toxicological profiles of each individual enantiomer of drugs or potential biologically active compound.
- Provide analytical support during the development and preparation of a single enantiomer.

Except the analysis of enantiomers (optically active isomers), it is necessary to include the same requirements for the analysis of stereoisomers, which include (except others) geometrical isomers (cis/trans) and diastereomers (isomers with more than one optically active center being mirror images of each other) [1, 2].

There are a lot of cases in the past when pure enantiomers exhibited completely different pharmacological and toxicological properties and effects on humans. A typical example is L-propranolol, an antagonist of β-adrenergic receptors (a β-blocker). D-propranolol does not have these pharmacological effects; D-carnitine causes myasthenia gravis symptoms, but L-carnitine has no side effects and thus it is possible to use it as an additive to energy drinks. Another example is (*S*)-warfarin, which has five times higher anticoagulant properties compared to (*R*)-warfarin.

On the other hand, both enantiomers of an anti-inflammatory agent (*R,S*)-ibuprofen have the same effect as a painkiller. In addition, enantiomers of quinolones and *β*-lactams exhibit antibacterial activity. The relation between drug chirality and the pharmacological, toxicological, and other physiological properties of drugs is summarized in the review by Nguyne et al. [3].

5.2 Basic Concepts of Enantioseparation by Capillary Electrophoresis

CE is one of the most frequently used techniques for the separation of optically active compounds. The major mechanism applied in the enantioseparation by CE is based on the formation of diastereomeric complexes between the separated enantiomers and the chiral selector (CS) in background electrolyte (BGE) during the separation process.

This concept is well known as direct enantioseparation, where no covalent bond is formed between the separated enantiomers and CS and the discrimination is based on non-covalent intermolecular interactions. The huge advantage of CE compared with other chromatographic methods is almost unlimited selection of the CSs and their significantly low consumption than those necessary in the case of chromatographic methods. CE also allows excellent resolution of enantiomers due to a high number of benefits. The most important ones can be summarized as follows:

- High separation efficiency, which enables the separation of enantiomers even if the mobility difference is very low.
- The ability to easily change the concentration of CS in BGE as well as the fast changing of CSs during the optimization process.
- Besides the chromatographic techniques, it allows the utilization of a very wide range of CSs and also a combination of them to reach the required and sufficient resolution.

- Very low sample and BGE consumption and hence a low consumption of the CS (this fact is very useful when using rare or very expensive CS).

These advantages highlight CE techniques as the favored method for the chiral separation of diverse enantiomers. The vast majority of enantioseparation techniques using CE is realized by capillary zone electrophoresis (CZE) and micellar electrokinetic chromatography (MEKC). Among others, capillary electrochromatography (CEC) is also gaining interest.

The successful separation of a pair of enantiomers is strongly dependent on the specifically used separation system selectivity. In an achiral environment, the enantiomers cannot be distinguished because they do not differ in their electrophoretic mobilities. To realize the chiral separation, the interactions between CS and the enantiomer must be mutual and this interaction must differ in at least one physicochemical property.

A typical condition for a successful chiral separation is the difference of the interaction constants for interactions of the individual enantiomers and CS. Another essential condition for chiral separation is a sufficient rate of the dynamic equilibrium during the chiral interaction. Such a situation is similar to dissociation equilibrium where the rate of the dynamics does not allow separation of dissociated and undissociated forms of a compound. The difference of enantiomer mobilities is given by creating transient diastereomeric complexes. These diastereomers have the same static ratio of the charge and size (mass) and thus they should have the same values of mobilities. The difference of the mobilities is caused by the discrepant effective densities of the charge, which are given by the different spatial orientation of the diastereomers or by specific intermolecular interactions between them [4–7].

A certain disadvantage of CE in spectrophotometric detection (as the most used detection for CE) is the poor concentration sensitivity due to a very short optical length of capillary detection window. From this point of view, the hyphenation of CE with more sensitive detection systems such as mass spectrometry (MS) can overcome this problem. Moreover, the combination of CE and MS can provide additional structural information of the separated enantiomers [8].

5.3 Capillary Electrophoresis with Mass Spectrometry

Capillary electrophoresis with mass spectrometry represents a very interesting combination of two highly sophisticated analytical techniques. Connecting CE and MS utilizes advantages of both the analytical methods:

- Rapid separation (CE)
- Very low sample consumption (CE)
- High resolution capability (CE)
- Selectivity (CE and MS)
- Detection sensitivity (MS)
- Obtaining structural information about each separated components of the sample (MS)

These advantages make it possible to perform analysis of very complex samples together with reliable identification and quantification of the individual sample components (including enantiomers).

Online connection of CE and MS requires quite difficult instrumentation design. The first application of CE-MS was realized at the end of 1980s [9–11]. Since that time, the hyphenation of CE with MS has undergone a very significant improvement, and at present, CE-MS systems are commercially available. Even if the hyphenation of liquid chromatography (LC) and MS is more spread, CE-MS represents a very notable complementary technique, which can outdo LC-MS in many cases. LC-MS separation of certain groups of analytes could be very complicated especially when the analytes are very polar or have a charged group.

The critical point of the CE-MS connection realization is the transfer of the separated ions (analytes, in general) from the separation capillary into the ion source of MS. When using online combination of CE-MS, this transfer is provided by a special interface. The separation itself is performed in a fused-silica capillary in predominantly aqueous BGEs, whereas the detection of the ions produced in the MS ion source is carried out in a mass analyzer situated in the MS under a high vacuum.

For the realization of the online CE-MS technique, only a limited number of available ionization techniques are suitable and usable.

These ion sources belong to the atmospheric-pressure ionization techniques—electrospray ionization (ESI), chemical ionization (APCI), photoionization (APPI), inductively coupled plasma (ICP), and matrix-assisted laser desorption ionization (MALDI). In the case of CE-MS, the most commonly used ion source is ESI, which allows relatively simple transfer of the ions in the liquid phase from the separation capillary to the gas phase in the MS ion source. The connection of CE to MS accomplishes a direct transfer of the separated analytes from the CE step to the MS detector. The application of CE-ESI-MS covers more than 99% of currently published works, and this is similar when CE-ESI-MS is used for the analysis of chiral compounds.

5.3.1 Electrospray Ionization in Hyphenation of CE with MS

The ESI belongs to the soft ionization techniques, which makes possible the ionization of a wide range of compounds. These include especially polar and charged analytes with very low-molecular-mass and high-molecular-weight compounds and biopolymers. The high-mass capability of ESI for the analysis of biopolymers is based on the formation of multiple charged ions, and the final m/z ratio is much lower and thus applicable for the mass detector [12, 13].

The principle of ESI is based on a flow of liquid through a metal-spraying capillary, and a high direct voltage (2–4 kV) is applied at the end of it. The electrically formed droplets generated at the end of the capillary expand and evaporate forming ions, which are focused into the mass analyzer. The spraying capillary is heated to the temperature ranging from 100–300°C, which supports the evaporation of the solvent. Finally, the ion formation is also supported by a stream of sheath gas (usually nitrogen). The ions of low-molecular-weight compounds (in the range of m/z 100–1500) created in the ESI are usually singly charged, forming ions of $[M+H]^+$ for positive ionization mode and $[M-H]^-$ for negative ionization mode, respectively. When multiple charged ions are formed, the scheme is different and strongly depends on the compound. Thus, ions of $[M+nH]^{n+}$ are generated (n represents the value of positively charged ions).

Except the typical formation of $[M+H]^+$ or $[M-H]^-$, the formation of cluster ions is also possible by creating ions of $[nM+H]^+$, etc. In addition, the formation of adducts with alkali metals (Na^+, Li^+, K^+, etc.) or ammonia is also possible, especially when the source of these ions is present (e.g., in the sample matrix or as counterions of CSs). For further and detailed reading summarizing the principle of ESI-MS, Refs. [14–16] are recommended.

The ions formed in the ESI during the ionization process usually do not undergo fragmentation. In certain cases, an ion source decay can occur. This could be caused by the setting of the ESI parameters (most often temperature). In such cases, some labile molecules can undergo a neutral loss, forming ions with lower m/z ratio than the original compounds. The typical neutral losses include cleavage of H_2O ($\Delta m/z = 18$), CO_2 ($\Delta m/z = 44$), and NH_3 ($\Delta m/z = 17$) or alkylated amines. From the aforementioned principle of ESI, it is clear that this ionization technique allows ion formation of simple inorganic or organic compounds as well as complicated structures and biopolymers.

On the other hand, CE-ESI-MS has certain disadvantages and limitations. The first significant limitation refers to the necessity of creating a stable electrospray, which requires sufficient flow of the liquid, typically in the range of units to hundreds of µL/min. The BGE flow from the CE separation capillary is generated particularly by the electroosmotic flow (EOF) or by applying a gentle pressure (usually up to 10 mbar) at the inlet part of the CE during the separation (pressure-assisted CE). The EOF velocity in the silica capillary (with the inner diameter of 50–75 µm) is in the order of tens to hundreds nL/min. This very low flow is insufficient for the generation of a stable electrospray. This problem can be instrumentally solved by the following approaches:

- By using sheath-spraying liquid that increases the total flow of the liquid phase entering the spraying capillary. The arrangement of this solution can be done using various modes such as coaxial sheath-flow or liquid-junction interface.
- By replacing the commercial interface with a sheathless arrangement (e.g., nano-ESI, which allows stable spraying with extremely low flows of the liquid phase).

The next issue with the online CE-ESI-MS connection consists of a limited number of BGEs to be used for the CE separation. The non-volatile BGEs commonly used in CE are not suitable for ESI because they significantly decrease the ionization efficiency and thus negatively affect the sensitivity of the MS detection. Furthermore, they increase the signal-to-chemical noise (increased signal-to-noise ratio, S/N) and can distinctly cause a contamination of the ion source.

These are particularly the non-volatile components of BGEs such as phosphate, borate, or citrate as well as other constituents of the BGE, namely surfactants (both ionic and non ionic), CSs, and other additives. A review dealing with the issues of the compatibility of CE and ESI-MS considering the latest trends was published by Pantůčková et al. [17], Zhong et al. [18], and Kleparník [19]. Volatility evaluation of various types of BGE additives for (LC) CE-ESI-MS was also published by Petritis et al. [20]. This evaluation was done using data obtained from evaporative light scattering detector (ELSD), which has the same requirements on the volatility of the mobile phase components such as ESI-MS. An ideal construction of the interface for the direct connection of CE and MS must fulfil these criteria:

- Provide sensitive, reproducible, and linear response.
- Should not decrease the efficiency and resolution of the separation.
- Electrical circuit should be secured for applying the separation voltage.
- Ionization of all eluting from the separation capillary analytes must be allowed.
- Ionization and stable electrospray formation even at low liquids should be enabled.

5.4 Interfacing of CE-ESI-MS

The interface that provides the direct connection between CE and ESI-MS is a modified interface used for LC-ESI-MS. Several construction solutions for the direct interfacing of CE and ESI-MS were developed during the last two decades. All of these solutions must carry out the conditions mentioned earlier. There are three possible construction solutions:

1. An interface without any additional flow of the sheath liquid, which is especially suitable and designed for nano-ESI (sheathless interface).
2. An interface with an auxiliary coaxial flow of the spraying (sheath) liquid—one of the most frequently used solutions, which is also available commercially (sheath liquid interface).
3. An interface with the addition of the spraying liquid flow realized by capillary connection (liquid-junction interface).

5.4.1 Sheathless Interface

Sheathless interface is largely handled by a metal coating of the separation capillary tip with the inner diameter of 100 μm and thus provides a conductive electrical connection, which is necessary for the particular separation in CE and ESI ionization.

However, the metal coating has only a limited shelf life (usually a few hours only), which considerably limits the measurements. The interface without the additional flow of the liquid was also used in the first construction of CE-ESI-MS [9].

In comparison to the sheath-flow interface and the liquid-junction interface, the dilution of the eluent from the capillary does not occur, which leads to a high sensitivity of detection (approximately 10 to 20 times higher). Most of the developed sheathless interfaces for the CE-ESI-MS still use the modification of the original solution. To increase the stability of the metal coating, procedures based on the utilization of conductive coating agents (e.g., silver, gold, nickel, graphite, conductive epoxy polymers, etc.) have been developed.

Moreover, the end of the capillary can be tapered to increase the resulting electric field intensity, which is applied at the end of the capillary (e.g., from 360 μm OD and 20 μm ID to 40 μm OD and 3 μm ID). This will significantly increase the sensitivity and efficiency of the ionization. The targeted narrowing of the spraying tip end often leads to its plugging. This interface has one huge disadvantage: The ionization efficiency and thereby the sensitivity of MS detection strongly depend on the EOF velocity, which affects the speed of the BGE through the capillary tip. All the electrolyte composition plays a key role in the ionization process. An example of one possible construction of the mentioned interface is depicted in Fig. 5.1.

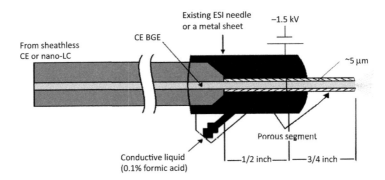

Figure 5.1 Construction of sheathless interface. Reprinted with permission from Ref. [21], Copyright 2007, American Chemical Society.

Nowadays, this type of interface is being intensely developed and studied, in particular, to achieve the best possible sensitivity of the MS detection as well as for connection of microseparation techniques with ESI-MS [22–27]. The main problem with this interface lies in the production of emitter tips, which are not sufficiently reproducible [8].

5.4.2 Sheath Liquid Interface

This CE-ESI-MS interface construction is the most widely used solution and also commercially available, mainly because of the simplicity of working with this interface. The interface with a coaxial additional flow was, in the case of CE-ESI-MS, first used by Smith et al. [11]. A general scheme of the coaxial sheath liquid interface is shown in Fig. 5.2. The ESI ion source functions optimally when the liquid flow is approximately 5 μL/min.

Since the capillary flows of the electrolyte are about 100 times lower, it is possible to increase the flow to the desired rate by adding an auxiliary liquid. This liquid is usually known as "sheath liquid" and its main role is to increase the volume of the flow at the end of spraying capillary and thus increasing the stability of the electrospray. The spraying capillary must also support ionization of analytes in the electrospray. It is, therefore, necessary to optimize not only the flow rate, but also the sheath liquid composition regarding the ionization of the target analytes.

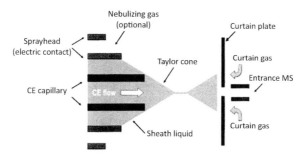

Figure 5.2 Schematic illustration of the sheath liquid interface. Reprinted from Ref. [28], Copyright 2003, with permission from Elsevier.

Furthermore, the auxiliary spraying capillary allows a conductive connection (electric circuit) as the spraying liquid must be conductive and in contact with the metal-spraying tip (usually electrically grounded) and the electrolyte eluting from the silica separation capillary. The sheath liquid is usually supplied coaxially around the metal-spraying capillary in which the separation silica capillary of CE is inserted. To achieve a stable electrospray, it is also necessary to coaxially supply the nebulizing gas (usually nitrogen), which supports the evaporation of the solvent from the BGE and the sheath liquid. This ensures the formation of a stable aerosol. An important benefit of this interface is that the CE and MS processes are separated from each other. This fact allows to employ a wide range of BGEs and separation capillaries.

While BGE affects the quality of the separation, the sheath liquid together with the BGE affects the efficiency of transfer of the analytes into the gas phase. In the case of the sheath liquid interface, the ionization efficiency is affected by a combination of many factors. Consequently, it is useful for optimizing the composition and the flow rate of the sheath liquid to use one of the chemometric methods [29].

Among the highlighted advantages of this construction solution is the electrospray formation robustness. Ionization in this case does not depend on the EOF velocity. Beyond that, the disadvantage is dilution of the analyte by the sheath liquid prior to the ionization as well as band broadening of the sprayed analyte zones. This leads to a reduced sensitivity of the MS detection [30–32]. Usually, the higher the flow rates of the sheath liquid, the lower the signal intensity of the analytes in MS.

Apart from the flow rate and composition of the sheath liquid, it is also necessary to optimize the flow rate and pressure of the nebulizing gas, temperature of the spraying capillary, and the polarity and value of the high voltage applied on the spraying capillary. The inner diameter as well as the position of the silica capillary in the metal interface should be taken into account as well. When the enantioseparation is performed, one of the crucial parameters determining the quality of the separation is resolution.

The resolution strongly depends on the nebulizing gas pressure, and this parameter must be thoroughly optimized. Unfortunately, due to the many parameters to be optimized, the final conditions of analytes ionization is usually a compromise of the aforementioned parameters [32, 33].

Generally, it is possible to define several experimental suggestions on how to achieve high ionization efficiency and MS detection sensitivity:

- The silica-spraying capillary must be adjusted in the metal interface, and the silica-spraying tip length should be approximately two-thirds of the capillary outer diameter.
- The tip of the separation silica capillary must be properly cut at both inlet and spraying positions.
- The polyimide coating of the silica capillary should be removed at the injection side as well as at the end entering the MS interface.
- The sheath liquid should be conductive to ensure and keep the electrical circuit for CE separation and ESI ionization. High ionic strength of the sheath liquid leads to instable and discontinuous electrospray.
- The nebulizing gas and the sheath liquid flow rates should be as low as possible to keep the formation of stable electrospray and also to minimize dilution of the sprayed analytes by the sheath liquid. Also a siphoning effect can occur when the flow rates are not properly optimized. Moreover, the pressure of the nebulizing gas should be relatively low (typically between 2 and 10 psi).

A certain disadvantageous effect of this interface is the fact that migration of counterions from the sheath liquid into the separation capillary can occur. In such cases, a moving boundary can be formed inside the separation capillary, which affects the migration order of analytes and causes change or loss of the resolution in comparison to the offline CE separations. These effects are more frequent and more significant in the BGEs with a low rate of EOF [34].

5.4.3 Liquid-Junction Interface

The liquid-junction interface was developed to reduce the dilution of the eluent by the sheath liquid. The silica separation capillary is introduced into a reservoir filled with the sheath-spraying liquid [35, 36]. The electrode is also set in the reservoir and closes the circuit, which allows applying the spraying voltage. The interface design with an additional flow is in Fig. 5.3.

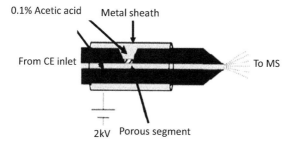

Figure 5.3 Liquid-junction interface. Reprinted with permission from Ref. [37], Copyright (2003), American Chemical Society.

In this case, it is difficult to regulate the flow of the sheath liquid. Another problem causes the connection of the CE capillary and the spraying capillary, which leads to a decrease in the peak efficiency and sensitivity [38, 39].

5.5 Compatibility of Capillary Electrophoresis and Mass Spectrometry

Separation in CE often uses BGE components such as phosphate, borate, alkali metals, surfactants, and CSs. All of these favorite components of the electrolytes and additives are non-volatile and, therefore, not applicable in the case of CE-ESI-MS analyses. When using non-volatile components, the resulting contamination of the ion source reduces the sensitivity of the MS detection and leads to significant deterioration of migration time and peak area repeatability. For this reason, it is necessary to replace the non-volatile components and additives by volatile components such as formate, acetate, and ammonium hydroxide. Similarly, the volatile components must be also present in the spraying sheath liquid.

Another approach to utilize the non-volatile additives (e.g., surfactants or CSs) for CE-ESI-MS separations is the use of partial filling technique (PFT), where the BGE containing the non-volatile components is filled to a defined length of the separation capillary only. During the separation itself, it is necessary to ensure such conditions when the ions of the non-volatile additives migrate toward the inlet electrode, preventing any contamination of the ion source. A more detailed discussion of the partial filling method can be found in Section 5.5.1.

Due to these limitations, also the separation selectivity manipulation is considerably reduced and a transfer of CE methods employing a spectrophotometric detector to the CE-ESI-MS system is difficult compared to more simple method transfer of LC-UV-VIS instrumentation to the LC-ESI-MS system. The usage of PFT for CE-ESI-MS enantioseparation can lead to lower resolution compared to CE with the spectrophotometric detection, owing to the lower interaction time between separated enantiomers and CS in PFT CE-ESI-MS analysis. On the other hand, a certain advantage in this case is the fact that MS allows identification and quantification of the components with insufficient resolution obtained during the CE separation.

Despite these limitations, the CE-ESI-MS instrument is an efficient and powerful separation technique with a higher sensitivity (lower limits of detection, LODs) in comparison to the commonly used CE equipped with a UV-VIS detector only. Moreover, it is also possible to detect substances that do not absorb UV-VIS radiation, which makes CE-ESI-MS more interesting because instrumental analysis of non-absorbing compounds (such as low-molecular carboxylic acids or inorganic ions) is still an issue in analytical chemistry. The most commercially available CE-ESI-MS interface with the coaxially supplied sheath liquid has concentration LODs around $\sim 10^{-7}$ to 10^{-9} mol/L (depending on the mass analyzer).

5.5.1 Basic Principles of Partial Filling Technique

The principle of PFT consists of filling the capillary with the BGE containing a non-volatile component (CS, surfactant, polymer, etc.) to a defined part of the total length of the capillary (e.g., 90% of the total capillary length). Thanks to this, the non-volatile BGE component does not enter the MS ion source. The next step is an injection of the sample zone after the BGE containing the non-volatile components.

This is followed by the separation voltage application at which separation occurs. Simultaneously, the migration of the non-volatile components must be opposite to the migration of the separating analytes (through self-electrophoretic mobility of the non-volatile components or by the EOF in the case of uncharged non-volatile components). This means that the non-volatiles migrate toward the inlet electrode at the CE side and the analyte ions migrates directly to the ion source at the MS side.

This arrangement when two interacting components migrate toward each other is referred to as counter-current mode. Such conditions could be easily achieved in a covalently coated capillary with eliminated EOF, and the used non-volatile component must be oppositely charged to the separated analytes. Other option is that the non-volatile component in the partially filled zone does not bear an electric charge (e.g., neutral CSs), and thus its effective mobility is zero. The analytes being separated then migrate through the immobile zone containing non-volatile additives toward the ion source. In this case, it is necessary to use the covalently coated capillary or to utilize very acidic buffers with pH about 2.5 or lower when the EOF is sufficiently suppressed and thus there is no elution of the non-volatile components into the ion source. A model showing analysis using PFT-CE-ESI-MS and a scheme of the counter-current migration is shown in Fig. 5.4.

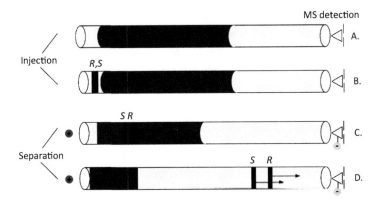

Figure 5.4 Scheme of PFT with counter-current migration for enantioseparation.

Separation of enantiomers using CE-ESI-MS employing non-volatile CS (cyclodextrins and their derivatives, bile acids, chiral surfactants, crown ethers, chiral antibiotics, etc.) is often performed by PFT. Even if PFT allows enantioseparation with the direct ESI-MS detection and identification, it has a number of disadvantages and limitations:

- The method optimization is difficult, and it is necessary to properly optimize the concentration of the non-volatile additive as well as the partial filling length.
- The different composition of the electrolytes in the two sections of the capillary results in undesired changes of the selectivity during the separation. PFT combines enantioselective mechanism in one part of the capillary with the CZE mechanism in the remaining part. This fact leads to a decrease in the analysis repeatability.
- At the boundary of the zones containing non-volatile additive and additive-free BGE, the peak broadening occurs. This phenomenon reduces the efficiency of the separation.
- The so-called "migration window" in which the interactions of the analytes with the non-volatile additive occur is shorter compared to a situation where the non-volatile additive is present throughout the whole separation in the capillary. The time when the analytes and additives are in their mutual interaction is shorter compared to the conventional electrolyte arrangement.

As evident from the drawbacks described earlier, also the total peak capacity of PF-CE-ESI-MS separation is lower compared to the CE separation when the non-volatile additive is present in the separation capillary during the whole analysis. This leads to a different selectivity and poorer separation resolution. Other possibility is to use high molecular weight chiral polymers as the CS, which forms ions with the m/z ratio higher than the mass analyzer range.

Although ESI requires volatile components of the BGEs for a sensitive and repeatable ionization, other types of ion sources could be used as well. Atmospheric-pressure chemical ionization (APCI) allows the utilization of BGEs based on non-volatile components. Nevertheless, the availability of CE-APCI-MS connection is limited and commercial instruments are rare. The majority of works dealing with the CE-MS utilize electrospray as the ion source of choice.

The PFT is applied wherever it is necessary to prevent the entry of any additive of the BGE to the detection window (in the case of an optical detection) or into the ion source of MS. The first work employing the PFT was published by Nilsson et al. [40]. A part of the separation capillary was filled with a cross-linked bovine serum albumin (BSA) for enantioseparation of tryptophan. Later, Valtcheva et al. [41] used non-immobilized cellulose for enantioseparations of β-blockers using the CZE mechanism. PFT can be used for enantioseparation using both uncharged CSs, which in this case can only migrate with EOF (unless suppressed by dynamic or covalent capillary coating), and charged CSs, which have their own electrophoretic mobility. A combination of PFT with the counter-current migration is favorable to achieve a sufficient resolution and to prevent a possible detector contamination.

In such cases, the experimental conditions allow migration of the CS zone to the opposite side from the ion source, whereas the separated analytes migrate toward the detector and interact with the CS during their migration. This experimental design was first described by Chankvetadze et al. [42] for the separation of basic enantiomers, and negatively charged sulfobutylether-β-CD (SBE-β-CD) was employed as the CS. The PFT is also preferably used for separation of achiral compounds by MEKC-ESI-MS, where the non-volatile surfactant must be thoroughly separated from the ion source [42–45]. Utilization of the PFT for enantioseparation was then documented in many works using different types of CS (e.g., vancomycin, charged derivatives of CD, uncharged CD, avidin, chiral crown ethers, pepsin, transferrin, cellulase, etc.) with UV-VIS or MS detection.

The next advantage of the PFT (except preventing the ion source contamination by CS) is the very low consumption of CS compared to the classical arrangement where the CS is a part of the BGE, which fills all the separation system compartments (inlet vial–capillary–outlet vial). The injected volumes of the BGE-containing CS are in the range of tens to hundreds nL in the case of the PFT method. This fact makes it possible to study rare or newly synthesized CSs whose amounts are very limited.

A special case is a technique referred to as complete partial filling (CFT) where the whole separation capillary is filled by the BGE-containing CS. The other compartments (inlet and outlet vials) contain the electrolyte without the addition of CS.

A certain benefit of the CFT compared to the PFT is the fact that the whole capillary is used for the enantioseparation, which makes the experiment less time consuming, because the capillary can be quickly flushed by the separation electrolyte containing CS. The injection of the partial zone in PFT requires an application of preciously defined low pressures (typically 3–5 kPa), which causes prolongation of the analyses.

The CFT method was described by Amini et al. [46] for the enantioseparation of (R,S)-ropivacaine using methyl-β-CD as CS. It should be noted that during the flushing step with the BGE-containing CS, the separation capillary must be put out of the ion source of the MS. The relation between resolution and efficiency of the enantioseparation employed both charged and uncharged CS in PFT was discussed in a review by Amini et al. [47].

The differences in migration velocities of the analyte in the zone containing the CS and in the CS-free zone lead to several significant specificities compared to the classical experimental design. When the analytes and the CS migrate both to the opposite direction, the applied zone length (PL_{app}) is higher than the effective zone length (PL_{eff}). Otherwise, if the analyte and CS migrate toward each other, then the PL_{app} is lower than PL_{eff} [48–51]. The electroosmotic mobility inside the zone with CS is different from the mobility in the zone with no CS. This can lead to the generation of a laminar flow inside the capillary and can negatively affect the separation efficiency. Thus, it is recommended to use as low amount of the CS as possible for the enantioresolution. Another contribution to the dispersion of the separated enantiomers is changing of the migration velocity of the enantiomers connected with their migration from the zone with CS to the CS-free zone.

These effects are always more apparent on the slowly migrating enantiomer because its mobility difference in the zones with and without CS is always greater compared to a faster migrating enantiomer [52, 53]. As a general rule, the higher the length of the separation zone in which the interaction between enantiomers and the CS takes place, the higher resolution and the higher enantioselectivity is achieved. During the PFT method development, a proper optimization of the BGE composition and CS concentration is necessary. Also the length of the zone containing CS must be thoroughly optimized taking into account the resolution, efficiency of the separation as well as the analysis duration. In the case of a quantitative analysis, it is suitable to evaluate the analyses using corrected peak areas to migration times [54].

5.6 Cyclodextrin as CS for CE-MS Enantioseparation

Native cyclodextrins (CDs) and their neutral or charged derivatives are non-volatile CSs, and thus it is important to avoid contamination of the MS ion source by such additives. The advantage of CDs and their derivatives consists of their broad application potential being among the widely used CS in CE.

Another advantage is their solubility in water as well as in aqueous-organic mixtures or purely organic BGEs [55–57]. Moreover, the fast kinetics of the transient complexes of CDs with the separated enantiomers highlights their use as CSs for highly efficient enantioseparation in CE. The basic mechanism of the enantioseparation using CDs and their miscellaneous derivatives is mainly based on the inclusion of a hydrophobic part of the analyte molecule (guest) to the relatively hydrophobic cavity of the CD (host). Aromatic rings are typical hydrophobic parts of such molecules, which can interact with the inner cavity of CDs [58]. Hydroxyl groups that are bonded on the outer cavity margin can be modified by various chemical reactions in order to prepare derivatized CDs with required properties leading, in particular, to:

- Increase complexing ability
- Affect the solubility
- Change CD electrophoretic mobility
- Detection compatibility

The trend in recent years is the preparation of well-defined derivatives of CDs. Different degrees of substitution of such derivatives might negatively affect the complexation ability of the CD derivatives [59–61].

For example, by replacing the hydroxyl groups by an alkyl or hydroxyalkyl substituent, a modification of the depth and shape of the inner cavity may be achieved, which leads to a better interaction of the analyte with the hydrophobic cavity of the CD derivative. This results in an increase in the stability constant of the resulting inclusive complex [58, 62].

Charged CD derivatives (positively or negatively) possess increased stability of the formed complexes with the separated enantiomers compared to the native or neutral ones. This can be explained by Coulombic (electrostatic) interactions between oppositely charged CDs and enantiomers. The stability of the complex is, among others, dependent on the size of the CD's effective charge, in particular when the functional groups are subject to dissociation/protonation, which depend on the BGE pH [61, 63]. Bifunctional reagents (such as diepoxydes or diisocyanates) could be used for CD polymerization [64] forming polymerized CDs with different properties such as increased solubility and rigidity, which can lead to a better enantioselectivity [65, 66].

5.6.1 Experimental Setup without Restrictions of the CD Entering MS Ion Source

The first separation using CD as CS for CE-ESI-MS/MS(QqQ) enantioseparation was published in 1995 [67]. Heptakis-2,6-di-O-methyl-β-cyclodextrin (DM-β-CD) of 5 mM concentration was used as CS for the separation of ephedrine and terbutaline enantiomers. The BGE consisted of 5 mM sodium phosphate of pH 2.50. Detection of the separated enantiomers was performed in SIM and SRM mode of MS, and no approach to prevent the contamination of the MS ion source was used during the separation. The authors reported only a negligible impact of the non-volatile BGE on the ionization efficiency of the separated enantiomers. Although the achieved sensitivity was 1000 times higher than in the case of the CE-UV detection (expressed as S/N ratio), the resulting sensitivity was not sufficient for the determination of ephedrine and terbutaline enantiomers in urine samples.

A special experimental approach based on a system of connected capillaries known as "voltage switching" during the various stages of the separation was chosen to prevent the entry of DM-β-CD into the ion source for the separation of (R,S)-ropivacaine using CE-ESI-MS(Q) [68]. The BGE consisted of 1 mM formic acid of pH 2.85 with an addition of 50 mM DM-β-CD. In the first capillary, the analytes are enantioseparated followed by the next step, where the analyte zones are subjected to heartcut and transferred to the second capillary with a CD-free BGE. The analytes in the second capillary then migrate directly to the MS ion source. Also a dissociation of the enantiomer-DM-β-CD complex takes place in the heartcut step. Even if this method prevents the entry and contamination of the ion source by CS, the setup is complicated and not useful for routine analyses. The aforementioned method shows the separation of (R,S)-ropivacaine with the enantiomer ratio of 1:20, which is insufficient for pharmaceutical control of enantiopurity, where a ratio of 1:100 is necessary in the case of ropivacaine.

CE-ESI-MS(Q) enantioseparation of terbutaline and ketamine employing DM-β-CD for enantioseparation of propranolol using 2-hydroxypropyl-β-CD (HP-β-CD) in water-organic BGE-containing 0.8 M of acetic acid and 5 mM ammonium acetate dissolved in methanol–water mixture (80:20, v/v) was published by Cole et al. [69]. Even in this case, no experimental approach was used to prevent the entry of the used CSs into the MS ion source. Furthermore, the influence of the CSs on the MS signal intensity (suppression of analyte signal) was evaluated. Significant reduction in the signal intensity of the studied enantiomers occurred at concentrations of the mentioned CD derivatives of 2 mM and higher.

A dual system of two CDs (3 mM β-CD, a 10 mM DM-β-CD) was used for the enantioseparation of methamphetamine, amphetamine, dimethylamphetamine, and p-hydroxymethamphetamine using CE-ESI-MS(Q) [70]. The BGE consisted of 1 M formic acid of pH 2.2, and the developed method was applied on the analysis of the selected amphetamine-based drugs in urine after an SPE extraction procedure. No special approach to avoid the ion source contamination was used in this case. Limits of detection for all studied amphetamines were 3 ng/mL (spiked urine samples).

5.6.2 Counter-Current Mode for CD-Mediated Enantioseparation by CE-ESI-MS

To reach a sufficient sensitivity of the MS detection, it is necessary to avoid entry of the CD into the MS ion source. The simplest way to avoid that is using a counter-current migration of charged CD derivative in conditions with a suppressed EOF or in strongly acidic pH.

The selected derivative of CD must be oppositely charged to the separated analyte. In such experimental approach, it is possible to utilize the electrostatic interactions between analyte and CS, which (compared to native CDs) allows and supports further selectivity manipulation. On the other hand, charged derivatives of CDs contribute to the total current of the electrophoretic separation and can lead to generation of excess Joule heating in the capillary. Except that, a sorption of the charged CDs on the inner capillary wall can occur. This can lead to decrease in migration time reproducibility and peaks resolutions and areas. When the charged CD is used as CS, the conditions must be set up so that the charged CD could migrate by self-electrophoretic mobility to the inlet part of CE and the analytes toward the MS ion source.

In the work published by Schulte et al. [71], a counter-current migration was used for the separation of basic drugs (etilefrine, mianserine, dimethindene a chlorpheniramine) in 10 mM acetic acid–ammonium acetate buffer of pH 3.5 employing 0.2–0.3 mg/mL of carboxymethyl-β-CD (CM-β-CD). Mianserine was also enantiosep-arated in the same conditions with 0.1 mg/mL of sulfobutylether-β-CD (SBE-β-CD). A chiral separation of ($\pm$) tropic acid was also described in the same publication using polyacrylamide-coated capillary filled with BGE based on ammonium acetate pH 5.0 with

8.0 mg/mL of tetramethylammonium-β-CD (TMA-β-CD) as CS with reversed voltage polarity.

Lio et al. [72] described a CE-ESI-MS(IT) method for the enantioseparation of methamphetamine and its metabolites in urine using uncoated silica capillary in strongly acidic electrolyte (1 M ammonium formate pH 2.0) with an addition of 1.5 mM heptakis(2,3-di-O-acetyl-6-O-sulfo)-β-cyclodextrin (HDAS-β-CD).

Another approach for the chiral separation of nine amphetamine-type stimulants was published by Iwata et al. [73]. The stimulants were separated using negatively charged highly sulfated γ-CD (HS-γ-CD) in the reversed separation polarity mode, which means that the HS-γ-CD entered the MS ion source. The amphetamines were detected using positive ESI ionization. Amphetamine forms a negatively charged transient diastereomeric complex with HS-γ-CD, which migrates as anions toward the MS ion source. Due to the positive ESI ionization process, a dissociation of the diastereomeric complexes between CS and amphetamine enantiomers occurs and negatively charged HS-γ-CD does not enter the ion source. No significant decrease in the MS detection sensitivity was observed.

The experimental setup that limits the entry of the non-volatile selector into the ion source using non-continuous flow CE-ESI-MS(Q-TOF) for enantioseparation of (R,S)-omeprazol was published by Olsson et al. [74]. During the enantioseparation in the CE step, the spraying voltage and the sheath liquid flow were switched off. Before the entry of the separated enantiomers to the MS interface, the spraying voltage and sheath liquid flow were switched on, resulting in the detection of the individual enantiomers without significant ion source contamination.

5.6.3 Partial Filling Technique Employing Neutral Derivatives of CDs

The first experiments concerning the PFT were demonstrated by Jäverfalk et al. [75] on enantioseparation of local anesthetics R,S-ropivacaine and (R,S)-bupivacaine. Polyacrylamide-coated capillary was used to eliminate the EOF, and the separation was performed in 50 mM acetic acid of pH 3.0 with the addition of 100 mg/mL of methyl-β-CD.

The polyacrylamide-coated capillary was used not only to suppress the EOF but also to avoid migration of the neutral β-CD derivative together with the EOF into the MS ion source. For this concrete case, the PFT enables to perform 7300 analyses with 1 g of CS. Other published works dealing with the PFT employed neutral CDs with suppressed EOF are summarized in Table 5.1. Another important point, which must be optimized, is the composition of the BGE together with proper selection of the CS and its amount (concentration). In addition, the parameters of MS detection can have a major impact on the enantioseparation process. The nebulizing gas pressure can cause resolution decrease during the enantioseparation with eliminated EOF (e.g., in acidic pH or coated capillary). This fact can be particularly critical especially when several structurally similar compounds are separated and their resolution is about 1.5 or lower. In such cases, it is possible to use the MS interface without the nebulizing gas as documented in Ref. [77].

Table 5.1 Counter-current method combined with PFT for CE-ESI-MS enantioseparation of drugs employed neutral derivatives of CDs

Chiral analyte	CE conditions	MS	Sample/ preparation	Ref.
Clenbuterol and salbutamol	10 mM ammonium acetate (pH 2.0), 20% (v/v) MeOH, 40 mM DM-β-CD, U = +30 kV, +10 mbar, fused-silica capillary, PFT 66%	ESI (QqQ)	Plasma/ SPE	[76]
Amphetamine, methamphetamine, MDA, MDMA, MDEA, MDPA, venlafaxine and its phase I metabolites, hyoscyamine, homoatropine, methadone	Ammonium acetate (various pH) HP-β-CD or CM-β-CD or S-β-CD, U = +30 kV, PVA coated capillary, PFT 90%	ESI (Q)	Standard serum (methadone/ LLE)	[77]

Chiral analyte	CE conditions	MS	Sample/ preparation	Ref.
Camphorsulfonic acid	40 mM ammonium formate (pH 4.0), 50 mM DM-β-CD, U = -30 kV, PAA-coated capillary, PFT 70%	ESI (QqQ)	Standard	[78]
(+)-2-(4,5-Dihydro-1H-imidazo-2-yl-)-2'-methoxy-5-fluorobenzo-[1,4]-dioxan (andrenoreceptor antagonist)	50 mM ammonium formate (pH 4.0), +10 mM HP-β-CD, U = +20 kV, fused-silica capillary, PFT 80%	+ESI (QqQ)	Standard	[79]
(±) Methadone	20 mM ammonium acetate (pH 4.0), 18 mg/mL HP-β-CD, U = +20 kV, PVA capillary, PFT 90% SBE- β-CD, CM-β-CD also studied; Optimization by DOEs	+ESI (Q)	Standard	[80]

5.6.4 Combination of PFT and Counter-Current Migration for CD-Mediated CE-ESI-MS Enantioseparation

This combination gives an advantageous possibility to utilize separation with CS bearing an opposite charge to the analyte when the EOF is eliminated (as mentioned in the previous sections). Tanaka et al. [78] published a work dealing with enantioseparation of (±) tropic acid using quaternary ammonium-β-CD (QA-β-CD) in a polyacrylamide-coated capillary. The separation was carried out in 40 mM ammonium acetate with pH of 5.0 where the tropic acid enantiomers were negatively charged whereas the used CS was charged positively. Both the PFT method and the counter-current migration method were combined as the capillary was filled with the BGE-containing CS to the 70% of its total length. Finally, the separation was realized using negative (reversed) polarity.

Enantioseparation of (R,S)-methadone using SBE-β-CD and CM-β-CD in 20 mM ammonium acetate of pH 4.0 in a poly(vinyl alcohol)-coated capillary was published by Rudaz et al. [80]. The PFT using 90% of the capillary's total length was used in the case of SBE-β-CD, while the amount of SBE-β-CD required for successful enantioseparation was higher (0.4 mg/mL) compared to CM-β-CD, where only 70% of the capillary was filled with the BGE and the amount of CM-β-CD was 0.25 mg/mL only. After a modification, the method was used for enantioseparation and determination of (R,S)-methadone and its metabolites in serum samples.

Enantiomers of basic anesthetics ketamine, prilocaine, and mepivacaine were separated by CE-ESI-MS(Q) in a poly(vinyl alcohol)-coated capillary using a combination of PFT (90%) and counter-current migration in 20 mM ammonium acetate of pH 3.0 with the additions of S-β-CD [81]. Ketamine was separated using BGE with 6 mg/mL of S-β-CD, whereas satisfactory resolution of prilocaine and mepivacaine was achieved with 10 mg/mL of S-β-CD in the BGE. The enantioresolution capability of S-β-CD was much higher compared to SBE-β-CD and CM-β-CD, which were evaluated as well.

A CE-ESI-MS(Q) enantioseparation of methadone, tramadol, venlafaxine, and fluoxetine in an uncoated silica capillary with strongly acidic BGE was published by Rudaz et al. [82]. The separation was performed in 30 mM ammonium formate of pH 2.5 with HS-γ-CD in the range of 0.1% to 0.85% (w/v) using PFT to 83% of the capillary total length. An infinite resolution was described in this publication due to the different effective charges of the resulting diastereomeric complexes between negatively charged CS and positively charged analytes.

Desiderio et al. published a CE-ESI-MS(QqQ) enantioseparation of (R,S)-baclofen employing negatively charged SBE-β-CD (1.75 mM) in strongly acidic BGE consisting of 0.25 M formic acid using the PFT (88%) [83]. The authors pointed out the effect of the spraying tip adjustment to the final resolution. This method of (R,S)-baclofen enantioseparation was validated and used for the analysis of pills and pharmaceutical materials with a different ratio of the individual enantiomers using L-tyrosine as internal standard.

Several publications dealing with HS-γ-CD for CE-ESI-MS(Q) enantioseparation of amphetamine-type stimulants (derivatives of amphetamine, 3,4-methylenedioxy-methamphetamine, methadone) in plasma after liquid–liquid extraction (LLE) in strongly acidic BGEs and an uncoated silica capillary using the PFT methods were published [84–86]. The use of the coated capillary has a benefit consisting in the possibility of using a longer injected zone of the BGE (longer PFT zone), which leads to higher resolution of the enantiomers.

Marina et al. published a work dealing with the determination of D-carnitine as an optical impurity of L-carnitine using CE-ESI-MS(IT) in pharmaceutical products. The enantioseparation was performed in an uncoated silica capillary using succinyl-γ-CD (SUC-γ-CD) in 0.5 M ammonium formate pH 2.5 [87]. The developed method allowed determination of the enantiomeric impurity (D-carnitine) in the concentration as low as 0.002% (to L-carnitine). The LOD for carnitine determination was 10 ng/mL.

Monii et al. used a sheathless arrangement of the ESI ionization for CE-ESI-MS(Q) enantioseparation of positional and optical isomers of amphetamine and cathinone derivatives (amphetamine, methamphetamine, 3-fluoromethcathinone, 4-fluoromethcathinone, pentedrone, 4-methylethcathinone, methylone, and pentylone) using HS-γ-CD in 0.5% formic acid (v/v) in an uncoated silica capillary with the inner diameter of 20 μm using a connection with a porous tip interface [88]. The resolution of the separated cathinone derivatives was higher when a dual selector system consisted of HS-γ-CD and (+)-18-crown-6-tetracarboxylic acid ($C_{18}H_6$) was studied. It was verified that there is no detection sensitivity decrease when the sheathless mode is used. This could be explained by the very low liquid flows in the separation capillary (nL/min) and thus the influence of the non-volatile CS on the signal intensity is negligible.

Enantioseparation of (R,S)-duloxetine by CE-ESI-MS(IT) was published by Sánchez-López et al. [89]. The separation was carried out in an uncoated capillary and acidic BGE based on 150 mM ammonium formate pH 3.0 using the PFT (38%). Uncharged HP-β-CD in the concentration of 0.5% (w/v) was used as CS. Compared to the CE-UV method, the capillary used for PFT had to be longer to reach the same resolution of the separated enantiomers. The LOD for

(R)-duloxetine was 20 ng/mL, and the method was applied for the determination of its enantiomers in pharmaceutical formulations (Fig. 5.5).

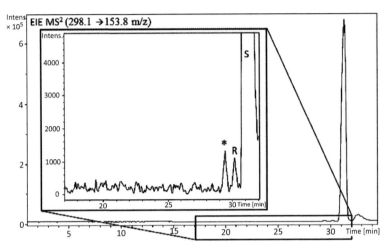

Figure 5.5 CE-MS2 EIE corresponding to the LOD of (R)-duloxetine (0.02 μg/mL) in the presence of 100 μg/mL (S)-duloxetine. Experimental conditions: BGE, 150 mM ammonium formate buffer (pH 3.0); PFT 0.5% (w/v) of HP-β-CD in BGE applying 1 bar during 1 min; uncoated fused-silica capillary, 104 cm × 50 μm I.D. Reprinted from Ref. [89], Copyright 2014, with permission from Elsevier.

Another CE-ESI-MS work dealing with enantioseparation of 12 cathinone derivatives using sodium phosphate-based BGE was published [90]. The BGEs based on volatile components (ammonium acetate and formate) were inadequate for the enantioseparation of the cathinone derivatives because the obtained resolutions were insufficient. These BGEs were replaced by 50 mM sodium phosphate with an addition of 0.5% (w/v) HS-γ-CD using PFT (70%). The authors also emphasized that by using the sheath liquid flow, the eluting BGE was diluted, the ESI ion source contamination was negligible, and the sensitivity decrease caused by the sodium phosphate was insignificant.

An ultra-trace determination of pheniramine in urine using multidimensional techniques ITP-CE-ESI-MS(QqQ) was published by Piešťanský et al. [91]. The used CS was CE-β-CD in the concentration of 5 mg/mL and the LOD for pheniramine enantiomers reached a level of tens pg/mL due to the online izotachophoretic preconcentration step. The published method allows determination of pheniramine in urine without any sample pre-treatment. Notable advantages of this method are injection of a long zone of the sample, online preconcentration, counter-current migration of the analytes, and a sensitive MS/MS detection.

Online sample preconcentration with the PFT for enantioseparation of chlorpheniramine using HP-β-CD as CS was published in 2014 [92]. A combination of field-enhanced sample injection (FESI) and micelle-to-solvent stacking (MSS) was employed to reach 500 times lower LOD compared to the method without the online preconcentration. The enantioseparation of chlorpheniramine was carried out using the PFT (40%) in 50 mM ammonium acetate of pH 3.5 with the addition of 20 mM HP-β-CD. The sample solvent suitable for the online preconcentration was 20 times diluted BGE with 60% (v/v) of acetonitrile. A zone of 10 mM ammonium lauryl sulfate was injected before the sample zone.

5.7 Enantioseparation using MEKC-MS

Like in the case of CDs and their derivatives, the surfactant micelles (both chiral and achiral) are usually non-volatile, which causes significant decrease in the MS detection sensitivity as well as contamination of the MS ion source by the surfactant monomers. In addition, in this case, it is possible to optimize the conditions of enantioseparation so that micelles of the surfactants do not enter the MS ion source. Another approach is to use high-molecular-mass micelle polymers as chiral pseudostationary phase. The use of these chiral polymeric surfactants as CSs enables to work in classical separation system arrangement for CE-ESI-MS (the surfactant is present in the inlet part as well as in the whole capillary; the outlet vial is usually replaced by a flow of a conductive sheath liquid).

The potential entrance of the polymeric surfactant into the ion source (ESI, APPI, etc.) does not lead to a significant suppression of the analyte signal in MS. Micelles of chiral surfactants derived from cholic acids or micelles of chiral polymers with high molecular weight and very low CMC values can be utilized as CS for MEKC-MS methods. The physicochemical properties of the polymeric chiral surfactants are compatible with the most frequently used ESI-MS detection. These are polymeric micelles with very low CMC (close to zero) and thus can be used also in combination with organic solvents. Moreover, they are difficult to ionize in the ESI source due to their high molecular weight. An important advantage that results from the mentioned physicochemical properties is the high MS detection sensitivity, when the chiral polymeric surfactant does not affect the ionization efficiency.

Polymeric chiral surfactants have lower droplet surface affinity or surface activity than the analytes when performing detection of chiral compounds with ESI-MS. Unlike the micelles of low-molecular surfactants, the polymeric surfactants are not subject to dissociation to the monomeric units in the ESI ion source. The presence of monomeric units formed by the dissociation during the spraying decreases the intensity of the MS signal as well as the electrospray stability. A disadvantage of these chiral polymeric surfactants is the fact that they are usually not commercially available. The monomeric units as well as the resulting polymer must be synthesized in the laboratory according to a procedure published by Wang et al. [93]. A list of works dealing with the polymeric chiral surfactants for MEKC-MS enantioseparation is shown in Table 5.2.

The first publication that utilized MEKC-ESI-MS for enantioseparation was published in 2001 by Shamsi et al. The micellar pseudostationary phase was composed of 0.2% (w/v) polysodium N-undecanoyl-L-valinate (poly-L-SUV) and used for enantioseparation of 1,1'-bi-2-naphthol (BINOL) atropisomers [94]. A crucial influence on the sensitivity of the MS detection was the concentration of the chiral pseudostationary phase. In this case, the separation was employed using basic pH BGE where a strong EOF is generated (20 mM ammonium acetate of pH 9.2 with the addition of 0.2% poly-L-SUV). To ensure the highest detection sensitivity, it was necessary to use the lowest possible concentration of the chiral polymer so that an acceptable resolution could be achieved. The nebulizer gas pressure and the sheath liquid flow rate significantly influenced the resolution of BINOL atropisomers.

Table 5.2 MEKC-MS methods employing chiral polymer micelles as CSs

Chiral analyte	MEKC conditions	MS	Sample/ preparation	Ref.
Atropisomers of 1,1′-bi-2-naphthol (BINOL)	20 mM ammonium acetate (pH 9.2), 0.20% (w/v) poly-L-SUV	ESI (Q)	Standard	[94]
Beta-blockers (atenolol, metoprolol, carteolol, pindolol, oxprenolol, talinolol, alprenolol, propranolol)	25 mM ammonium acetate/TEA (pH 8.8), 15 mM poly-L-SUCL	ESI (Q)	Standard	[95]
Oxazepam, lorazepam, nefopam	Comparison of six polymeric surfacants (poly-L-SUL, poly-L-SUV, poly-L,L-SULV, poly-L-SUCL, poly-L-SUCV, poly-L,L-SULV) 25 mM ammonium acetate (pH 8.0)	ESI (Q)	Standard	[96]
Warfarine and coumachlor	25 mM ammonium acetate (pH 6.0), with 25 mM poly-L,L-SULV	ESI (Q)	Plasma/SPE	[97]
Phenylethylamines, beta-blockers, hydrobenzoin, benzoin and derivatives, benzodiazepines, PTH-derivatives of amino acids, chlorophenoxypropionic acid	Poly-L-SUCLS, poly-L-SUCILS, poly-L-SUCVS, poly-L-SUCAAS at acidic, neutral, and basic pH	ESI-Q	Standard/ human urine pseudoephedrine analysis	[98]
Ephedrine, pseudoephedrine, norephedrine, N-methyl-ephedrine	15 mM ammonium acetate (pH 6.0), 30% ACN with 35 mM poly-L-SUCL	ESI-Q	Standard	[99]

(Continued)

Table 5.2 (*Continued*)

Chiral analyte	MEKC conditions	MS	Sample/ preparation	Ref.
Ephedrine, pseudoephedrine, norephedrine, *N*-methyl-ephedrine, norpseudoephedrine, *N*-methylpseudoephedrine, phenylephrine (IS)	15 mM ammonium acetate (pH 6.0), 30% ACN with 35 mM poly-L-SUCL	ESI-Q	Dietary supplements	[100]
1,1′-binaphtol (BINOL) a 1,1′-binaphthyl-2,2′-dihydrogenphosphate (BNP)	35 mM ammonium acetate (pH 10.8), 27 mM poly-L-SUCL/poly-L-SULV (1:1)	ESI-Q	Standard	[101]
Venlafaxine and *O*-desmethylvenlafaxine	20 mM ammonium acetate/25 mM TEA (pH 8.5), 25 mM poly-L,L-SULA	ESI-QqQ	Plasma (pharmako-kinetics and drug–drug interaction)	[102]
Warfarine and hydroxy-metabolites	25 mM ammonium acetate (pH 5.0), 25 poly-L,L-SULV, 15% MeOH	ESI-Q ESI-QqQ	Plasma/SPE	[104]
Benzoin ethyl ether, benzoin methyl ether, benzoin, hydrobenzoin	55 mM ammonium acetate (pH 8.0), 15 mM poly-L-SUCL, 50 mM poly-L,L-SULV	APPI-Q	Standard	[105]
Mephobarbital, pentobarbital, secobarbital	25 mM ammonium acetate (pH 7.0), 39.7 mM poly-L-SUCIL	ESI-Q	Serum	[106]

Polysodium *N*-undecenoxycarbonyl-L-leucinate (Poly-L-SUCL) was used as a pseudostationary phase for the enantioseparation of eight structurally related β-blockers. In this work, the influence of the polymerized and unpolymerized surfactant was evaluated in terms of MS detection signal (as S/N ratio) and the polymerized surfactant showed higher S/N ratio. The presented method has an advantage that all of the eight β-blockers were separated together using one CS under the same conditions [95].

Six other chiral polymers (three carbamates and three amide-based) were studied for MEKC-ESI-MS(Q) enantioseparation of two benzodiazepines (oxazepam and lorazepam) and nefopam (non-opioid drug of the benzoxazocine group) [96]. The studied polymeric micelles showed different selectivity to the separated enantiomers. A simultaneous chiral separation of all three analytes (oxazepam, lorazepam, and nefopam) was achieved using 15 mM polysodium undecanoyl-L-leucinate (poly-L-SUL) in 25 mM ammonium acetate of pH 8.0 containing 10% (v/v) of acetonitrile. Longer retention times were obtained using the carbamate polymeric surfactants, which also showed excellent enantioselectivity for oxazepam and lorazepam but no stereoselectivity for nefopam.

Hou et al. [97] published enantioseparation of warfarine and coumachlor using 25 mM poly-L,L-SULV in 25 mM ammonium acetate of pH 6.0. The authors described the influence of the buffer pH on the resolution of the separated enantiomers using polysodium *N*-undecanoyl-L,L-leucylvalinate (poly-L,L-SULV). In the range of pH 5.5 to 8.0, the resolution was decreasing with increasing pH. This was explained by a higher electrostatic repulsion between the polymer micelles due to a higher effective charge of the analyte and surfactant in less acidic pHs.

Several polymeric micelles based on amino acids were studied for MEKC-ESI-MS enantioseparation of various drugs and metabolites using acidic, neutral, and basic electrolytes [98].

Designs of experiment (DOE) approaches were utilized for the optimization of MEKC-ESI-MS enantioseparations of eight stereoisomers of ephedrine and related compounds using poly-L-SUCL [99, 100]. This approach was also used in a study dealing with the enantioseparation of BINOL and 1,1′-binaphthyl-2,2′-dihydrogenphosphate (BNP) atropisomers employing a mixture of poly-L-SUCL and poly-L,L-SULV [101].

A (polysodium *N*-undecenoyl-L,L-leucylalaninate) (poly-L,L-SULA) was used as CS for the enantioseparation of venlafaxine and its structurally related metabolite *O*-desmethylvenlafaxine by MEKC-ESI-MS [102]. The authors evaluated the influence of three types of chiral surfactants, namely, poly-L,L-SULA, poly(sodium *N*-undecenoyl-L,L-leucylvalinate) (poly-L,L-SULV), and poly(sodium *N*-undecenoyl-L,L-leucyl-leucinate) (poly-L,L-SULL) on the enantioresolution of venlafaxine and *O*-desmethylvenlafaxine. The obtained results showed that the increasing hydrophobicity of the surfactant chain on which the chiral amino acid is bonded (at the C-terminal end of the dipeptide surfactant head group) led to prolongation of migration time of the separated analytes. The resolution of both studied compounds was decreasing in the following order: poly-L,L-SULA > poly-L,L-SULV > poly-L,L-SULL (Fig. 5.6).

The polymeric surfactants with the lowest hydrophobicity (poly-L,L-SULA) gave the shortest migration times together with the best resolution, and thus it was chosen as the most appropriate chiral surfactant. When compared to previously published method dealing with the chiral separation of venlafaxine and its metabolites using charged CDs [103], the LODs obtained using the poly-L,L-SULA were significantly lower (10.5 ng/mL for venlafaxine enantiomers and a 31 ng/mL for *O*-desmethylvenlafaxine enantiomers, respectively).

Wang et al. [104] published enantioseparation of warfarine and its hydroxylated metabolites (enantiomers and positional isomers) by MEKC-MS(MS) with poly-L,L-SULV as CS. As well as other chiral polymeric surfactants on the same base, poly-L,L-SULV has an important benefit consisting of a wide elution window and no interferences with the separated enantiomers during the ionization in MS. The authors also studied the effects of pH and organic modifiers on the resolution. The developed method was compared with CEC-ESI-MS employing a packed column with vancomycine covalently bonded on the stationary phase. The MEKC-MS method with the chiral polymer proved to be about half faster than the CEC-ESI-MS and the poly-L,L-SULV as CS allowed enantioseparation of all isomers of warfarin and its metabolites (except 8-OH-warfarin). Finally, the developed MEKC-ESI-MS method was used for profiling enantiomers of warfarin and its metabolites in plasma after an SPE extraction by mixed-mode anion exchange cartridges.

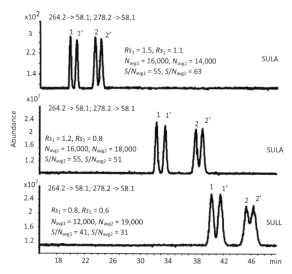

Figure 5.6 Effect of polymeric dipeptide surfactant head groups on the simultaneous enantioseparation of O-desmethylvenlafaxine (1,1′) and venlafaxine (2,2′). Conditions: 60 cm long (375 μm O.D., 50 μm I.D.) fused-silica capillary. Buffer: 20 mM NH$_4$OAc + 25 mM TEA, pH 8.5, 25 mM polydipeptide surfactant. Applied voltage, +20 kV, injection, 5 mbar, 100 s. Spray chamber parameters: nebulizer pressure: 3 psi, drying gas temperature: 200°C, drying gas flow: 8 L/min; capillary voltage: +3000 V; fragmentor voltage, 113 V for O-DVX and 117V for VX; collision energy: 17 eV; MRM transition: ODVX: 264.2→58.1; VX: 278.2→58.1. Sheath liquid: MeOH/H$_2$O (80/20, v/v), 5 mM NH$_4$OAc, pH 6.8 with flow rate of 0.5 mL/min; sample concentration: 5 μg/mL in MeOH/H$_2$O (10/90, v/v). Reprinted from Ref. [102], Copyright 2015, with permission from Elsevier.

Another application of MEKC-APPI-MS was developed for the enantioseparation of photoinitiators (benzoin, hydrobenzoin, benzoin methyl ether, benzoin ethyl ether) as potential drugs for cancer therapy. This enantioseparation was based on a mixture of two chiral surfactants (poly-L-SUCL a poly-L,L-SULV) [105]. The APPI was selected as the ion source because the studied benzoins are uncharged and thus ESI was not suitable for their ionization. A multivariate central composite design of experiment was used to optimize the conditions of benzoins enantioseparation. It should be noted that this work was the first one using MEKC-APPI-MS for the enantioseparation and detection of chiral compounds. Due to this fact, a very strong potential of the APPI-MS was demonstrated as a detection mode, which showed high sensitivity and the possibility of utilizing non-volatile buffers.

Enantioseparation of three chiral barbiturates by MEKC-ESI-MS was published by Wang et al. [106]. In this work, 11 chiral polymers were tested using the multivariate optimization. Using computer optimization, the migration times, resolution, and S/N ratio were estimated and then proved experimentally. The values predicted by the computed optimization were in good agreement with the experimentally obtained parameters. Poly(sodium N-undecenoxy-carbonyl-isoleucinate) (poly-L-SUCIL) was the most suitable CS for the separation of three barbiturates, and separation was achieved within 32 min.

5.8 Crown Ethers Mediated Enantioseparations by CE-MS

Crown ethers are macrocyclic CSs that have relatively limited stereoselectivity, especially for primary amines. From a wide spectrum of chiral crown ethers, only 18-crown-6-tetracarboxylic acid (18C6H4) is commonly used. The polyether cycle of 18C6H4 forms a cavity, which can interact with alkali metals, alkaline earth metals, ammonium cation, and protonated primary amines [107]. Diastereomeric complexes with different electrophoretic mobility are formed by the interaction of 18C6H4 with the primary amines. The first application of crown ethers was published by Kuhn et al. [108, 109], who also formulated the essential rules that must apply in the structure of the separated enantiomers [110]. Crown ethers are often used in a combination with other CSs (especially from the CD group) to manipulate the resulting resolution. A complex review dealing with applications of crown ethers in CE was published recently [111]. In addition to other mentioned CSs, the crown ethers also belong to the non-volatile additives and thus their entrance to the MS ion source should be avoided.

Tanaka et al. published a CE-ESI-MS(Q) method for the enantioseparation of 3-aminopyrrolidine and α-amino-ε-caprolactam [112]. By using the PFT, the enantioseparations were performed with very high sensitivities. The authors pointed out to a relation between the concentrations of 18C6H4 and the PFT zone length. The higher the concentration of 18C6H4 as well as the longer the PFT zone, the higher resolution was achieved. A similar method was used by Zhou et al. [113] for the enantioseparation of 18 pharmaceutical products by CE-ESI-MS(Q) and 18C6H4 as the CS.

Xia et al. [114] developed a method of chiral dipeptides enantioseparation (D-Ala-D-Ala, L-Ala-L-Ala, D,L-Leu-D,L-Leu, Gly-L-Phe, Gly-D-Phe). The method employed CE-ESI-MS and PFT with (+)-(18-crown-6)-2,3,11,12-tetracarboxylic acid as CS. The separation was carried out in an uncoated silica capillary using 2 M acetic acid of pH 2.15 as the BGE. The PFT method used 30.6% of the total capillary length, and the injected zone with the CS consisted of 5 mM 18C6H4 in 3 M acetic acid of pH 3.0. The dissociation of the CS is strongly suppressed in the acidic pH, and thus the CS migrates with the same velocity as the residual EOF and the combination with PFT does not lead to ion source contamination. The developed method was also applied for the enantioseparation of the dipeptides in spiked serum samples. Even if the dipeptides served as model compounds only, the presented method can be an example for further application on the enantioseparation of peptide-based drugs.

5.9 Macrocyclic Antibiotics as CS for CE-MS

Glycopeptide macrocyclic antibiotics (ATB) are CSs produced by bacterial fermentation. They are used as CS either binded to silicagel-forming CSP particles for LC or CEC, or dissolved in a BGE for CE. Commercially available macrocyclic ATBs that are used as CS are especially the following: vancomycine, ristocetin R, teicoplanin, rifamycin, and teicoplanin aglycone. Due to many functional groups (e.g., amino, carboxyl, and hydroxyl), these ATBs can be charged depending on the electrolyte's pH. Glycopeptide ATBs usually have one or more saccharide units linked to the macrocyclic rings. These saccharide units increase the solubility of the aglycone in water. On the other hand, an absence of the saccharide units does not cause loss of antibiotic stereoselective properties [115].

The most widely used CS from this group is vancomycin chloride (VC), which is produced by bacterial strains of *Streptomyces orientalis*. The spatial arrangement of the VC molecule is similar to a basketball hoop shape. VC contains several functional groups in its structure (hydroxyl, amino-, amido-, carboxy-, aromatic rings, and saccharide units) and has 18 stereogenic centers, which play the key role during the enantioseparation process. VC is well soluble in water and other polar organic solvents. So far, VC is the only one CS from the macrocyclic ATB group, which has been used for enantioseparation using CE-ESI-MS. A summary of the physicochemical properties of VC is given in Table 5.3.

Table 5.3   Physicochemical properties of vancomycin

Vancomycin chloride	
Molecular weight	1449
pK_a	2.9 (carboxyl group), 6.8 (phenol group), 7.7 (phenol group), 8.5 (phenol group), 10.4 (primary amine), 11.7 (primary amine)
Isoelectric point	7.2
Stereogenic centers	18
Number of macrocycles	3
Number of –OH groups	9 (three groups are phenolic)
Number of –NH$_2$ groups	1
Number of secondary amino groups	1
Number of –COOH groups	1
Aromatic rings	5
Amido linkages	7
Sugar moieties	2

VC is positively charged in pH lower than its isoelectric point (pI = 7.2) and negatively charged in pH higher than isolelectric point. The positive charge in the acidic environment is given by the protonation of the amino groups. The carboxyl groups are dissociated, forming anions in the pH ranging from 4 to 13. In neutral pH, the positive charge of the protonated amino groups is compensated by the negative charges of the carboxyl groups as well as by the negative charges of the phenolic groups. The dissociation of the carboxyl and phenolic groups increases with an increase in the pH of the BGE. In the pH range from 10 to 12, all the carboxyl and phenolic groups are in anionic form and the amino groups are deprotonated and neutral. The resulting effective charge of VC in the pH above 12 is −4. The aforementioned properties indicate that the electrostatic interactions are one of the crucial stereoselective interactions driving the enantioseparation in CE. Other types of interaction (π–π, dipole–dipole, induced dipole, hydrogen bonds, steric hindrance, and van der Waals forces) also play an important role in enantiorecogntion.

The main disadvantages of VC are sorption on the inner silica capillary wall and also the fact that VC absorbs UV radiation under 250 nm. Thus, the typical mode of separation is usually performed in coated capillaries using the PFT method, where the risk of sorption is minimized and the interferences caused by the UV radiation absorption are avoided. The discussed sorption of VC can also influence a decrease in migration times and can lead to the broadening of enantiomer zones caused by retarded EOF, which is given by the VC sorption [116].

For the determination of the optimal conditions for enantioseparations employing vancomycin as CS, the proper selection of BGE composition and pH is necessary. A suitable pH for VC-based enantioseparation is under the pI value (pH less than 7.2), when the VC is charged positively. Furthermore, VC solutions are stable in the pH range from 4 to 7.

The first enantioseparation using VC as CS for CE-UV was published by Armstrong et al. [117]. A review summarizing the utilization of various types of ATBs for CE enantioseparations was published later by Prokhorova et al. [118].

The VC properties and the issues mentioned earlier predetermine the usage of the PFT method for CE-ESI-MS because VC is not volatile. Because of the basic character of vancomycin, this CS is suitable especially for enantioseparations of negatively charged analytes (carboxylic and hydroxy carboxylic acid, amino acids derivatives, etc.).

Fanali et al. [119] published the enantioseparation of propionic acid-based NSAIDs (namely, carprofen, flurbiprofen, ketoprofen, naproxen, etodolac and its metabolites, 2′-hydroxyibuprofen, and carboxyibuprofen) performed by CE-ESI-MS(Q) in the negative ionization mode using VC as CS. Vancomycin was positively charged in the acidic environment of pH 4.8. By using the PFT with polyacrylamide-coated capillary, it is possible to realize a separation where the non-volatile CS (VC) migrates toward the inlet part of the instrument, whereas the analytes migrate into the MS ion source. The CE-ESI-MS enantioseparation of the selected NSAIDs represents a cheap and fast method compared to chiral HPLC separation. No other publications dealing with other ATBs for the enantioseparation of optically active pharmaceuticals have been available so far.

5.10 Non-Aqueous Capillary Electrophoresis-Mass Spectrometry for Enantioseparation

The electrophoretic separation of enantiomers is usually performed in aqueous or organic-aqueous BGEs with the addition of a suitable CS. In certain cases, the solubility of the analytes can be limited, which can lead to complications during the separation. This problem can be overcome by using non-aqueous capillary electrophoresis (NACE). The utilization of non-aqueous electrolytes has another advantage of using CS with a lack of solubility in aqueous BGEs. NACE has several other advantages, besides affecting the solubility of the separated enantiomers and CS. NACE enables to influence the separation selectivity and, therefore, allows affecting the final resolution. This can be suitable especially in the case of separation of enantiomeric impurities in pharmaceutical products. A review describing enantioseparations of pharmaceuticals using NACE was published in the past [120]. This review also contains descriptions of the enantioseparation mechanisms in non-aqueous environment. Two other reviews dealing with the fundamentals of NACE were published by Sarmini et al. [121] and Kenndler [122, 123].

A comparison of salbutamol enantioseparation by NACE equipped with UV and MS detection was published by Servais et al. [124]. HDAS-β-CD was employed as CS for the separation of (R,S)-salbutamol. The BGE consisted of 0.75 M formic acid in methanol, and the concentration of the CS was 15 mM. The selected composition of the BGE provides appropriate conditions for the connection of

NACE with ESI-MS detection. Negatively charged CD derivative migrates under these conditions as anion toward the inlet part of the instrument, and the separated enantiomers migrate directly into the ion source of MS. Methanol causes EOF suppression, which was negligible in this case. Very low detection limits were achieved by combining an SPE method with NACE-MS for the separation of salbutamol enantiomers (8 and 14 ng/mL for the first and second enantiomers, respectively). The developed method was used for the determination of salbutamol enantiomers in urine.

Mebeverine, its derivatives, and salbutamol were separated in non-aqueous BGE based on 0.75 M formic acid, 30 mM potassium camphor sulfonate, and 30 mM heptakis(2,3-di-O-methyl-6-O-sulfo)-β-cyclodextrin (HDMS-β-CD) in methanol. The EOF is strongly suppressed in methanolic BGEs, so the contamination of the ion source by the negatively charged CS does not occur [125]. The authors also demonstrated a certain disadvantage of the charged HDMS-β-CD, which is usually employed as sodium salt. The sodium ions migrate toward the ion source and cause suppression of the ionization efficiency.

Chiral separation of several nonsteroidal anti-inflammatory drugs (NSAIDs), namely, ibuprofen, fenoprofen, ketoprofen, indoprofen, and flurbiprofen, was published by Mol et al. [126]. These chiral drugs were enantioseparated using positively charged single-isomer CDs (6-monodeoxy-6-mono-(3-hydroxypropyl) propylamino-β-cyclodextrin, PA-β-CD and 6-monodeoxy-6-mono(2-hydroxy)propylamino-β-cyclodextrin, IPA-β-CD) in 20 mM ammonium acetate in methanol employing PAA-coated capillary. In this case, a suppression of the separated enantiomers by chloride anions (co-ions of the CSs) was described. This phenomenon was explained by the fact that chlorides migrate in the same direction as the separated enantiomers.

(–)-2,3:4,6-di-O-isopropylidene-2-keto-L-gulonic acid [(–)DI-KGA] was used as CS for CE-ESI-MS(Q-TOF) enantioseparation of pronethalol employed PFT and counter-current migration [127]. The optimal composition of the BGE was 20 mM ammonium acetate/20 mM acetic acid in MeOH:2-PrOH (75:25) mixture with the addition of 80 mM (–)DIKGA.

5.11 Enantioseparation by CEC-MS

Binding of CS to a stationary phase (SP) is another approach to prevent the entry of non-volatile CS into the MS ion source. The CS is an integral part of the SP, which is bound to the capillary wall (column), and thus no decrease in MS signal occurs due to contamination. Several possible capillary columns are used for CEC separations: (i) packed, (ii) monolithic, and (iii) open tubular (OT). The packed columns are not frequently used for CEC-MS; the frits that retent the chiral stationary phase (CSP) in the separation column can contribute to bubble formation in the mobile phase. The bubble formation could be potentially eliminated by the pressurization of the capillary at its both ends (pCEC). The outlet part of the CEC capillary, which is connected by the interface of the ESI (or other atmospheric-pressure ionization technique), cannot be pressurized. This issue can be solved by tapering the capillary ends when no additional pressurization is necessary.

Zheng et al. [128] used pCEC-ESI-MS to demonstrate a positive effect on the repeatability of (R,S)-warfarin enantioseparation by replacing the untapered with a tapered column.

A sulfonated polysaccharide-based CSP packed in the capillary with the tapered end was used for the CEC-MS separation of (R,S)-aminogluthetimide employing mobile phase composed of ACN/water (7:3, v/v) containing 5 mM ammonium formate pH 3.5 [129].

Monolithic SPs represent another type of CSPs for CEC enantioseparation. There are three possible methods for their preparation: (i) post-polymerization modification, (ii) utilization of chiral monomers in the reaction mixture, and (iii) molecular imprinting using chiral templates (MIP).

A chiral monomer glycidyl methacrylate-bonded CD was polymerized for CSP preparation, which was then used for the enantioseparation of hexobarbital, catechine, pseudoephedrine, and other enantiomers [130]. The prepared CEC column with the CSP showed very good repeatability as well as selectivity together with a homogenous microflow during the CEC-ESI-MS(Q) analyses.

The first enantioseparation using OT-CEC-MS was published by Schurig et al. [131]. In this publication, permethyl-β-CD binded to polysiloxane (Chirasil-Dex) immobilized on the capillary wall was used for the enantioseparation of hexobarbital enantiomers in urine. The separation was performed in 10 mM ammonium acetate of pH 7.0, and after an LLE, the LODs of the enantiomers were lower than the therapeutic levels of hexobarbital in urine (20–50 ng/mL). Except that the analysis time was only 2 min. At the spraying end, the CSP-containing capillary was connected with a fused-silica capillary by a PTFE tube. Furthermore, the authors also detected a presence of hexobarbital with sodium adducts, $[M+Na]^+$ a $[M+2Na]^+$, which was explained as the main mechanism of the ionization process.

Eight β-blockers (oxprenolol, alprenolol, pindolol, metoprolol, propranolol, talinolol, atenolol a carteolol) were enantioseparated by CEC-ESI-MS employed internally tapered column with covalently bound vancomycine and teicoplanin [132]. The enantioselectivity of duplex columns (a combination of two columns with different CS bound) and mixed-mode columns was compared in this work.

In another work, silicagel modified with permethylated-β-CD (5 µm, 300 Å) was used as CSP for the enantiosepration of hexobarbital, mephobarbital, and chlorinated alkyl phenoxy propanoate enantiomers [133]. The separation was done using pCEC connected with coordination ion spray mass spectrometry (CIS-MS), when selected metal ions [silver(I), cobalt(II), copper(II), and lithium(I)] were added to the sheath liquid serving as complexing ions, which influence the ionization efficiency of the separated enantiomers. The enantiomers formed complexes with the metal ions, which led to ionization efficiency improvement. For a sensitive determination of hexobarbital, a presence of silver(I), cobalt(II), and copper(II) in the sheath liquid was applicable, whereas a presence of lithium(I) ions led to an increase in the sensitivity of the alkyl phenoxy propanoate. The mobile phase was based on 0.5 mM ammonium acetate in a mixture of water/methanol (40:60, v/v) with a pH of 6.6. The developed CEC-CIS-MS approach was finally compared to an HPLC-CIS-MS method, and it was demonstrated that the CEC enantioseparation is six times faster with higher separation efficiency.

Packed capillary with different CSP was utilized for the enantioseparation of two anticoagulant drugs, warfarin and coumachlor. Zheng et al. used a commercial CSP (3R, 4S)-Whelk-O1 CSP (0.5 µm) [134] and evaluated three types of packed CEC columns: (i) untapered, (ii) externally tapered, and (iii) internally tapered. The tapered column showed the highest repeatability of retention times compared to the untampered column. The final developed method was used for the determination of warfarin isomers in plasma samples.

5.12 Specifics of Quantitative Analysis of Enantiomers by CE-MS

Nowadays, the methods of enantioseparation do not serve as simple separations of individual enantiomers. The methods discussed in this chapter can be primarily used for the sensitive determination of low concentrations of one enantiomer in a mixture with a high excess of the second one (pharmaceutical control of impurities) as well as for the determination of enantiomers in complicated matrices such as crude or final pharmaceutical products.

Quantitation of individual enantiomers is also required in other samples, e.g., from the environment, foodstuff, or biological materials (urine, blood derivatives, cerebrospinal liquid, etc.). The influence of the matrix on the final resolution and the signal intensity is very significant in the case of CE-ESI-MS.

Even if it is possible to increase and improve the selectivity and detection sensitivity by MS in the selected ion monitoring (SIM) mode or by using tandem mass spectrometry (MS/MS or MS^n), it is still necessary to thoroughly optimize the sample pre-treatment. Stable conditions of separation and detection are indispensable to achieve acceptable repeatabilities of the separation. The technical solution of CE-MS hyphenation is still the major weakness, which has the greatest impact on the poorer repeatability and reproducibility of quantitation.

Besides the optimization of the enantioseparation conditions at the CE side, it is also necessary to properly optimize the conditions of the ESI-MS detection. Except the parameters that affect the sensitivity of the MS detection (sheath liquid composition and flow rate, temperature, nebulizing gas pressure, and flow rate), it is also required to optimize the spraying tip position. Finding the optimal position of the spraying tip is necessary to avoid the formation of a corona discharge and to ensure a stable electrical current in the ion source. As it is apparent from the number of parameters to be optimized, it is recommended to employ a method of experimental design [135]. Monitoring of the current in the ion source is very important not only for the stability of the ionization, but also if the charged CS enters the ion source the total current will increase [136].

In the case of biological samples, it is necessary to evaluate the matrix effects like in the case of LC-(ESI)-MS. For a precious quantitative analysis of the real samples, it is better to use internal standards (IS), which can minimize the variability of migration times and peak areas. When the enantioseparation is carried out by CE-(ESI)-MS, the IS should migrate close to the separated enantiomers to assure the most similar conditions of the IS ionization as possible. It is clear that the ideal IS is an isotopically labeled standard (usually deuterated or labeled by ^{13}C) of the separated compound, which will migrate as the non-labeled analyte and thus will be ionized under identical conditions. By using the isotopically labeled standard, it is also possible to eliminate further variabilities of the whole analytical process (sample treatment, injection into the capillary, matrix effects, EOF and temperature variability during the migration, etc.). It must be noted that the CE-ESI-MS usually requires a certain part of the separation capillary to put outside the thermostated cassette of the CE instrument, which complicates the temperature control. This part of the separation capillary makes a connection of two independent instruments—CE and the interface of the MS detection system.

When charged CSs that contain co-ions are used (usually in combination of PFT and counter-current migration), the fact that the co-ions (e.g., sodium or chloride ions) migrate toward the MS ion source should be taken into account. These co-ions can negatively influence the resulting ionization efficiency. Also an adduct formation can occur with the separated analytes, which is often especially for sodium or chloride ions. The formed adducts can complicate the identification of the separated analytes as well as the interpretation of the mass spectra. A possible solution for the future is the use of electrokinetic partial filling combined with counter-current migration and suppressed EOF. By using such an approach, it is possible to inject cations or anions of the CS into the capillary without the corresponding co-ions [137]. The utilization of the PFT method and counter-current migration is currently the most frequent approach for enantioseparation and sensitive detection by CE-MS.

Acknowledgment

The authors gratefully acknowledge the support by the project LO1305 of the Ministry of Education, Youth and Sports of the Czech Republic.

Appendix

List of abbreviations

Abbreviation	Definition
BGE	Background electrolyte
BINOL	1,1′-Bi-2-naphthol
BNP	1,1′-Binaphthyl-2,2′-Dihydrogenphosphate
CM-β-CD	Carboxymethyl-β-CD
DM-β-CD	Dimethyl-β-cyclodextrin
DOEs	Design of experiments
HDMS-β-CD	Heptakis(2,3-di-O-methyl-6-O-sulfo)-β-cyclodextrin
HP-β-CD	Heptakis-2,6-di-O-methyl-β-cyklodextrin
HS-γ-CD	Highly sulfated γ-cyclodextrin
LLE	Liquid–liquid extraction
MDA	Methylenedioxyamphetamine
MDEA	3,4-Methylenedioxyethylamphetamine
MDMA	3,4-Methylenedioxymethylamphetamine
MDPA	3,4- Methylenedioxypropylamphetamine
PFT	partial filling technique
Poly-L,L-SULA	(Polysodium N-undecenoyl-L,L-leucylalaninate)
Poly-L,L-SULL	Poly(sodium N-undecenoyl-L,L-leucyl-leucinate)
Poly-L,L-SULV	Poly(sodium N-undecenoyl-L,L-leucylvalinate)
Poly-L-SUCASS	Polysodium N-undecenoxycarbonyl-L-amino acid sulfates
Poly-L-SUCIL	Poly(sodium N-undecenoxy carbonyl-L-isoleucinate)
Poly-L-SUCILS	Polysodium N-undecenoyl-L-isoleucine sulfate
Poly-L-SUCLS	Polysodium N-undecenoxycarbonyl-L-leucine sulfate
Poly-L-SUL	Poly(sodium undecenoyl-L-leucinate)
Poly-L-SULA	Sodium poly(N-undecanoyl-L-leucyl-alaninate)

Abbreviation	Definition
Poly-L-SULV	Sodium poly(N-undecanoyl-L-leucyl-valinate)
QA-β-CD	Quaternary ammonium-β-cyclodextrin
QA-β-CD	Quaternary ammonium-β-CD
SBE-β-CD	Sulfobutylether-β-CD
SUC-γ-CD	Succinyl-γ-cyclodextrin
TMA-β-CD	Tetramethylammonium-β-CD

References

1. Caner, H., Groner, E., Levy, L., and Agranat, I. (2004). Trends in the development of chiral drugs, *Drug Discov. Today.*, **9**, pp. 105–110.

2. US Food and Drug Administration website. http://www.fda.gov/drugs/GuidanceComplianceRegulatoryInformation/Guidances/ucm122883.htm. Published May 1, 1992. Accessed January 6, 2013. See more at: http://www.ajmc.com/journals/issue/2014/2014-vol20-n3/assessing-the-chiral-switch-approval-and-use-of-single-enantiomer-drugs-2001-to 2011/P-3#sthash.qPM23kDY.dpuf.

3. Nguyen, L. A., He, H., and Pham-Huy, C. (2006). Chiral drugs: An overview, *Int. J. Biomed. Sci.*, **2**, pp. 85–100.

4. Gübitz, G. and Schmid, M. G. (1997). Chiral separation principles in capillary electrophoresis, *J. Chromatogr. A*, **792**, pp. 179–225.

5. Vespalec, R. and Boček, P. (2000). Chiral separation in capillary electrophoresis, *Chem. Rev.*, **100**, pp. 3715–3754.

6. Stalcup, A. M. (2010). Chiral separations, *Ann. Rev. Anal. Chem.*, **3**, pp. 341–363.

7. Van Eeckhaut, A. and Michotte, Y. (eds). (2009). *Chiral Separations by Capillary Electrophoresis*, CRC Press, Boca Raton.

8. Maxwell, E. J. and Chen, D. D. Y. (2008). Twenty years of interface development for capillary electrophoresis-electrospray ionization-mass spectrometry, *Anal. Chem. Acta*, **627**, pp. 25–33.

9. Olivares, J. A., Nguyen, N. T., Yonker, C. R., and Smith, R. D. (1987). On-line mass spectrometry detection for capillary zone electrophoresis, *Anal. Chem.*, **59**, pp. 1230–1232.

10. Smith, R. D., Olivares, J. A., Nguyen, C. R., and Udseth, H. R. (1988). Capillary zone electrophoresis-mass spectrometry using an electrospray ionization interface, *Anal. Chem.*, **60**, pp. 436–441.

11. Smith, R. D., Baringa, C. J., and Udseth, H. R. (1988). Improved electrospray ionization interface for capillary zone electrophoresis-mass spectrometry, *Anal. Chem.*, **60**, pp. 1948–1952.

12. Dole, M., Mack, L. L., Hines, R. L., Mobley, R. C., Ferguson, L. D., and Alice, M. B. (1984). Molecular beams of macroions, *J. Phys. Chem.*, **49**, pp. 2240–2249.

13. Yamashita, M. and Fenn, J. B. (1984). Electrospray ion source. Another variation on the free-jet theme, *J. Phys. Chem.*, **88**, pp. 4451–4459.

14. Yamashita, M. and Fenn, J. B. (1984). Negative ion production with the electrospray ion source, *J. Phys. Chem.*, **88**, pp. 4671–4675.

15. Wilm, M. (2011). Principles of electrospray ionization, *Mol. Cel. Proteomics*, **10**, pp. M111.009407.

16. Gaskell, S. J. (1997). Electrospray: Principles and practice, *J. Mass Spec.*, **32**, pp. 677–688.

17. Pantůčková, P., Gebauer, P., Boček, P., and Křivánková, L. (2011). Recent advances in CE-MS: Synergy wet chemistry and instrumentation innovations, *Electrophoresis*, **32**, pp. 43–51.

18. Zhong, X., Zhang, Z., Jiang, S., and Li, L. (2014). Recent advances in coupling capillary electrophoresis based separation techniques to ESI and MALDI, *Electrophoresis*, **35**, pp. 1214–1225.

19. Klepárník, K. (2015). Recent advances in combination of capillary electrophoresis with mass spectrometry: Methodology and theory, *Electrophoresis*, **36**, pp. 159–178.

20. Petritis, K., Dessans, H., Elfakir, C., and Dreux, M. (2002). Volatility evaluation of mobile-phase/electrolyte additives for mass spectrometry, *LC-GC Europe*, **15**, pp. 98–102.

21. Moini, M. (2007). Simplifying CE–MS operation. 2. Interfacing low flow separation techniques to mass spectrometry using a porous tip, *Anal. Chem.*, **79**, pp. 4241–4246.

22. Huang, J. L., Hsu, R. Y., and Her, G. R. (2012). The development of a sheathless capillary electrophoresis electrospray ionization–mass spectrometry interface based on thin conducting liquid film, *J. Chromatogr. A,* **1267**, pp. 131–137.

23. Wang, C. W. and Her, G. R. (2013). Sheathless capillary electrophoresis electrospray ionization-mass spectrometry interface based on poly(dimethylsiloxane) membrane emitter and thin conducting liquid film, *Electrophoresis,* **34**, pp. 2538–2545.

24. Wang, C. W. and Her, G. R. (2014). The development of a counterflow-assisted preconcentration technique in capillary electrophoresis

electrospray-ionization mass spectrometry, *Electrophoresis,* **35**, pp. 1251–1258.

25. Moini, M. and Martinez, B. (2014). Ultrafast capillary electrophoresis/mass spectrometry with adjustable porous tip for a rapid analysis of protein digest in about a minute, *Rapid. Commun. Mass Spec.,* **28**, pp. 305–310.

26. Moiny, M., Klauenberg, K., and Ballard, M. (2011). Dating silk by capillary electrophoresis mass spectrometry, *Anal. Chem.,* **83**, pp. 7577–7581.

27. Moini, M., Rollman, C. M., and France, C. A. M. (2013). Dating human bone: Is racemization dating species-specific? *Anal. Chem.,* **85**, pp. 11211–11215.

28. Brenner-Weiss, G., Kirschhofer, F., Kuhl, B., Nusser, M., and Obst, U. (2003). Analysis of non-covalent protein complexes by capillary electrophoresis-time of flight mass spectrometry, *J. Chromatogr. A,* **1009**, pp. 147–153

29. Varesio, E., Cherkaou, S., and Veuthey, J. L. (1998). Optimization of CE-ESI-MS parameters for the analysis of ecstasy and derivatives in urine, *J. High Resol. Chromatogr.,* **21**, pp. 653–657.

30. Moini, M. (2002). Capillary electrophoresis mass spectrometry and its application to the analysis of biological mixtures, *Anal. Bioanal. Chem.,* **373**, pp. 466–480.

31. Grunmann, M. and Matysik, M. (2011). Fast capillary electrophoresis-time-of-flight mass spectrometry using capillaries with inner diameters ranging from 75 to 5 μm, *Anal. Bioanal. Chem.,* **400**, pp. 269–278.

32. Huiko, K., Kotiaho, T., and Kostianien, R. (2002). Effects of nebulizing and drying gas flow on capillary electrophoresis/mass spectrometry, *Rapid Commun. Mass Spetrom.,* **16**, pp. 1562–1568.

33. Ross, G. A. (2001). Electrophoresis-mass spectrometry: Practical implementation and applications, *LC-GC Europe,* **1**, pp. 2–6.

34. Foret, F., Thomson, T. J., Vouros, P., Karger, B. L., Gebauer, P., and Boček, P. (1994). Liquid sheath effects on the separation of proteins in capillary electrophoresis/electrospray mass spectrometry, *Anal. Chem.,* **66**, pp. 4450–4458.

35. Lee, E. D., Muck, W., Henion, J. D., and Covey, T. R. (1988). On-line capillary zone electrophoresis-ion spray tandem mass spectrometry for the determination of dynorphins, *J. Chromatogr.,* **458**, pp. 313–321.

36. Lee, E. D., Muck, W., Henion, J. D., and Covey, T. R. (1989). Liquid junction coupling for capillary zone electrophoresis/ion spray mass spectrometry, *Biomed. Environ. Mass Spectrom.*, **18**, pp. 844–850.

37. Whitt, J. T. and Moini, M. (2003). Capillary electrophoresis to mass spectrometry interface using a porous junction, *Anal. Chem.*, **75**, pp. 2188–2191.

38. Pleasance, S., Thibault, P., and Kelly, J. (1992). Comparison of liquid-junction and coaxial interfaces for capillary electrophoresis-mass spectrometry with application to compounds of concern to the aquaculture industry, *J. Chromatogr.*, **591**, pp. 325–339.

39. Cai, J. and Henion, J. (1992). Capillary electrophoresis-mass spectrometry, *J. Chromatogr. A*, **703**, pp. 667–692.

40. Birnbaum, S. and Nilsson, S. (1992). Protein-based capillary affinity gel electrophoresis for the separation of optical isomers, *Anal. Chem.*, **64**, pp. 2872–2874.

41. Valtcheva, L., Mohammad, J., Petterson, G., and Hjertén, S. (1993). Chiral separation of β-blockers by high-performance capillary electrophoresis based on non-immobilized cellulase as enantioselective protein, *J. Chromatogr.*, **638**, pp. 263–267.

42. Chankvetadze, B., Endresz, G., and Blaschke, G. (1994). Neutral and anionic cyclodextrins in capillary zone electrophoresis enantiomeric separation of ephedrine and related compounds, *Electrophoresis*, **15**, pp. 804–807.

43. Nelson, M. W., Tang, Q., Harrata, A. K., and Lee, C. S. (1996), On-line partial filling micelllar electrokinetic chromatography-electrospray ionization mass spectrometry, *J. Chromatogr.*, **749**, pp. 219–226.

44. Wiedmer, S. K., Jussila, M., and Riekkola, M. L. (1998). On-line partial filling micellar electrokinetic capillary chromatography electrospray ionization mass spectrometry of corticosteroids, *Electrophoresis*, **19**, pp. 1711–1718.

45. Yang, L., Harrata, A. K., and Lee, C. S. (1997). On-line micellar electrokinetic chromatography–electrospray ionization mass spectrometry using anodically migrating micelles, *Anal. Chem.*, **69**, pp. 1820–1826.

46. Amini, A., Wiersma, B., Westerlund, D., and Paulsen-Sörman, U. (1999). Determination of the enantiomeric purity of S-ropivacaine by capillary electrophoresis with methyl-β-cyclodextrin as chiral selector using conventional and complete filling techniques, *European J. Pharm. Sci.*, **9**, pp. 17–24.

47. Amini, A., Paulsen-Sorman, U., and Westerlund, D. (1999). Principle and applications of the partial filling technique in capillary electrophoresis, *Chromatographia*, **50**, pp. 497–506.

48. Amini, A., Petersson, C., and Westerlund, D. (1997). Enantioresolution of disopyramide by capillary affinity electrokinetic chromatography with human α1-acid glycoprotein (AGP) as chiral selector applying a partial filling technique, *Electrophoresis*, **18**, pp. 950–957.

49. Amini, A. and Westerlund, D. (1998). Evaluation of association constants between drug enantiomers and human α1-acid glycoprotein by applying a partial-filling technique in affinity capillary electrophoresis, *Anal. Chem.*, **70**, pp. 1425–1430.

50. Amini, A. and Paulson-Sorman, U. (1997). Enantioseparation of local anaesthetic drugs by capillary zone electrophoresis with cyclodextrins as chiral selectors using a partial filling technique, *Electrophoresis*, **18**, pp. 1019–1025.

51. Amini, A., Merclin, N., Bastamin, S., and Westerlund, D. (1999). Determination of association constants between enantiomers of orciprenaline and methyl-β-cyclodextrin as chiral selector by capillary zone electrophoresis using a partial filling technique, *Electrophoresis*, **20**, pp. 180–188.

52. Nelson, W. M. and Cheng, S. L. (1996). Mechanistic studies of partial-filling micellar electrokinetic chromatography, *Anal. Chem.*, **68**, pp. 3265–3269.

53. Hjertén, S. and Kubo, K. (1993). A new type of pH- and detergent-stable coating for elimination of electroendosmosis and adsorption in (capillary) electrophoresis, *Electrophoresis*, **14**, pp. 390–395.

54. Nilsson, L. B. (1990). Determination of remoxipride in plasma and urine by reversed-phase column liquid chromatography, *J. Chromatogr.*, **526**, pp. 139–150.

55. Scriba, G. K. E., (2008). Cyclodextrins in capillary electrophoresis enantioseparations: Recent developments and applications, *J. Sep. Sci.*, **31**, pp. 1991–2011.

56. Juvancz, Z., Kendrovics, R. B., Ivanyi, R., and Szente, L. (2008). The role of cyclodextrins in chiral capillary electrophoresis, *Electrophoresis*, **29**, pp. 1701–1712.

57. Cserhati, T. (2008). New applications of cyclodextrins in electrically driven chromatographic systems: A review, *Biomed. Chromatogr.*, **22**, pp. 563–571.

58. Vespalec, R. and Boček, P. (2000). Chiral separations in capillary electrophoresis, *Chem. Rev.*, **100**, pp. 3715–3753.

59. Vespalec, R. and Boček, P. (1999). Chiral separations in capillary electrophoresis, *Electrophoresis*, **20**, pp. 2579–2591.

60. Mikuš, P., Kaniansky, D., Šebesta, R., and Sališová, M. (1999). Analytical characterizations of purities of alkyl- and arylamino derivatives of beta-cyclodextrin by capillary zone electrophoresis with conductivity detection, *Enantiomer*, **4**, pp. 279–287.

61. Mikuš, P. and Kaniansky, D. (2007). Capillary zone electrophoresis resolutions of 2,4-dinitrophenyl labeled amino acids enantiomers by n-methylatedamino-β-cyclodextrins, *Anal. Lett.*, **40**, pp. 335–347.

62. Blanco, M. and Valverde, I. (2003). Choice of chiral selector for enantioseparation by capillary electrophoresis, *TrAC Trends Anal. Chem.*, **22**, pp. 428–439.

63. Wenz, G., Strassnig, C., Thiele, C., Angelke, A., Morgenstern, B., and Hegetschweiler, K. (2008). Recognition of ionic guests by ionic β-cyclodextrin derivatives, *Chem. Eur. J.*, **14**, pp. 7202–7211.

64. Fenyvesi, E. (1998). Cyclodextrin polymers in the pharmaceutical industry, *J. Incl. Phenom. Macrocycl. Chem.*, **6**, pp. 537–545.

65. Ingelse, B., Everaerts, F. M., Desiderio, C., and Fanali, S. (1995). Enantiomeric separation by capillary electrophoresis using a soluble neutral beta-cyclodextrin polymer, *J. Chromatogr. A*, **709**, pp. 89–98.

66. Ševčík, J., Stránský, Z., Ingelse, B., and Lemr, K. (1996). Capillary electrophoretic enantioseparation of selegiline, methamphetamine and ephedrine using a neutral beta-cyclodextrin epichlorhydrin polymer, *J. Pharm. Biomed. Anal.*, **14**, pp. 1089–1094.

67. Sheppard, R. L., Tong, X., Cai, J., and Henion, J. (1995). Chiral separation and detection of terbutaline and ephedrine by capillary electrophoresis coupled with ion spray mass spectrometry, *Anal. Chem.*, **67**, pp. 2054–2058.

68. Lamoree, M. H., Sprang, A. F. H., Tjaden, U. R., and van der Greef, J. (1996). Use of heptakis(2,6-di-O-methyl)-beta-cyclodextrin in on-line capillary zone electrophoresis-mass spectrometry for the chiral separation of ropivacaine, *J. Chromatogr. A*, **742**, pp. 235–242.

69. Lu, W. and Cole, R. B. (1998). Determination of chiral pharmaceutical compounds, terbutaline, ketamine and propranolol, by on-line capillary electrophoresis–electrospray ionization mass spectrometry, *J. Chromatogr. B*, **714**, pp. 69–75.

70. Lio, R., Chinaka, S, Tanaka, S., Takayama, N., and Haykawa, K. (2003). Simultaneous chiral determination of methamphetamine and its metabolites in urine by capillary electrophoresis-mass spectrometry, *Analyst*, **128**, pp. 646–650.

71. Schulte, G., Heitmeier, S., Chankvetadze, B., and Blaschke, G. (1998). Chiral capillary electrophoresis-electrospray mass spectrometry coupling with charged cyclodextrin derivatives as chiral selectors, *J. Chromatogr. A*, **800**, pp. 77–82.

72. Lio, R., Chinaka, S., Takayama, N., and Hayakawa, K. (2005). Chiral capillary electrophoresis of amphetamine-type stimulants, *J. Health. Sci.*, **51**, pp. 693–701.

73. Iwata, Y. T., Kanamori, T., Ohmae, Y., Tsujikawa, K., Inoue, H., and Kishi, T. (2003). Chiral analysis of amphetamine-type stimulants using reversed-polarity capillary electrophoresis/positive ion electrospray ionization tandem mass spectrometry, *Electrophoresis*, **24**, pp. 1770–1776.

74. Olsson, J., Marlin, N. D., and Blomberg, L. G. (2007). Enantiomeric separation of omeperazole enantiomers by aqueous CE using UV and MS detection, *Chromatographia*, **66**, pp. 421–425.

75. Jäverfalk, E. M., Amini, A., Westerlund, D., and Per Andrén, E. (1998). Chiral separation of local anaesthetics by a capillary electrophoresis/ partial filling technique coupled on-line to micro-electrospray mass spectrometry, *J. Mass. Spetrom.*, **33**, pp. 183–186.

76. Toussaint, B., Palmer, M., Chiap, P., Hubert, P., and Crommen, J., (2001). On-line coupling of partial filling-capillary zone electrophoresis with mass spectrometry for the separation of clenbuterol enantiomers, *Electrophoresis*, **22**, pp. 1363–1372.

77. Cherkaoui, S., Rudaz, S., Varesio, E., and Veuthey, J. L. (2001). On-line capillary electrophoresis-electrospray mass spectrometry for the stereoselective analysis of drugs and metabolites, *Electrophoresis*, **22**, pp. 3308–3315.

78. Tanaka, Y., Kishimoto, Y., and Terabe, S. (1998). Separation of acidic enantiomers by capillary electrophoresis–mass spectrometry employing a partial filling technique, *J. Chromatogr. A*, **802**, pp. 83–88.

79. Grard, S., Morin, Ph., Dreux, M., and Ribet, J. P. (2001). Efficient applications of capillary electrophoresis–tandem mass spectrometry to the analysis of adrenoreceptor antagonist enantiomers using a partial filling technique, *J. Chromatogr. A*, **926**, pp. 3–10.

80. Rudaz, S., Cherkaoui, S., Gauvrit, J., Lantéri, P., and Veuthey J. L. (2001). Experimental designs to investigate capillary electrophoresis-electrospray ionization-mass spectrometry enantioseparation with the partial-filling technique, *Electrophoresis*, **22**, pp. 3316–3326.

81. Cherkaoui, S. and Veuthey, J. L. (2002). Use of negatively charged cyclodextrins for the simultaneous enantioseparation of selected anesthetic drugs by capillary electrophoresis–mass spectrometry, *J. Pharm. Biomed. Anal.*, **27**, pp. 615–626.

82. Rudaz, S., Calleri, E., Geiser, L., Cherkaoui, S., Prat, J., and Veuthey, J. L. (2003). Infinite enantiomeric resolution of basic compounds using highly sulfated cyclodextrin as chiral selector in capillary electrophoresis, *Electrophoresis*, **24**, pp. 2633–2641.

83. Desiderio, C., Rossetti, D. V., Perri, F., Giardina, B., Messana, I., and Castagnola, M. (2008). Enantiomeric separation of baclofen by capillary electrophoresis tandem mass spectrometry with sulfobutylether-β-cyclodextrin as chiral selector in partial filling mode, *J. Chromatogr. B*, **875**, pp. 280–287.

84. Rudaz, S., Geiser, L., Souverain, S., Prat, J., and Veuthey, J. L. (2005). Rapid stereoselective separations of amphetamine derivatives with highly sulfated cyclodextrin, *Electrophoresis*, **26**, pp. 3910–3920.

85. Schappler, J., Guillarme, D., Prat, J., Veuthey, J. L., and Rudaz, S. (2006). Enhanced method performances for conventional and chiral CE-ESI/MS analyses in plasma, *Electrophoresis*, **27**, pp. 1537–1546.

86. Schappler, J., Guillarme, D., Prat, J., and Veuthey, J. L. (2008). Validation of chiral capillary electrophoresis-electrospray ionization-mass spectrometry methods for ecstasy and methadone in plasma, *Electrophoresis*, **29**, pp. 2193–2202.

87. Sanchez-Hernández, L., García-Ruiz, C., Crego, A. L., and Marina, M. L. (2010). Sensitive determination of D-carnitine as enantiomeric impurity of levo-carnitine in pharmaceutical formulations by capillary electrophoresis–tandem mass spectrometry, *J. Pharm. Biomed. Anal.*, **53**, pp. 1217–1223.

88. Moini, M. and Rollman, C. M. (2015). Compatibility of highly sulfated cyclodextrin with electrospray ionization at low nanoliter/minute flow rates and its application to capillary electrophoresis/electrospray ionization mass spectrometric analysis of cathinone derivatives and their optical isomers, *Rapid. Comm. Mass Spec.*, **29**, pp. 304–310.

89. Sánchez-López, E., Montealegre, C., Marina, M. L., and Crego, A. L. (2014). Development of chiral methodologies by capillary electrophoresis with

ultraviolet and mass spectrometry detection for duloxetine analysis in pharmaceutical formulations, *J. Chromatogr. A*, **1363**, pp. 256–362.

90. Merola, G., Fu, H., Tagliaro, F., Macchia, T., and McCord, B. R. (2014). Chiral separation of 12 cathinone analogs by cyclodextrin-assisted capillary electrophoresis with UV and mass spectrometry detection, *Electrophoresis*, **35**, pp. 3231–3241.

91. Piešťanský, J., Maráková, K., Koval, M., Havránek, E., and Mikuš, P. (2015). Enantioselective column coupled electrophoresis employing large bore capillaries hyphenated with tandem mass spectrometry for ultra-trace determination of chiral compounds in complex real samples, *Electrophoresis*, **36**, pp. 3069–3079.

92. Wuethrich, A., Haddad, P. R., and Quirino, J. P. (2014). Online sample concentration in partial-filling chiral electrokinetic chromatography–mass spectrometry, *Chirality*, **26**, pp. 734–738.

93. Wang, J. and Warner, I. M. (1994). Chiral separations using micellar electrokinetic capillary chromatography and a polymerized chiral micelle, *Anal. Chem.*, **66**, pp. 3773–3376.

94. Shamsi, S. A. (2001). Micellar electrokinetic chromatography–mass spectrometry using a polymerized chiral surfactant, *Anal. Chem.*, **73**, pp. 5103–5108.

95. Akbay, C., Rizvi, S. A. A., and Shamsi, S. A. (2010). Simultaneous enantioseparation and tandem UV–MS detection of eight β-blockers in micellar electrokinetic chromatography using a chiral molecular micelle, *Anal. Chem.*, **77**, pp. 1672–1683.

96. Hou, J., Rizvi, S. A. A., Zheng, J., and Shamsi, S. A. (2006). Application of polymeric surfactants in micellar electrokinetic chromatography-electrospray ionization mass spectrometry of benzodiazepines and benzoxazocine chiral drugs, *Electrophoresis*, **27**, pp. 1263–1275.

97. Hou, J., Zheng, J., and Shamsi, S. A. (2007). Separation and determination of warfarin enantiomers in human plasma using a novel polymeric surfactant for micellar electrokinetic chromatography–mass spectrometry, *J. Chromatogr. A*, **1159**, pp. 208–216.

98. Rizvi, S. A. A., Zheng, J., Apkarian, R. P., Dublin, S. N., and Shamsi, S. A. (2007). Polymeric sulfated amino acid surfactants: A class of versatile chiral selectors for micellar electrokinetic chromatography (MEKC) and MEKC-MS, *Anal. Chem.*, **79**, pp. 879–898.

99. Hou, J., Zheng, J., Rizvi, S. A. A., and Shamsi, S. A. (2007). Simultaneous chiral separation and determination of ephedrine alkaloids by

MEKC-ESI-MS using polymeric surfactant I: Method development, *Electrophoresis*, **28**, pp. 1352–1363.

100. Hou, J., Zheng, J., and Shamsi, S. A. (2007). Simultaneous chiral separation of ephedrine alkaloids by MEKC-ESI-MS using polymeric surfactant II: Application in dietary supplements, *Electrophoresis*, **28**, pp. 1426–1434.

101. He, J. and Shamsi, S. A. (2009). Multivariate approach for the enantioselective analysis in micellar electrokinetic chromatography-mass spectrometry: I. Simultaneous optimization of binaphthyl derivatives in negative ion mode, *J. Chromatogr. A*, **1216**, pp. 845–856.

102. Liu Y., Jann, M., Vandenberg, C., Eap, C. B., and Shamsi, S. A. (2015). Development of an enantioselective assay for simultaneous separation of venlafaxine and O-desmethylvenlafaxine by micellar electrokinetic chromatography-tandem mass spectrometry: Application to the analysis of drug-drug interaction, *J. Pharm. Biomed. Anal.*, **1420**, pp. 119–128.

103. Rudaz, S., Stella, C., Balant-Gorgia, A. E., Fanali, S., and Veuthey, J. L. (2000). Simultaneous stereoselective analysis of venlafaxine and O-desmethylvenlafaxine enantiomers in clinical samples by capillary electrophoresis using charged cyclodextrins, *J. Pharm. Biomed. Anal.*, **23**, pp. 107–115.

104. Wang, X., Hou, J., Jann, M., Hon, Y. Y., and Shamsi, S. A. (2013). Development of a chiral micellar electrokinetic chromatography-tandem mass spectrometry assay for simultaneous analysis of warfarin and hydroxywarfarin metabolites: Application to the analysis of patients serum samples, *J. Chromatogr. A*, **1271**, pp. 207–216.

105. He, J. and Shamsi, S. A. (2011). Chiral micellar electrokinetic chromatography-atmospheric pressure photoionization of benzoin derivatives using mixed molecular micelles, *Electrophoresis*, **32**, pp. 1164–1175.

106. Wang, B., He, J., and Shamsi, S. A. (2010). A high-throughput multivariate optimization for the simultaneous enantioseparation and detection of barbiturates in micellar electrokinetic chromatography—mass spectrometry, *J. Chromatogr. Sci.*, **48**, pp. 572–583.

107. Behr, J. P., Lehn, J. M., and Vierling, P. (1982). Molecular receptors. Structural effects and substrate recognition in binding of organic and biogenic ammonium ions by chiral polyfunctional macrocyclic polyethers bearing amino acid and other side-chains, *Helv. Chim. Acta*, **65**, pp. 1853–1867.

References | 221

108. Kuhn, R., Sreinmetz, C., Bereuter, T., Haas, P., and Erni, F. (1994). Enantiomeric separations in capillary zone electrophoresis using a chiral crown ether, *J. Chromatogr. A*, **666**, pp. 367–373.

109. Kuhn, R., Stoeklin, F., and Erni, F. (1992). Chiral separations by host-guest complexation with cyclodextrin and crown ether in capillary zone electrophoresis, *Chromatographia*, **33**, pp. 32–36.

110. Kuhn, R. (1995). Enantiomerentrennung mittels chiralem kronenether in der kapillarzonenelektrophorese, *GIT Fachz. Lab.*, **11**, pp. 1031–1034.

111. Mohammadzadeh Kakhki, R. and Assadi, H. (2015). Capillary electrophoresis analysis based on crown ethers, *J. Incl. Phenom. Macrocycl. Chem.*, **81**, pp. 1–12.

112. Tanaka, Y., Otsuka, K., and Terabe, S. (2000). Separation of enantiomers by capillary electrophoresis–mass spectrometry employing a partial filling technique with a chiral crown ether, *J. Chromatogr. A*, **875**, pp. 323–330.

113. Zhou, L., Lin, Z., Reamer, R. A., Mao, B., and Ge, Z. (2007). Stereoisomeric separation of pharmaceutical compounds using CE with a chiral crown ether, *Electrophoresis*, **28**, pp. 2658–2666.

114. Xia, S., Zhang, L., Lu, M., Qiu, B., Chi, Y., and Chen, G. (2009). Enantiomeric separation of chiral dipeptides by CE-ESI-MS employing a partial filling technique with chiral crown ether, *Electrophoresis*, **30**, pp. 2837–2844.

115. Kaplan, J. (2001). The role of sugar residues in molecular recognition by vancomycin, *J. Med. Chem.*, **44**, pp. 1837–1840.

116. Gasper, M. (1996). Comparison and modeling study of vancomycin, ristocetin A, and teicoplanin for CE enantioseparations, *Anal. Chem.*, **68**, pp. 2501–2514.

117. Armstrong, D. W., Rundlett, K. L., and Chen, J. R. (1994). Evaluation of the macrocyclic antibiotic vancomycin as a chiral selector for capillary electrophoresis, *Chirality*, **6**, pp. 496–509.

118. Prokhorova, A. F., Shapalovova, E. N., and Shpigun, O. A. (2010). Chiral analysis of pharmaceuticals by capillary electrophoresis using antibiotics as chiral selectors, *J. Pharm. Biomed. Anal.*, **53**, pp. 1170–1179.

119. Fanali, S., Desiderio, C., Schulte, G., Heitmeier, S., Strickmann, D., Chankvetadze, B., and Blaschke, G. (1998). Chiral capillary electrophoresis–electrospray mass spectrometry coupling using vancomycin as chiral selector, *J. Chromatogr. A*, **800**, pp. 69–76.

120. Ali, I., Sanagi, M. M., and Aboul-Enein, H. Y. (2014). Advances in chiral separations by nonaqueous capillary electrophoresis in pharmaceutical and biomedical analysis, *Electrophoresis*, **35**, pp. 926–936.

121. Sarmini, K. and Kenndler, E. (1997). Influence of organic solvents on the separation selectivity in capillary electrophoresis, *J. Chromatogr. A.*, **792**, pp. 3–11.

122. Kenndler, E. (2014). A critical overview of non-aqueous capillary electrophoresis. Part I: Mobility and separation selectivity, *J. Chromatogr. A*, **1335**, pp. 16–30.

123. Kenndler, E. (2014). A critical overview of non-aqueous capillary electrophoresis. Part II: Separation efficiency and analysis time, *J. Chromatogr. A*, **1335**, pp. 31–40.

124. Servais, A. C., Fillet, M., Mol, R., Somsen, G. W., Chiap, P., de Jong, G. J., and Crommen, J. (2006). On-line coupling of cyclodextrin mediated nonaqueous capillary electrophoresis to mass spectrometry for the determination of salbutamol enantiomers in urine, *J. Pharm. Biomed. Anal.*, **40**, pp. 752–757.

125. Mol, R., Servais, A. C., Fillet, M., Crommen, J., de Jong, G. J., and Somsen, G. W. (2007). Nonaqueous electrokinetic chromatography–electrospray ionization mass spectrometry using anionic cyclodextrins, *J. Chromatogr. A*, **1159**, pp. 51–57.

126. Mol, R., de Jong, G. J., and Somsen, G. W. (2008). Coupling of non-aqueous electrokinetic chromatography using cationic cyclodextrins with electrospray ionization mass spectrometry, *Rapid. Comm. Mass Spec.*, **22**, pp. 790–796.

127. Lode, H., Hedeland, Y., Hedeland, M., Bondenson, U., and Petterson, C. (2003). Development of a chiral non-aqueous capillary electrophoretic system using the partially filling technique with UV and mass spectrometric detection, *J. Chromatogr. A*, **986**, pp. 143–152.

128. Zheng, J. and Shamsi, S. A. (2003). Combination of chiral capillary electrochromatography with electrospray ionization mass spectrometry: Method development and assay of warfarin enantiomers in human plasma, *Anal. Chem.*, **75**, pp. 6295–6305.

129. Bragg, W. and Shamsi, S. A. (2011). Development of a fritless packed column for capillary electrochromatography-mass spectrometry, *J. Chromatogr. A*, **1218**, pp. 8691–8700.

130. Gu, C. and Shamsi, S. A. (2011). Evaluation of a methacrylate bonded cyclodextrins as a monolithic chiral stationary phase for

capillary electrochromatography (CEC)-UV and CEC coupled to mass spectrometry, *Electrophoresis*, **32**, pp. 2727–2737.

131. Schurig, V. and Mayer, S. (2001). Separation of enantiomers by open capillary electrochromatography on polysiloxane-bonded permethyl-β-cyclodextrin, *J. Biochem. Biophys. Methods,* **28**, pp. 117–141.

132. Zheng, J. and Shamsi, S. A. (2006). Simultaneous enantioseparation and sensitive detection of eight β-blockers using capillary electrochromatography-electrospray ionization-mass spectrometry, *Electrophoresis*, **27**, pp. 2139–2151.

133. Van Brocke, A., Wistuba, D., Gfrorer, P., Stahl, M., Schurig, V., and Bayer, E. (2002). On-line coupling of packed capillary electrochromatography with coordination ion spray-mass spectrometry for the separation of enantiomers, *Electrophoresis*, **23**, pp. 963–2972.

134. Zheng, J. and Shamsi, S. A. (2003). Combination of chiral capillary electrochromatography with electrospray ionization mass spectrometry: Method development and assay of warfarin enantiomers in human plasma, *Anal. Chem.*, **75**, pp. 6259–6305.

135. Nilsson, S. L., Bylund, D., Joertnten-Karlsson, M., Petersson, P., and Markides, K. E. (2004). A chemometric study of active parameters and their interaction effects in a nebulized sheath-liquid electrospray interface for capillary electrophoresis-mass spectrometry, *Electrophoresis,* **25**, pp. 2100–2107.

136. Geiser, L., Rudaz, S., and Veuthey, J. L. (2003). Validation of capillary electrophoresis–mass spectrometry methods for the analysis of a pharmaceutical formulation, *Electrophoresis*, **24**, pp. 3049–3056.

137. Maier, V., Petr, J., Knob, R., Horáková, J., and Ševčík, J. (2007). Electrokinetic partial filling technique as a powerful tool for enantiomeric separation of D,L-lactic acid by CE with contactless conductivity detection, *Electrophoresis*, **28**, pp. 1815–1822.

Chapter 6

Enantioselective Drug–Plasma Protein-Binding Studies by Capillary Electrophoresis

Laura Escuder-Gilabert,[a] Yolanda Martín-Biosca,[a]
Salvador Sagrado,[a,b] and María José Medina-Hernández[a]

[a]Department of Analytical Chemistry, University of Valencia, C/ Vicente A. Estellés s/n
E-46100, Burjassot, Valencia, Spain
[b]Instituto Interuniversitario de Investigación de Reconocimiento Molecular y
Desarrollo Tecnológico (IDM), Universitat Politècnica de València, Universitat de
València, C/ Vicente A. Estellés s/n E-46100, Burjassot, Valencia, Spain
lescuder@uv.es

6.1 Introduction

Drug action in living organisms is the result of a large number of pharmacological processes. In this sense, the interactions between drugs and biomembranes, plasma proteins, enzymes, or receptors are decisive features of the final biological activity of drugs. Most of the pharmacological processes responsible of drug action present a high degree of stereoselectivity, resulting in a difference between the activities of the enantiomers of drugs. In fact, very often one of them is the most active, while the other may produce side effects and

---

Capillary Electrophoresis: Trends and Developments in Pharmaceutical Research
Edited by Suvardhan Kanchi, Salvador Sagrado, Myalowenkosi Sabela, and Krishna Bisetty
Copyright © 2017 Pan Stanford Publishing Pte. Ltd.
ISBN 978-981-4774-12-3 (Hardcover), 978-1-315-22538-8 (eBook)
www.panstanford.com

even toxicity in some cases [1]. Pharmacological differentiation of enantiomers may occur at pharmacodynamic and pharmacokinetic levels. The term pharmacokinetics involves all the processes that take place since a drug enters the human body and until the moment it is eliminated by different pathways. Pharmacokinetic processes are drug absorption, distribution, metabolism, and excretion, and all of them can be affected by enantioselectivity. The main prerequisite of a biological process for being enantioselective is stereospecific recognition of the enantiomers of drugs by a chiral biomolecule responsible for the process [2]. Examples of enantioselective processes include active transport, protein binding, or enzymatic reactions such as metabolism.

The distribution of a drug involves its penetration, movement, and storage in different tissue compartments. The distribution properties of a drug are of vital importance, since they determine if the drug can reach the active site and be effective. Distribution processes of chiral drugs can be enantioselective. Protein binding is an important distribution process [3]. Drugs usually do not travel through the circulatory system themselves, but they bind to different plasma proteins. This can affect their disposition for other processes such as first-pass metabolism, metabolic and renal clearance, and distribution to other compartments. Plasma proteins facilitate drug transport throughout the body but also can limit drug availability for reaching target organs, since only the free fraction of the drug can leave plasma and penetrate membranes. When proteins and chiral drugs interact in an enantioselective way, two diastereomeric adducts are formed with potential differences in the protein binding that may result in different pharmacokinetic profiles for the individual enantiomers [4]. For drugs that are highly protein bound, small changes in protein binding result in large changes in the free active fraction and thus the therapeutic benefit of such compounds can be markedly affected by stereoselective protein binding. So the investigation of enantioselectivity of drugs in their binding with human plasma proteins and the identification of the molecular mechanisms involved in the stereodiscrimination by the proteins represents a great challenge for clinical pharmacology.

Methodologies able to perform such studies play an important role in the development of new drugs and new formulations as well as in investigations on bioavailability and metabolism. Furthermore, it has often become the basis of therapeutic drug monitoring and drug management in patients [3]. This chapter is focused on the applications of capillary electrophoresis (CE) in the study of the enantioselective binding of chiral drugs with human plasma proteins.

6.2 Plasma Proteins

Human plasma contains more than 60 proteins, with albumin (HSA), α_1-acid glycoprotein (AGP), lipoproteins, and globulins being the most important from the point of view of binding to drugs [3]. HSA is the major plasma protein in the circulatory system since it represents 60% of the protein content of the plasma, with a concentration of 35–45 g/L (550–600 µM) [5]. It presents a high degree of enantioselective interaction with chiral drugs [6]. HSA is a single peptide chain of 585 amino acid residues. Its 3D structure, shown in Fig. 6.1, consists of a monomeric globular protein made out of three domains (I, II, and III), with a similar helical pattern [7, 8].

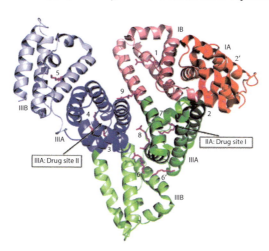

Figure 6.1 Structure of HSA with its two main binding sites for drugs.

HSA molecule has at least six selective binding sites and a great number of low-affinity binding sites. Each drug preferably binds to a binding site, although it can also bind to another site with minor affinity. Several studies suggest that the binding sites are formed due to conformational changes that take place during the binding process of drug to HSA. Up to now, the regions of binding best defined in HSA molecule are the sites warfarin-azapropazone (site I) and indol-benzodiazepine (site II) [9, 10]. Site I is in the sub-domain IIA and is formed by six helices that make up the sub-domain and a helix bound to the sub-domain IA (Fig. 6.1); the interior of the cavity is predominantly apolar though it contains two groups of polar residues, one at the end and the other one at the entry of the cavity. Site I seems to be specific for anionic or electronegative drugs such as warfarin, fenilbutazone, and valproic acid. The interactions are hydrophobic, and hydrogen bonds with the hydroxyl group of the polar group Y150 [10]. Site II is in the sub-domain IIIA; it is apolar and presents major selectivity for benzodiazepines and their derivatives, as well as for different carboxylic acids such as non-steroidal anti-inflammatory drugs.

In spite of sites I and II being responsible for a great number of interactions between drugs and HSA, some crystalographic [7, 8] and chromatographic studies [11, 12] confirm the existence of other regions of binding in the HSA molecule of minor importance such as the binding site of digitoxine (site III) and the binding sites of bilirubin and tamoxifen. Nowadays, the locations of the binding sites of tamoxifen and bilirubin and site III are not exactly known, but there are evidences that confirm that site III is far from sites I and II, that there are allosteric effects between site I and the binding site of tamoxifen, and that the sites of bilirubin and tamoxifen are overlapped. The second plasma protein in abundance is AGP, also called orosomucoid, ORM. AGP is a α_1-globulin protein characterized by its high content in carbohydrates. It is the plasmatic protein of minor molecular mass (41,000 g/mol), and its content in sialic acid awards an isoelectric point of 2.7 [9, 13]. Its plasmatic levels are between 0.4–1.0 g/L, with a plasmatic concentration of 20 μM.

The AGP molecule consists of a single polypeptide chain of 183 amino acids and five *N*-glycan chains, which have di-, tri-, and tetra-antennary structures, with sialic acids as the terminal group. AGP has three polymorphic variants (F1, S, and A) with different primary structure. Its 3D structure and physiological function are still not solved. AGP presents a selective binding site and other sites with low affinity [14]. AGP has affinity for basic drugs such as tricyclic antidepressants, local anesthetics, phenothiazines and β-blockers, and in minor extent for some neutral and acid drugs. In general, the drug–AGP interactions are of hydrophobic nature. The contribution of AGP to the total plasma protein binding of drugs is much lower than HSA contribution, due to its lower concentration in plasma [15]. The binding of drugs to AGP can be also enantioselective.

Lipoproteins are spherical pseudo-micellar particles soluble in water with a hydrophobic nucleus composed principally of triglycerides, esters of cholesterol, sphingolipids, and A, D, E, and K vitamins surrounded with a hydrophilic layer of phospholipids, non-esterified cholesterol, and apoproteins. Plasma lipoproteins are classified into several subclasses according to their density, such as high-density lipoprotein (HDL), low-density lipoprotein (LDL), very low-density lipoprotein (VLDL), and chylomicron (CM). Among these, HDL and LDL are the most important drug-transporting proteins because of their higher plasma concentrations than others. Oxidized LDL resulting from LDL in vivo conversion has also reported to be a high-affinity drug-binding protein [3, 16]. These proteins principally bind to liposoluble drugs, with a high volume of distribution and generally of basic nature such as imipramine and cyclosporine [5]. Since apolipoproteins and lipid constituents such as free cholesterol, cholesterol ester, and some phospholipids are chiral compounds, the binding of a racemic drug to lipoproteins may be enantioselective.

The term globulin refers to all plasmatic proteins except HSA and prealbumin. There are principally five types of globulins: α_1-, α_2-, β_1-, β_2-, and γ-globulin. α- and β-globulins have great affinity for different endogenous and exogenous substances of similar structure such as steroids (prednisone and transcortin), whereas γ-globulins selectively interact with antigens and their interaction with the majority of drugs is inappreciable [5].

6.3 Capillary Electrophoresis for Enantioselective Protein-Binding Experiments

The in vitro measurement of drug binding to plasma proteins is used to calculate pharmacokinetic parameters and make predictions on the pharmacokinetic behavior of the drug. Generally, drug–plasma protein binding is a reversible and kinetically rapid interaction; therefore, it should be analyzed without disturbing binding equilibrium.

Different approaches have been proposed for the evaluation of protein binding of drugs that can be classified into two main groups [17–21]. The first group involves the separation of the free ligand (drug) from the bound species and the determination of the drug in one fraction or both of them by means of separation techniques such as equilibrium dialysis (ED), ultrafiltration (UF), liquid chromatography (HPLC), and CE among others. The second group is based on the detection of a change in some physicochemical properties of the drug or the protein due to the binding. Examples of these methodologies include nuclear magnetic resonance spectroscopy, surface plasmon resonance, and calorimetry among others. Among the separation techniques, CE offers some advantages in enantioselective drug–protein binding studies such as short analysis times, high separation efficiencies, versatility of conditions and performances that can be employed, easy conditioning of the capillary, low sample requirements, and compatibility with different detection systems. CE has been used in this kind of studies in two ways: combined with other separation techniques for the chiral analysis of the bound or unbound fraction and for direct evaluation of enantioselective protein binding of drugs (Fig. 6.2).

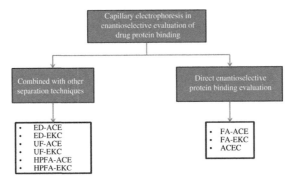

Figure 6.2 Role of capilary electrophoresis in enantioselective protein binding studies. ED: equilibrium dialysis, UF: ultrafiltration, HPFA: high performance frontal analysis, FA: frontal analysis, ACE: affinity capillary electrophoresis, EKC: electrokinetic chromatography, ACEC: affinity capillary electrochromatography.

6.3.1 Combination of CE with other Separation Techniques

In the bibliography, the majority of the methodologies for the investigation of drug enantioselective binding are conducted by combining two separation techniques. The methodology involved requires three steps: (i) equilibration of drug and protein (or plasma) mixtures, (ii) separation of the free drug (unbound fraction) and drug–protein complex (bound fraction), and (iii) determination of the enantiomer concentration in one of the fractions (unbound or bound) by means of an achiral or chiral separation technique depending on the availability of individual enantiomers or racemic compounds, respectively.

The first and second steps have been mainly performed by means of classical membrane-based separation methodologies, such as the traditional equilibrium dialysis, ultracentrifugation, and ultrafiltration, but also by means of chromatographic methods.

Among the different methodologies used for drug–protein binding studies, ED and UF are undoubtedly the most widely used because of their simplicity and general applicability to many different systems in vitro and ex vivo. Moreover, commercial high-throughput devices in the standard 96-well plate format have been developed, such as the Equilibrium Dialyzer-96^{TM} from Harvard Bioscience [22] and the MultiScreen^{TM} filter assembly with Ultracel^{TM}-PPB membrane from the Millipore Corporation specifically optimized for in vitro plasma protein-binding assays [23].

ED is based on the establishment of an equilibrium state between a protein compartment and a buffer compartment (usually containing the drug or ligand), which are separated by a membrane only permeable for a low-molecular-weight ligand. Also the use of immobilized protein onto membranes instead of protein solutions has been proposed [24]. This method has many problems, including the time needed to reach equilibrium (3–24 h), volume shifts, Donnan effects, hindering of the passage of free ligand, non-specific adsorption to dialysis device and dialysis membrane particularly for highly lipophilic drugs, among others [20]. Many researchers have used ultrafiltration centrifugal devices for protein-binding measurements since they offer significant advantages represented by short analysis time (usually up to 30 min), simplicity, lack of dilution effects and volume shifts. UF is a simple and rapid method in which centrifugation forces the buffer containing free drugs through the size-exclusion membrane and achieves a fast separation of free from protein-bound drug. The major controversy involves the stability of the binding equilibrium during the separation process, especially in the case of low-affinity interactions [20] and non-specific binding of drugs on filter membranes and plastic devices, which can be avoided by means of a pre-treatment of the filter membranes [25].

Although in a lesser extent than membrane-based separation techniques, high-performance size-exclusion chromatographic techniques in the modality of frontal analysis (HPFA) have also been used for the separation of unbound/bound fractions. HPFA was first developed by using a restricted access type HPLC column, which excludes a large molecule of plasma protein but retains a drug of small molecular size [26]. In HPFA, an excess volume of sample is injected directly using mobile phase compositions that guarantee not to modify the binding equilibrium. For drugs with high affinity toward the protein, the online preconcentration of unbound fraction has been proposed. Once the unbound/bound fractions have been separated, the last step consists in the enantioselective analysis of one of the fractions, usually the unbound fraction. If pure enantiomers have been used for the study, traditional achiral chromatographic or electrophoretic techniques can be used. For racemic mixtures, the determination of enantiomers using different techniques such as CE and different chiral selectors is required.

Enantioseparations with CE can be performed by adding a chiral selector to the background electrolyte (BGE) in the so-called electrokinetic chromatography (EKC) modality. Different chiral selectors can be used, and cyclodextrines [27] and proteins are the most used in these studies. When proteins are added to BGE, the modality is called affinity capillary electrophoresis (ACE). An alternative to the addition of chiral selector to the BGE is the use of partial (PFT) and complete (CFT) filling techniques in which the capillary is (partially or totally, respectively) filled with chiral selector solution prior to injection of the analyte, while the BGE remains without chiral agent. PFT allows using chiral selectors to which given detectors have a significant response, as well as to avoid contamination of the ion-source when CE is coupled to mass spectrometry. The most important property of PFT and CFT, compared with the conventional technique, is its reduced cost due to the drastic reduction of the chiral selector consumption [28]. A special form of PFT and CFT, when the chiral selector and analyte possess opposite charges and move in the capillary in the opposite direction, has been called counter-current separation.

6.3.2 CE for Direct Enantioselective Protein-Binding Evaluation

CE offers some advantages in direct drug–protein binding studies, such as low sample requirements, short analysis times, high separation efficiencies and sample throughput. Several CE approaches have been developed for the direct quantitative assessment of drug–protein interactions, which include ACE methods or the immobilization of proteins into the capillary in the modality of affinity capillary electrochromatography (ACEC).

ACE offers several possibilities for studying ligand–protein binding interaction depending on the disposition of protein and drug in the electrophoretic system, such as frontal analysis (FA), Hummel–Dreyer method, vacancy peak and vacancy affinity CE [29]. All these approaches are complementary rather than competitive since the information that can be obtained from each kind of experiments is different [29]. Among these methodologies, for enantioselective drug–protein binding studies, frontal analysis is concluded to be superior to the other because of its greater simplicity, speed, and versatility to study multiple equilibria [29]. FA is based on the injection of a relatively large sample plug (100–200 nL), which consists of a pre-equilibrated mixture of drug and protein and, therefore, some complex may also be present [30]. In FA, both hydrodynamic and electrokinetic injection are possible. In FA with hydrodynamic injection, a plug of drug–protein mixed solution is introduced into the capillary containing the running buffer (usually at pH 7.4), and a positive voltage is applied. In these conditions, basic drugs are positively charged, while plasma proteins such as HSA and AGP have a net negative charge. Therefore, the unbound drug can be separated from the protein and the drug–protein complex.

By the addition of a chiral selector in the running buffer (FA/EKC) or by the action of the protein itself as chiral selector (FA/ACE), the enantiomers present in the unbound fraction are separated and their concentrations are calculated. Figure 6.3 shows the chiral separation of enantiomers of imazalil in the unbound fraction after incubation with HSA in the FA/EKC modality using highly sulfated β-cyclodextrin (HS-β-CD) as chiral selector [31].

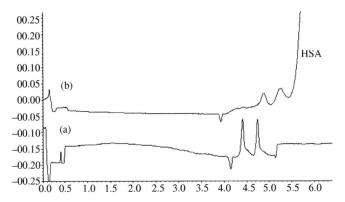

Figure 6.3 Electropherograms corresponding to the chiral separation of (a) a 200 μM standard solution of imazalil and (b) a mixture of imazalil (200 μM)–human serum albumin (HSA; 675 μM). Reprinted with permission from Ref. [30], Copyright 2004, John Wiley and Sons.

In FA with electrokinetic injection, only the unbound drug present in the pre-equilibrated mixture is introduced selectively into the capillary by applying positive voltage (for positively charged drugs) at the sample injection side. Using this strategy, the negatively charged protein and bound drug are not introduced. The unbound drug zone migrates through the capillary toward the cathodic end and can be separated into two zones of the enantiomers by the addition into the running buffer of an appropriate chiral selector in optimized experimental conditions [20].

The sample injection volume in FA methodologies (100–200 nL) is smaller by more than two orders of magnitude than the sample volume required by the ultrafiltration method. The current major disadvantage of FA is the relatively high detection limit (μM range). Limitations of this method also include the need to minimize protein–wall adsorption.

In ACEC, silica particles with immobilized AGP and HSA have been packed into capillaries. Assuming that the immobilization procedure does not influence the binding properties of protein, the studies allow the direct evaluation of enantioselective binding of a chiral drug with a particular protein. Other important applications in the context of enantioselective drug–protein binding of ACEC are related with the study of drug–drug and enantiomer–enantiomer interactions [32].

6.4 Experimental Design and Mathematical Models in Protein-Binding Studies

While the separation/analytical aspects related with enantioselective protein-binding estimations have been extensively investigated, the impact of the experimental design (in the incubation step) and the mathematical approach (involving several assumptions and model simplification) on the results has been underestimated in many studies, entailing a risk of lack of quality results, or an underestimation of the process uncertainty [33, 34]. The adequate experimental design and mathematical treatment of the experimental data are important to obtain accurate estimations. In recent papers concerning the enantioselective binding of the antidepressant fluoxetine [33] and catechin [34] to HSA, using ultrafiltration followed by chiral separation of the unbound fraction by EKC-CFT with HS-β-CD as chiral selector, some weak aspects of the experimental methodologies and mathematical models previously published for the estimation of binding parameters have been pointed out.

6.4.1 Mathematical Models and Deficiencies

The reversible binding interaction between a drug and a protein has been usually described by the following model:

$$r = \frac{b}{P} = \frac{D-d}{P} = \sum_{i=1}^{m} n_i \frac{K_i d}{1 + K_i d}$$

(6.1)

where r is the fraction of bound drug per molecule of protein, b is the bound concentration of drug after equilibrium, P is the total protein concentration, D is the total drug concentration (enantiomer in the case of a enantioselective study), d is the free concentration of drug after equilibrium, m is the number of classes of independent active sites in the protein, n is the number of binding sites of one class per protein molecule (apparent stoichiometry), and K is the binding constant of the drug–protein interaction. For example, for the molecule of albumin, m represents the different sites described in the molecule (sites I, II, III, ...). For instance, if two main binding sites are considered to be important ($m=2$; warfarin site or site I, and diazepam site or site II), the equation is:

$$r = n_1 \frac{K_1 d}{1 + K_1 d} + n_2 \frac{K_2 d}{1 + K_2 d} \qquad (6.2)$$

Equation (6.1) or (6.2) is used in achiral studies but also in enantioselective studies considering the corresponding enantiomer data. These equations are too complex to provide reliable estimations for the parameters, since several combinations of the parameters provide similar d values. However, HSA normally exhibits a single high-affinity site when binding to small molecules; thus, $m = 1$ is often adopted (without further attention [34]). This assumption reduces the problem to a two-parameter estimation, n_1 and K_1:

$$r = n_1 \frac{K_1 d}{1 + K_1 d} \qquad (6.3)$$

This equation still suffers the same problem, since more than one single set of suitable n_1 and K_1 values can be found [35].

Table 6.1 Classical linear models for protein-binding estimations

Model	Equation	
Klotz	$\dfrac{1}{r} = \dfrac{1}{n_1} + \dfrac{1}{n_1 K_1} \dfrac{1}{d}$	(6.4)
Scatchard	$\dfrac{r}{d} = n_1 K_1 - K_1 r$	(6.5)
y-reciprocal	$\dfrac{d}{r} = \dfrac{1}{n_1 K_1} + \dfrac{1}{n_1} d$	(6.6)

From this model, several linear equations have been derived (Table 6.1) in the past and have been mostly used. However, linear plots can provide mathematically inconsistent results, since r and *d* appear both as independent and dependent variables. These mathematical models lack robustness, so little differences in the experimental data will provide great changes in the estimations.

More recently, alternative equations derived from Eq. (6.3) have been used [34]:

$$K_1 = \frac{1}{d}\left(\frac{1}{n_1 - r}\right) \tag{6.7}$$

$$n_1 = r\frac{(1 + K_1 d)}{K_1 d} \tag{6.8}$$

These equations allow a different strategy to estimate the parameters. In order to avoid the simultaneous estimation of n_1 and K_1 from a single equation, most authors assume a 1:1 stoichiometry ($n_1 = 1$) without further verification. A particular case of Eq. (6.7), assuming $n_1 = 1$, is [33, 34]:

$$K_1 = \frac{1}{d}\frac{r}{1 - r} \tag{6.9}$$

This equation provides a direct approach for estimations of K_1 (note that each *D–P–d* experimental data provide an individual K_1 estimate; univariate strategy), avoiding the classical regression/fitting strategies (bivariate approaches).

6.4.2 Experimental Design and Verification of the Assumptions

The first consideration that should be taken into account is the in vivo situation. Estimations performed far from physiological conditions could provide results far from the real situation. For instance, the steady-state concentration of fluoxetine in human plasma (around 0.2 µM of racemic fluoxetine in a normal dosage situation) and the albumin plasma concentration (500–600 µM) show that in a physiological normal situation, the ratio *D/P* should be lower than 1 (*D* << *P* in most cases) [33]. Thus, the most probable situation is that the drug only binds to a kind of binding site in the protein (so *m* = 1). Furthermore, as there is a considerable excess of protein over the amount of drug in plasma, the stoichiometry of the binding will probably be 1:1 (only one molecule of drug per molecule of protein), thus $n_1 = 1$. If the in vitro model tries to assemble the in vivo conditions and keeps the D/P ratio low enough, these assumptions can also be done in vitro and simplify the calculations considerably without losing reliability.

One common experimental design in the literature uses a fixed P level and different D levels (i.e., P-constant design). In most papers, D covers several orders of magnitude and does not keep the D/P ratio < 1, so the aforementioned assumptions become invalid. In contrast, experimental designs defining short concentration ranges of D (fixing P and keeping the D/P ratio ≤ 0.5) have also been planned [33, 34]. It has been demonstrated that the impact of assuming $m = 1$ on log K_1 estimates (from Eq. 6.7) as a function of D/P (when an m = 2 situations exists) is low below this critical 0.5 ratio [34]. Also in this work, it has been demonstrated that a P-constant design is preferable to the opposite D-constant design; concretely, an ideal P-constant design with $P = 530$ µM and (at least) five spaced D levels from 50 to 250 µM has been recommended.

Using this suggested D-range, estimations of n_1 and K_1 ($m = 1$) should be satisfactory. However, to validate the direct estimation of K_1 by means of the simple direct univariate approach (Eq. 6.9), the assumption $n_1 = 1$ should be confirmed. Two approaches are possible. The first one is a linear equation derived from this equation:

$$\log \frac{r}{1-r} = \log K_1 + \log(d) \tag{6.10}$$

Plotting $\log(r/(1-r))$ versus $\log(d)$, a slope = 1 verifies the assumption $n_1 = 1$ [33]. The other possibility proposed for the confirmation of the $n_1 = 1$ assumption is the use of a nonlinear equation derived from Eq. (6.3) or (6.7), in which d is a dependent variable of D and P [30]:

$$d = \frac{-(1 - K_1 D + n_1 K_1 P) + \sqrt{(1 - K_1 D + n_1 K_1 P)^2 + 4K_1 D}}{2K_1} \tag{6.11}$$

n_1 and K_1 in Eq. 6.11 can be estimated using nonlinear approaches (e.g., SIMPLEX) to confirm if $n_1 = 1$ is a valid solution. However, several sets of estimated values can be found, an intrinsic problem of these classical equations/approaches.

More recently [33, 34], the "stoichiometry verification plot" has been proposed to confirm the $n_1 = 1$ assumption. The median value of K_1 estimates from Eq. (6.9) is used to obtain n_1 estimates from Eq. (6.8) and plotted against d values. This plot should provide a straight line with slope = 0 if the assumption of n_1 is valid. In practice, considering data imprecision and inaccuracies, the absence of upward/downward trends in the plot can be considered an indication of slope = 0.

On the other hand, prior to K_1 estimations, outliers identification and elimination could be necessary, since outlier values can considerably affect the K_1 estimates. The use of the direct approach facilitates to evaluate outliers directly on the K_1 estimates (e.g., typical outliers test from univariate statistics could be used). Anyway, in general, the use of robust statistics (e.g., the median) is recommendable [33, 34].

In summary, a protocol for the direct strategy, applied for the case of an enantioselective study (i.e., to estimate the constants for both, the first and second eluted enantiomers, K_{E1} and K_{E2}, respectively), could be:

- Elimination of K_{E1} and K_{E2} outliers (Eq. 6.9), and/or using the median as provisional K_{E1} and K_{E2} estimates.
- Verification of $n_1 = 1$ assumption (Eq. 6.10, Eq. 6.11, or the stoichiometry verification plot) and, in this case, acceptation of K_{E1} and K_{E2} estimates and $n_1 = 1$ for both enantiomers.

Note that this protocol assumes the estimation of K_{E1} and K_{E2} from an independent model. However, if a competitive model is suspected, the correct equations, instead of Eq. (6.11), should be [33]:

$$d_{E1} = \frac{-b + \sqrt{b^2 + 4ac}}{2a} \tag{6.12}$$

where $a = K_{E1}$, $\quad b = 1 - K_{E1} D + n_1 K_{E1} P + K_{E2} d_{E2}$, $\quad$ and $c = D$.

$$d_{E2} = \frac{-b + \sqrt{b^2 + 4ac}}{2a} \tag{6.13}$$

where $a = K_{E2}$, $\quad b = 1 - K_{E2} D + n_1 K_{E2} P + K_{E1} d_{E1}$, $\quad$ and $c = D$.

6.4.3 Examples Illustrating the Direct Approach Strategy

Figure 6.4 shows some results using the direct approach in the case of fluoxetine [33]. Figure 6.4(left) shows the experimental d values obtained using a P-constant design (P = 0.0004766 M and D values from 0.0000789 to 0.0001842 M corresponding to the E1 and E2 enantiomers). A D/P ratio under 0.35 (below the critic ratio 0.5) is used. Assuming m = 1 and n_1 = 1 and applying Eq. (6.9), 10 independent K_1 (or logK_1) values were obtained per enantiomer and tested for outliers. From their median values (logK_{E1} = 4.47 and logK_{E2} = 4.15), 10 n_1 values were estimated (Eq. 6.8) and plotted against the corresponding d values (stoichiometry verification plot; Fig. 6.4, right part). As can be observed in Fig. 6.4(right), no trends were found, indicating the validity of n_1 = 1.

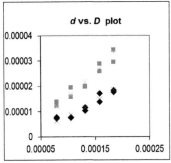 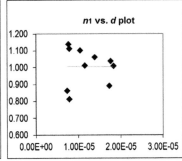

Figure 6.4 Protein binding studies for fluoxetine enantiomers. Left: Experimental d values obtained using a P-constant design (P = 0.0004766 M and D values from 0.0000789 to 0.0001842 M, values corresponding to enantiomers; E1: diamonds, E2: squares). Crosses represent the theoretical model (Eq. 6.9) using the estimated K_1 values for each enantiomer. Right: Stoichiometry verification plot, where n_1 are estimated using Eq. 6.8 from the median log K_1 values (log K_{E1} = 4.47 and log K_{E2} = 4.15). Reprinted with permission from Ref. [33], Copyright 2010, John Wiley and Sons.

The enantioselectivity degree (ES) is an important parameter of the enantiomer-binding studies. Enantioselectivity in drug–protein binding (nK or K ratio) is defined as the ratio between the affinity constant of enantiomer that presents the greatest affinity toward protein and the affinity constant of the other enantiomer [20]. For fluoxetine enantiomers, ES has been found to be equal to 2.20 (ratio K_{E1}/K_{E2}, assuming an independent model), so the binding of fluoxetine enantiomers to HSA can be considered enantioselective [33].

Another interesting parameter from a pharmacokinetic point of view is the percentage of protein binding (*PB*), which may be estimated at physiological conditions. It can be calculated as:

$$PB\% = \frac{D-d}{D} \times 100 \tag{6.14}$$

For fluoxetine, d was estimated via Eq. (6.11) using the D and P physiological levels found in the literature. The *PB* values obtained were 95.2% and 90.0% for E1 and E2, respectively [33]. These values are consistent with that reported for racemic fluoxetine (94.75%).

Figure 6.5 shows the final results using the direct approach in the case of catechin corresponding to a *P*-constant design (*D* values from 0.0001006 to 0.0002515 M corresponding to the enantiomers E1 and E2) [34]. Considering the *P* concentration (0.0004766 M), a *D/P* ratio under 0.47 is used. The stoichiometry verification plot suggested a valid $n_1 = 1$ assumption. The univariate format of the direct approach allows presenting the estimates as vectors; therefore, conventional univariate plots, such as the box-and-whisker plot, are possible, indicating the experimental parameters dispersion (uncertainty). The final estimates were $\log K_{E1} = 3.47 \pm 0.06$, $\log K_{E2} = 3.28 \pm 0.16$, $ES = 1.5 \pm 0.2$ (moderate enantioselectivity) and *PB* values of 64% and 53% calculated at $D = 5$ µM and $P = 600$ µM for E1 and E2 enantiomers, respectively.

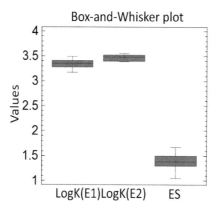

Figure 6.5 Box-and-whisker plot of estimated log K_{E1}, log K_{E2} and the corresponding enantioselectivity (ES) for catechin. Reprinted with permission from Ref. [34], Copyright 2012, Springer.

6.5 Studies on the Application of Capillary Electrophoresis for Evaluating Enantioselective Plasma Protein Binding of Chiral Drugs

Table 6.2 shows some binding parameters obtained using different electrophoretic techniques for evaluating the stereoselectivity in protein binding of chiral drugs and the methodology used for the evaluation of the enantioselective binding [31, 33, 34, 36–45]. These studies are devoted to evaluate the enantioselective binding of chiral drugs with a specific protein (HSA, bovine serum albumin (BSA), AGP, lipoproteins), genetic variants of proteins or whole human plasma, some of them are commented as follows.

Zopiclone is a hypnotic drug used for the treatment of insomnia. The hypnotic activity resides in the (S)-enantiomer, while (R)-zopiclone has no effect. Consequently, eszopiclone, a new drug consisting of the pure (S)-enantiomer, has been recently commercialized. Asensi Bernardi et al. [36] evaluated the enantioselective protein binding of zopiclone to HSA and whole plasma. Ultrafiltration followed by chiral separation of the unbound fraction by EKC-PFT with carboxymethyl-β-cyclodextrin (CM-β-CD) was used. For the evaluation of the enantioselective binding of zopiclone to HSA, an experimental design with a constant P value near the physiological concentration (450 µM) and five D levels in a short D-concentration interval (100–200 µM, three independent replicates of each level, total 15 individual samples) was used. The values obtained for the (S)- and (R)-zopiclone were: 1:1 stoichiometry, medium affinity constants ($\log K_{1S} = 3.38 \pm 0.14$ and $\log K_{1R} = 3.09 \pm 0.19$) and medium protein-binding percentage ($PB = 47\%$ and 36% for (S)- and (R)-enantiomers, respectively). The ES obtained as K_{1S}/K_{1R} was around 2. So (S)-zopiclone has higher affinity to the HSA molecule than its antipode. Finally, an estimation of the binding of zopiclone enantiomers to plasma proteins was performed. PB of 45 $\pm$ 3 and 49 $\pm$ 6 were found for (S)- and (R)-zopiclone, respectively, values which agree with the PB found in the literature for racemic zopiclone, 45% [46]. Comparing these results with those obtained for HSA, it can be concluded that the (S)-enantiomer mainly binds to HSA, while (R)-zopiclone binds also to other plasma proteins.

Table 6.2 Literature data of binding parameters for stereoselective protein-binding studies by capillary electrophoresis

Chiral drug	Protein/s	Binding parameters		nK (or K) ratio	Techniques (chiral selector)	Reference
		(R)-enantiomer	(S)-enantiomer			
(R,S)-Fluoxetine	HSA	$K = 3.23 \times 10^4 \, M^{-1}$ (R- or S-enantiomer)	$K = 1.48 \times 10^4 \, M^{-1}$ (R- or S-enantiomer)	2.2	UF-EKC (HS-β-CD)	[33]
$(\pm)$-Catechin	HSA	$K = 2.95 \times 10^3 \, M^{-1}$ (R- or S-enantiomer)	$K = 1.91 \times 10^3 \, M^{-1}$ (R- or S-enantiomer)	1.5	UF-EKC (HS-β-CD)	[34]
(R,S)-Zopiclone	HSA	$K = 1.23 \times 10^3 \, M^{-1}$	$K = 2.40 \times 10^3 \, M^{-1}$	1.9 (S/R)	UF-EKC (CM-β-CD)	[36]
(R,S)-Nomifensine	HSA	$K = 1.69 \times 10^3 \, M^{-1}$ (R- or S-enantiomer)	$K = 4.90 \times 10^3 \, M^{-1}$ (R- or S-enantiomer)	2.7	UF-EKC (TM-β-CD)	[37]
(R,S)-Propanocaine	HSA	$K = 1.58 \times 10^3 \, M^{-1}$ (R- or S-enantiomer)	$K = 2.51 \times 10^3 \, M^{-1}$ (R- or S-enantiomer)	1.5	UF-ACE (HSA)	[38]
(R,S)-Imazalil	HSA	$K = 2.51 \times 10^3 \, M^{-1}$	$K = 1.26 \times 10^3 \, M^{-1}$	2.0	FA-ACE (HSA)	[31]
(R,S)-Brompheniramine (R,S)-Chlorpheniramine (R,S)-Hydroxyzine (R,S)-Orphenadrine	HSA	$K = 9.39 \times 10^2 \, M^{-1}$ $K = 9.20 \times 10^2 \, M^{-1}$ $K = 5.3 \times 10^3 \, M^{-1}$ (R- or S-enantiomer) $K = 1.26 \times 10^3 \, M^{-1}$ (R- or S-enantiomer)	$K = 2.60 \times 10^3 \, M^{-1}$ $K = 1.69 \times 10^3 \, M^{-1}$ $K = 6.3 \times 10^3 \, M^{-1}$ (R- or S-enantiomer) $K = 1.67 \times 10^4 \, M^{-1}$ (R- or S-enantiomer)	2.8 (S/R) 1.8 (S/R) 1.2 13.3	UF-ACE (HSA)	[39]

(*Continued*)

Table 6.2 (Continued)

Chiral drug	Protein/s	Binding parameters		nK (or K) ratio	Techniques (chiral selector)	Reference
		(R)-enantiomer	(S)-enantiomer			
(R,S)-Rotigotine	HSA	$K = (17.6\pm0.6) \times 10^3\,M^{-1}$	$K = (8.90\pm0.3) \times 10^3\,M^{-1}$,	2.0 (R/S)	FA-ACE	[40]
	BSA	$K = (9.3\pm0.3) \times 10^3\,M^{-1}$	$K = (7.3\pm0.2) \times 10^3\,M^{-1}$		(BSA, HSA)	
(R)- or (S)-Amlodipine	HSA	$K = 1.06 \times 10^5\,M^{-1}$	$K = 9.71 \times 10^4\,M^{-1}$	1.1 (R/S)	FA	[41]
(R,S)-Verapamil	HSA	$K = 2670\,M^{-1}$	$K = 850\,M^{-1}$	3.1 (R/S)	FA-EKC	[42]
(R,S)-Propranolol		K = not given	K = not given	1	(TM-β-CD)	
(R,S)-Acenocoumarol	ORM1	Affinity constants not given		2.61(S/R)	UF-EKC	[43]
(R,S)-Phenprocoumon	ORM2			1.1 (R/S)	(SP-β-CD)	
(R,S)-Warfarin	AGP			2.26 (S/R)	UF-EKC	
	ORM1			1.12 (R/S)	(α-CD)	
	ORM2			1.76 (R/S)	UF-EKC	
	AGP			1.21(S/R)	(M-β-CD)	
	ORM1			1.85(S/R)		
	ORM2			1.27 (S/R)		
	AGP			1.77(S/R)		
(R)- or (S)-Nilvadipine	HDL	$nK = 1.02 \times 10^6\,M^{-1}$	$nK = 1.02 \times 10^6\,M^{-1}$	1.00 (R/S)	FA	[16]
	Normal LDL	$nK = 7.66 \times 10^6\,M^{-1}$	$nK = 7.47 \times 10^6\,M^{-1}$	1.03 (R/S)	-	
	Oxidized LDL	$nK = 3.66 \times 10^7\,M^{-1}$	$nK = 3.57 \times 10^7\,M^{-1}$	1.03(R/S)		
(R)- or (S)-Verapamil	HDL	$nK = 2.75 \times 10^4\,M^{-1}$	$nK = 2.81 \times 10^4\,M^{-1}$	1.02(S/R)	FA	[44]
	Normal LDL	$nK = 1.99 \times 10^5\,M^{-1}$	$nK = 1.96 \times 10^5\,M^{-1}$	1.02(R/S)	-	
	Oxidized LDL	$nK = 2.03 \times 10^6\,M^{-1}$	$nK = 2.06 \times 10^6\,M^{-1}$	1.01(S/R)		
(R)- or (S)-Propranolol	HDL	$nK = 2.12 \times 10^4\,M^{-1}$	$nK = 2.43 \times 10^4\,M^{-1}$	1.02(S/R)	HPFA-CE	[45]
	Normal LDL	$nK = 4.01 \times 10^5\,M^{-1}$	$nK = 4.02 \times 10^5\,M^{-1}$	1.00(S/R)	-	

α-CD: α-cyclodextrin; CM-β-CD: carboxymethyl-β-cyclodextrin; HS-β-CD: highly sulfated-β-cyclodextrin; M-β-CD: methylated-β-cyclodextrin; SP-β-CD: sulfopropylated-β-cyclodextrin; TM-β-CD: heptakis-(2,3,6-O-methyl)-β-cyclodextrin.

Nomifensine is an isoquinoline derivative used in clinical Parkinson's disease studies and diagnosis. The enantioselective binding of nomifensine to HSA and total plasma proteins has been evaluated using the UF-EKC and the experimental design and the equations previously described [33, 34]. The separation of nomifensine enantiomers was performed by EKC-CFT using heptakis-(2,3,6-O-methyl)-β-cyclodextrin (TM-β-CD), a neutral cyclodextrin, as chiral selector [37]. Figure 6.6 shows the electropherograms obtained under the selected experimental conditions.

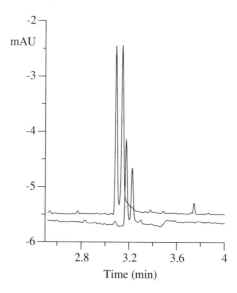

Figure 6.6 Enantioseparation of nomifensine enantiomers in a standard solution (upper part) and in an ultrafiltrate (unbound fraction, lower part) by EKC-CFT. Experimental conditions: 50 mM Tris buffer at pH 6.0 as BGE, 30 mM TM-β-CD in BGE injected at 10 psi/min, separation voltage 15 kV and capillary temperature 50°C. Reprinted with permission from Ref. [37], Copyright 2012, John Wiley and Sons.

A P-constant experimental design was used for the evaluation of nomifensine-HSA enantioselective binding, keeping the D/P ratio < 0.5. P was fixed near the physiological concentration, and D was varied between 70 and 200 µM of each enantiomer (5 concentration levels, 3 replicates per level). The logK_1 estimates were 3.23 ± 0.11 and 3.64 ± 0.09 for E1 and E2, respectively; these uncertainties were obtained under intermediate precision conditions (two independent working sessions). Enantioselectivity, defined as the ratio between affinity constants (K_{E2}/K_{E1}), was estimated as 2.7 ± 0.1, indicating higher affinity of the second eluted enantiomer for the HSA molecule. The estimations of enantioselective *PB* to the HSA molecule at the physiological level were 40 ± 5 and 63 ± 4 for E1 and E2, respectively. The enantioselective *PB* to total plasma proteins was calculated by incubating racemic nomifensine with human serum and applying the same methodology described for HSA. The results for the protein binding of nomifensine enantiomers to total plasma proteins were 58 ± 7% and 64 ± 4% for E1 and E2, respectively. From the *PB* estimates, it can be deduced that E1 binds not only to HSA but also to other plasma proteins; however, the role of HSA is the most important in its protein binding. For E2, the values of *PB* obtained for HSA and total plasma proteins were similar, indicating that this enantiomer binds a priori only to HSA.

Following similar experimental designs and equations, authors from the same research group have evaluated the enantioselective binding of propanocaine [38] as well as imazalil [31] enantiomers to HSA (see results in Table 6.2). Maddi et al. [47] studied the binding of the (*R*)- and (*S*)-enantiomers of amlodipine to BSA, HSA, AGP, and human plasma by equilibrium dialysis over the concentration range of 75–200 μM at a protein concentration of 150 μM. Unbound drug concentrations were determined by EKC using α-cyclodextrin as chiral selector. (*S*)-amlodipine was bound to a higher extend by HSA and human plasma compared with (*R*)-amlodipine, whereas the opposite binding of the enantiomers was observed for BSA and AGP. Data treatment was performed by Scatchard analysis (see Table 6.1).

In addition to the evaluation of affinity constants of enantiomers toward proteins, the study of the specific sites in the protein molecule where the enantiomers interact is of special interest. To identify the binding sites of enantiomer drugs in the HSA molecule, and elucidate if the binding of enantiomers follows an independent or competitive model, various site marker ligands are used. Warfarin and phenylbutazone, diazepam, and digitoxin were usually used as marker ligand representatives of sites I, II, and III, respectively. In these studies, the displacement of equilibrium between a racemic drug or an individual enantiomer and HSA in the presence of a marker ligand is evaluated. Additionally, other studies deal with drug–drug inhibitory interaction because of its great clinical significance. Based on the same principle, when two or more drugs that bind at the same binding site are administered together, an increase in their unbound drug concentrations can be observed. These interactions can also occur between enantiomers of different drugs.

Martínez-Gómez et al. [39] studied the enantioselective binding of the antihistamines brompheniramine, chlorpheniramine, hydroxyzine, and orphenadrine to HSA. Due to the high affinity of these drugs toward HSA, the authors proposed the treatment of the insoluble fraction obtained after ultrafiltration and the chiral separation and determination of enantiomers by ACE using HSA as chiral selector. The study performed to identify the binding sites of enantiomer drugs in the HSA molecule indicated that both enantiomers of brompheniramine and chlorpheniramine bind to site II and, therefore, they follow a competitive model. For orphenadrine, the least bound enantiomer binds to digitoxin site (site III), while for the most bound enantiomer, no specific site can be defined. For hydroxyzine, the least bound enantiomer does not bind to any of these three sites; however, the most bound enantiomer binds preferentially to site I. So for orphenadrine and hydroxyzine, an independent binding model of enantiomers to HSA could be considered. The affinity constants of drug enantiomer-HSA revealed the existence of enantioselective binding of antihistamines to HSA (Table 6.2). The binding of orphenadrine to HSA showed the highest enantioselectivity among the studied antihistamines.

The same authors studied the enantioselective binding of the antihistamines brompheniramine, chlorpheniramine, hydroxyzine, orphenadrine, and phenindamine; the phenothiazines promethazine and trimeprazine; and the local anaesthetic bupivacaine to whole plasma by determining the protein binding of drug enantiomers using UF-ACE [48]. The enantioselectivity values, defined as the ratio between the bound drug concentrations of E2 and E1 eluted enantiomers, were ranged from 1.01 to 2.51, indicating that a different degree of enantioselectivity in the binding of these compounds to plasma proteins exists. The observed order of enantioselectivity was: phenindamine > trimeprazine > promethazine ≈ orphenadrine > bupivacaine >> chlorpheniramine ≈ hydroxyzine ≈ brompheniramine. Orphenadrine and trimeprazine showed concentration-dependent enantioselectivity and extensive binding to human plasma. As it has been described previously, the enantioselective binding of (±)-catechin to HSA has been evaluated using UF-EKC with HS-β-CD as chiral selector [34]. In this paper, in order to identify the main active site in the HSA molecule, the effects of well-known site marker ligands on the binding of (±)-catechin enantiomers were examined. Warfarin and diazepam were used as marker representatives of sites I and II, respectively, in the HSA molecule. The results showed that the addition of warfarin produced an increase in the free concentration of both catechin enantiomers, independently of the total concentration of catechin used. In contrast, the addition of diazepam only produced an increase in the free catechin enantiomer concentrations at high total enantiomer concentration. These results suggest that both catechin enantiomers bind to site I in the HSA molecule at low concentrations as reported bibliography. Only at high concentrations (unexpected in vivo) could the catechin enantiomers also bind site II in some degree. Experimental enantioselectivity values were compared and agreed with estimations made using molecular docking.

Chu et al. [40] demonstrated the enantiospecific binding of antiparkinsonian drug rotigotine ((S)-enantiomer), and its antipode to HSA or BSA was using ACE-PF under near-physiological conditions. Based on the ACE-PF data, the following binding constants were obtained: $K_{HSA,S}$ = 8884 ± 255 M^{-1}, $K_{HSA,R}$ = 17648 ± 587 M^{-1}, $K_{BSA,S}$ = 7348 ± 237 M^{-1}, $K_{BSA,R}$ = 9353 ± 352 M^{-1}. The results revealed that rotigotine had low affinity for the two serum albumins, and both enantiomers showed stronger affinity for HSA than for BSA. The presence of warfarin or ketoprofen site markers had adverse effect on the enantioseparation due to the competitive binding, or even eliminated the enantioselective binding of the enantiomers to the albumin when the molar ratio of the site marker to the albumin was at certain level. The authors suggested that although there might be a synergistic binding between the drug and the albumin, sites II and I were the preferential binding site of the drug on HSA and BSA, respectively.

Liu et al. [41] applied a flow injection (FI)-FA to the study of stereoselectivity binding of amlodipine to HSA. The authors indicated that the combined flow injection (FI)-CE system can enhance sampling frequencies and improve reproducibility of classical CE systems. The stereoselectivity of amlodipine binding to HSA was proved by the different free fractions of the two enantiomers. In

Studies on the Application of Capillary Electrophoresis | **249**

physiological phosphate solution when 200 µM racemic amlodipine was equilibrated with 300 µM HSA, the concentration of unbound (R)-amlodipine was about 1.5 times higher than that of its antipode. The binding constants of two enantiomers were comprised between 9910 and 11,200 M^{-1} for (R)-amlodipine and between 90,200 and 104,000 M^{-1} for (S)-amlodipine. Hydroxypropyl-β-CD (HP-β-CD) was used as a chiral selector. L-tryptophan and ketoprofen were used as displacement reagents to investigate the binding sites of amlodipine to HSA. A binding synergism effect between hydrochlorothiazide and amlodipine was observed and the results suggested that hydrochlorothiazide can destroy binding equilibrium of (R)-amlodipine and (S)-amlodipine toward HSA and they can occupy the same binding site of HSA (site I).

The enantioselective binding of verapamil and propranolol to HSA using TM-β-CD as chiral selector and the FA-EKC technique was evaluated [42]. The authors indicated that the unbound concentration of (S)-verapamil was 1.7 times higher than that of (R)-isomer. Moreover, when racemic ibuprofen was added into verapamil-HSA solution, (R)-verapamil was partially displaced, while (S)-verapamil was not displaced at all. The authors indicated that the (R)-ibuprofen is the enantiomer responsible of the observed displacement. A binding synergism effect between bupivacaine and verapamil was observed, and the binding sites study suggested that verapamil and bupivacaine occupy different binding sites of HSA (site II and site III, respectively). No obvious stereoselective binding of propranolol to HSA was observed.

In spite of the fact that HSA is the major plasma protein responsible of the enantioselective binding of chiral drugs, AGP and lipoproteins can contribute to this behavior. Some reports are dedicated to this aim. In a paper by Hazai et al. [43], coumarin-type anticoagulants, warfarin, phenprocoumon, and acenocoumarol were tested for their stereoselective binding to native AGP and ORM 1 and ORM 2 genetic variants. For this study, ultrafiltration and enantiomeric separation of the unbound fraction of chiral drugs were used. Sulfopropylated-β-cyclodextrin, α-cyclodextrin, and random methylated-β-cyclodextrin were used as chiral selectors for the separation of the enantiomers acenocoumarol, phenprocoumon, and warfarin, respectively. The results showed that all investigated compounds bind stronger to ORM 1 variant than to ORM 2, and no significant enantioselectivity was observed in binding to ORM 2. Binding to native AGP (consisting of about 70% ORM 1 and 30% ORM 2) resembled binding to ORM 1 rather than to ORM 2. ORM 1 and human native AGP bind preferentially to (S)-enantiomers of warfarin and acenocoumarol, while slight enantioselectivity was observed in phenprocoumon binding. Acenocoumarol possessed the highest enantioselectivity in AGP binding due to the weak binding of its (R)-enantiomer.

Shiono et al. [49] studied the function of sialic acid groups at the terminal of AGP glycan chains with respect to chiral discrimination between enantiomers of propranolol and verapamil, using the FA-EKC method and TM-β-CD as chiral selector. The authors found that the unbound concentration of (S)-verapamil was 1.3 times higher than that of (R)-verapamil in native AGP solution, and this selectivity was not affected by desialylation of AGP. Further, enzymatic elimination of end-terminal galactose residues of the desialylated AGP did not change the binding of either isomer of verapamil. On the other hand, the unbound concentration of (R)-propranolol was 1.27 times higher than that of (S)-propranolol in native AGP solution. Desialylation did not change the unbound concentration of (R)-propranolol but caused the unbound concentration of (S)-propranolol to rise up to the same level of (R)-propranolol, resulting in the loss of enantioselectivity. The authors suggested that the sialic acid residues may be regarded as one origin of enantioselectivity in AGP-propranolol binding, while they are not responsible for the enantioselective AGP-verapamil binding.

The effect of pH on the disopyramide binding properties of genetic variants of AGP, A variant, and a mixture of F1S variants was investigated using the FA method [50]. The unbound concentrations of disopyramide in AGP genetic variants were estimated at pH 4.0, 5.0, 6.0, and 7.4 to evaluate binding constants. The binding between disopyramide and A variant decreased as the BGE pH decreased (from pH 7.4 to 4.0), while the binding between disopyramide and F1S variants decreased at first (from pH 7.4 to 6.0), and then gained (from pH 6.0 to 4.0). Consequently, disopyramide was more strongly bound to A variant than to F1S variants at pH 7.4, while at pH 4.0, disopyramide was more strongly bound to F1S variants. At any pH, (S)-disopyramide was bound more strongly than (R)-disopyramide, and the enantioselectivity of A variant was significantly higher than that of F1S variants. The authors suggested that the conformational change induced by acidification of the BGE differs between these genetic variants, and this causes the difference in disopyramide binding ability.

The enantioselective binding of individual nilvadipine, verapamil, and propranolol enantiomers to plasma lipoproteins has been studied by FA [16, 44, 45]. It was found that the binding of nilvadipine, verapamil, and propranolol to HDL, LDL, and oxidized LDL was non-specific and not enantioselective. Partition-like binding to the lipid part of these lipoproteins seemed to occur dominantly. The total binding affinities of nilvadipine [16] and verapamil [44] to LDL were about seven times stronger than those to HDL, and the oxidation of LDL enhanced the binding affinity significantly. For propranolol, the total binding affinity to LDL was 17 times higher than that of propranolol–HDL binding [45].

6.6 Conclusion

The development of new drugs is a long, complex, and expensive process, which involves the evaluation of pharmacokinetic and pharmacodynamic properties of new molecules in different stages. In the early stages of drug development, in vitro methods for the high-throughput screening of the properties of new molecules are commonly used to obtain preliminary data about their potential pharmacological activity and also their pharmacokinetics. When chiral molecules are used, both pharmacokinetic and pharmacodynamic properties can show a certain degree of enantioselectivity due to the interaction with optically active biomacromolecules, so the methods applied in these preclinical early stages must allow the evaluation of these properties in an enantioselective way. Stereoselectivity in protein binding can have a significant effect on the drug disposition such as first-pass metabolism, metabolic clearance, renal clearance, and tissue binding. The amount of the enantiomer drug that binds to plasma proteins, as well as the strength of this binding, influences considerably the availability of the drug for reaching target organs and also for being eliminated. Therefore, the development of fast and reliable procedures for the in vitro evaluation of enantioselective drug–protein binding is interesting for the pharmaceutical industry research.

Capillary electrophoresis is a convenient analytical technique to be applied for these preclinical studies. It provides advantageous possibilities for the chiral analysis of small molecules and also for the evaluation of some pharmacokinetic and pharmacodynamic processes, such as the binding of drugs to plasma proteins. Its main features, which are extremely interesting for these applications, are speediness of analysis, low consumption of chemicals and reagents, high peak efficiencies, and low environmental impact.

In this chapter, the strength and weakness of CE in these studies are pointed out. While the separation/analytical aspects related with enantioselective protein-binding estimations have been extensively investigated, the impact of the experimental design and the mathematical approach on the results has been underestimated in many studies, entailing a risk of lack of quality results, or an underestimation of the process uncertainty. Among the different techniques, ultrafiltration combined with electrokinetic chromatography using cyclodextrins as chiral selector is the preferred methodology in most studies. Results show clear evidences on the enantioselective binding of chiral drugs to plasma proteins.

Acknowledgments

The authors acknowledge the Spanish Ministry of Science and Technology (MCYT), and the European Regional Development Fund (ERDF) (Project CTQ2015-70904R) for the financial support.

References

1. Nguyen, L. A., He, H., and Pham-Huy, C. (2006). Chiral drugs: An overview, *Int. J. Biomed. Sci.*, **2**, pp. 85–100.

2. Brocks, D. R. (2006). Drug disposition in three dimensions: An update on stereoselectivity in pharmacokinetics, *Biopharm. Drug Dispos.*, **27**, pp. 387–406.

3. Howard, M. L., Hill, J. J., Galluppi, G. R., and McLean, M. A. (2010). Plasma protein binding in drug discovery and development, *Comb. Chem. High T. Scr.*, **13**, pp. 170–87.

4. Chuang, V. T. G. and Otagiri, M. (2006). Stereoselective binding of human serum albumin, *Chirality*, **18**, pp. 159–166.

5. González-Alonso, I. and Sánchez-Navarro, A. (1998). *Unión a proteínas. En Biofarmacia y Farmacocinética,* eds. Doménech Berrozpe, J., Martínez Lanao, J., and Plá Delfina, J. M. Volumen *2.* (Síntesis, Madrid).

6. Ascoli, G. A., Domenici, E., and Bertucci, C. (2006). Drug binding to human serum albumin: Abridged review of results obtained with high-performance liquid chromatography and circular dichroism, *Chirality,* **18**, pp. 667–679.

7. He, X. M. and Carter, D.C. (1992). Atomic structure and chemistry of human serum albumin, *Nature,* **358**, pp. 209–215.

8. Sugio, S., Kashima, A., Mochizuki, S., Noda, M., and Kobayashi, K. (1999). Cristal structure of human serum albumin at 2.5 Å resolution, *Protein Eng.,* **12**, pp. 439–446.

9. Otagiri, M. (2005). A molecular functional study on the interactions of drugs with plasma proteins, *Drug Metab. Pharmacokinet.,* **20**, pp. 309–323.

10. Ghuman, J., Zunszain, P. A., Petitpas, I., Bhattacharya, A. A., Otagiri, M., and Curry, S. (2005). Structural basis of the drug-binding specificity of human serum albumin, *J. Mol. Biol.,* **353**, pp. 38–52.

11. Hage, D. S. and Sengupta, A. (1999). Characterization of the binding of digitoxin and acetyldigitoxin to human serum albumin by high-performance affinity chromatography, *J. Chromatogr. B,* **724**, pp. 91–100.

12. Sengupta, A. and Hage D. S. (1999). Characterization of minor site probes for human serum albumin by high-performance affinity chromatography, *Anal. Chem.,* **71**, pp. 3821–3827.

13. Kopecky, Jr., V., Ettrich, R., Hofbauerova, K., and Baumruk, V. (2003). Structure of human alpha1-acid glycoprotein and its high-affinity binding site, *Biochem. Biophys. Res. Commun.,* **300**, pp. 41–46.

14. Zsila, F. and Iwao, Y. (2007). The drug binding site of α1-acid glycoprotein: Insight from induced circular dichroism and electronic absorption spectra, *Biochim. Biophys. Acta,* **1770**, pp. 797–809.

15. Israili, Z. H. and Dayton, P. G. (2001). Human alpha-1-glycoprotein and its interactions withdrugs, *Drug Metab. Rev.,* **33**, pp. 161–235.

16. Mohamed, N. A. L., Kuroda, Y., Shibukawa, A., Nakagawa, T., El Gizawy, S., Askal, H. F., and El Kommos, M. E. (1999). Binding analysis of nilvadipine to plasma lipoproteins by capillary electrophoresis-frontal analysis, *J. Pharm. Biomed. Anal.,* **21**, pp. 1037–1043.

17. Vuignier, K., Schappler, J., Veuthey, J. L., Carrupt, P. A., and Martel, S. (2010). Drug-protein binding: A critical review of analytical tools, *Anal. Bioanal. Chem.*, **398**, pp. 53–66.

18. Lambrinidis, G., Vallianatou, T., and Tsantili-Kakoulidou, A. (2015). In vitro, in silico and integrated strategies for the estimation of plasma protein binding. A review, *Adv. Drug Deliv. Rev.*, **86**, pp. 27–45.

19. Shen, Q., Wang, L., Zhou, H., Di Jiang, H., Yu, L. S., and Zeng, S. (2013). Stereoselective binding of chiral drugs to plasma proteins, *Acta Pharm. Sinica*, **34**, pp. 998–1006.

20. Escuder-Gilabert, L., Martínez-Gómez, M. A., Villanueva-Camañas, R. M., Sagrado, S., and Medina-Hernández, M. J. (2008). Microseparation techniques for the study of the enantioselectivity of drug-plasma protein binding, *Biomed. Chromatogr.*, **23**, pp. 225–238.

21. Chen, J., Fitos, I., and Hage, D. S. (2006). Chromatographic analysis of allosteric effects between ibuprofen and benzodiazepines on human serum albumin, *Chirality*, **18**, pp. 24–36.

22. Kariv, I., Cao, H., and Oldenburg, K. R. (2001). Development of a high throughput equilibrium dialysis method, *J. Pharm. Sci.*, **90**, pp. 580–587.

23. Lázaro, E., Lowe, P. J., Briand, X., and Faller, B. (2008). New approach to measure protein binding based on a parallel artificial membrane assay and human serum albumin, *J. Med. Chem.*, **51**, pp. 2009–2017.

24. Randon, J., Garnier, F., Rocca, J. L., and Maïsterrena, B. (2000). Optimization of the enantiomeric separation of tryptophan analogs by membrane processes, *J. Membr. Sci.*, **175**, pp. 111–117.

25. Lee, K.-J., Mower, R., Hollenbeck, T., Castelo, J., Johnson, N., Gordon, P., Sinko, P. J., Holme, K., and Lee, Y.-H. (2003). Modulation of nonspecific binding in ultrafiltration protein binding studies, *Pharm. Res.*, **20**, pp. 1015–1021.

26. Shibukawa, A., Kuroda, Y., and Nakagawa, T. (1999). Development of high-performance frontal analysis and the application to the study of drug-plasma protein binding, *Trends Anal. Chem.*, **18**, pp. 549–556.

27. Escuder-Gilabert, L., Martín-Biosca, Y., Medina-Hernández, M. J., and Sagrado, S. (2014). Cyclodextrins in capillary electrophoresis: Recent developments and new trends, *J. Chromatogr. A.*, **1357**, pp. 2–23.

28. Chankvetadze, B. (2007). Enantioseparations by using capillary electrophoretic techniques. The story of 20 and a few more, *J. Chromatogr. A.*, **1168**, pp. 45–70.

29. Busch, M. H. A., Kraak, K. C., and Poppe, H. (1997). Principles and limitations of methods available for the determination of binding constants with affinity capillary electrophoresis, *J. Chromatogr. A*, **777**, pp. 329–353.

30. Martínez-Pla, J. J., Martínez-Gómez, M. A., Martín-Biosca, Y., Sagrado, S., Villanueva-Camañas, R. M., and Medina-Hernández, M. J. (2004). High-throughput capillary electrophoresis frontal analysis method for the study of drug interactions with human serum albumin at near-physiological conditions, *Electrophoresis*, **25**, pp. 3176–3185.

31. Asensi-Bernardi, L., Martín-Biosca, Y., Escuder-Gilabert, L., Sagrado, S., and Medina-Hernández, M. J. (2015). Evaluation of the enantioselective binding of imazalil to human serum albumin by capillary electrophoresis, *Biomed. Cromatogr.*, **29**, pp. 1637–1642.

32. Ye, M., Zou, H., Liu, Z., Wu, R., Lei, Z., and Ni, J. (2002). Study of competitive binding of enantiomers to protein by affinity capillary electrochromatograpy, *J. Pharm. Biomed. Anal.*, **27**, pp. 651–660.

33. Asensi-Bernardi, L., Martín-Biosca, Y., Villanueva-Camañas, R. M., Medina-Hernández, M. J., and Sagrado, S. (2010). Evaluation of enantioselective binding of fluoxetine to human serum albumin by ultrafiltration and CE. Experimental design and quality considerations, *Electrophoresis*, **31**, pp. 3268–3280.

34. Sabela, M. I., Gumede, N. J., Escuder-Gilabert, L., Martín-Biosca, Y., Bisetty, K., Medina-Hernández, M. J., and Sagrado, S. (2012). Connecting simulated, bioanalytical and molecular docking data on the stereoselective binding of (±)-catechin to human serum albumin, *Anal. Bioanal. Chem.*, **402**, pp. 1899–1909, Corrected erratum 404 (2012). 285

35. Šoltés, L. and Mach, M. (2002). Estimation of drug–protein binding parameters on assuming the validity of thermodynamic equilibrium, *J. Chromatogr. B*, **768**, pp. 113–119.

36. Asensi-Bernardi, L., Martín-Biosca, Y., Medina-Hernández, M. J., and Sagrado, S. (2011). On the zopiclone enantioselective binding to human albumin and plasma proteins. An electrokinetic chromatography approach, *J. Chromatogr. A*, **1218**, pp. 3111–3117.

37. Asensi-Bernardi, L., Martín-Biosca, Y., Medina-Hernández, M. J., and Sagrado, S. (2012). Electrokinetic chromatographic estimation of the enantioselective binding of nomifensine to human serum albumin and total plasma proteins, *Biomed. Chromatogr.*, **26**, pp. 1357–1363.

38. Martínez-Gómez, M. A., Villanueva-Camañas, R. M., Escuder-Gilabert, L., Sagrado, S., and Medina-Hernández, M. J. (2012). Evaluation of

enantioselective binding of propanocaine to human serum albumin by ultrafiltration and electrokinetic chromatography under intermediate precision conditions, *J. Chromatogr. B,* **889–890**, pp. 87–94.

39. Martínez-Gómez, M. A., Villanueva-Camañas, R. M., Sagrado, S., and Medina-Hernández, M. J. (2007). Evaluation of enantioselective binding of antihistamines to human serum albumin by ACE, *Electrophoresis,* **28**, pp. 2635–2643.

40. Chu, B. L., Lin, J. M., Wang, Z., and Guo, B. (2009). Enantiospecific binding of Rotigotine and its antipode to serum albumins: Investigation of binding constants and binding sites by partial-filling ACE, *Electrophoresis,* **30**, pp. 2845–2852.

41. Liu, X., Song, Y., Yue, Y., Zhang, J., and Chen, X. (2008). Study of interaction between drug enantiomers and human serum albumin by flow injection-capillary electrophoresis frontal analysis, *Electrophoresis,* **29**, pp. 2876–2883.

42. Ding, Y., Zhu, X., and Lin, B. (1999). Study of interaction between drug enantiomers and serum albumin by capillary electrophoresis, *Electrophoresis,* **20**, pp. 1890–1894.

43. Hazai, E., Visy, J., Fitos, I., Bikádi, Z., and Simonyi, M. (2006). Selective binding of coumarin enantiomers to human α_1-acid glycoprotein genetic variants, *Bioorg. Med. Chem.,* **14**, pp. 1959–1965.

44. Mohamed, N. A. L., Kuroda, Y., Shibukawa, A., Nakagawa, T., El Gizawy, S., Askal, H. F., and El Kommos, M. E. (2000). Enantioselective binding analysis of verapamil to plasma lipoproteins by capillary electrophoresis–frontal analysis, *J. Chromatogr. A,* **875**, pp. 447–453.

45. Ohnishi, T., Mohamed, N. A. L., Shibukawa, A., Kuroda, Y., Nakagawa, T., El Gizawy, S., Askal, H. F, and El Kommos, M. E. (2002). Frontal analysis of drug-plasma lipoprotein binding using capillary electrophoresis, *J. Pharm. Biomed. Anal.,* **27**, pp. 607–614.

46. Gaillot, J., Houghton, G. W., Marc Aurele, J., and Dreyfus, J. F. (1983). Pharmacokinetics and metabolism of zopiclone, *Pharmacology,* **27**, pp. 76–91.

47. Maddi, S., Yamsani, M. R., Seeling, A., and Scriba, G. K. (2010). Stereoselective plasma protein binding of amlodipine, *Chirality,* **22**, pp. 262–266.

48. Martínez-Gómez, M. A., Villanueva-Camañas, R. M., Sagrado, S., and Medina-Hernández, M. J. (2007). Evaluation of enantioselective binding of basic drugs to plasma by ACE, *Electrophoresis,* **28**, pp. 3056–3063.

49. Shiono, H., Shibukawa, A., Kuroda, Y., and Nakagawa, T. (1997). Effect of sialic acid residues of human α1-acid glycoprotein on stereoselectivity in basic drug–protein binding, *Chirality*, **9**, pp. 291–296.

50. Kuroda, Y., Matsumoto, S., Shibukawa, A., and Nakagawa, T. (2003). Capillary electrophoretic study on pH dependence of enantioselective disopyramide binding to genetic variants of human α1-acid glycoprotein, *Analyst,* **128**, pp. 1023–1027.

Chapter 7

Clinical Use of Capillary Zone Electrophoresis: New Insights into Parkinson's Disease

Pedro Rada,[a,b] Luis Betancourt,[a] Sergio Sacchettoni,[c] Juan Félix del Corral,[c] Hilarión Araujo,[b] and Luis Hernández[a]

[a]*Laboratory of Behavioral Physiology, School of Medicine, University of Los Andes, Mérida, Venezuela*

[b]*Neurology Unit, University Hospital (IAHULA), University of Los Andes, Mérida, Venezuela*

[c]*Hospital and School of Medicine, JM Vargas, Universidad Central de Venezuela, Caracas, Venezuela*

radap@ula.ve

Capillary zone electrophoresis (CZE), in the last 25 years, has emerged as a robust analytical technique that has gained wide acceptance in the biomedical environment. In the present chapter, we will focus on the use of CZE, in its micellar electrokinetic chromatography modality (MEKC), to monitor the zwitterion gamma-aminobutyric acid (GABA) and a polyamine putrescine, two biomolecules involved in disparate aspects of Parkinson´s disease. The first study consists of the sampling of GABA with microdialysis

---

Capillary Electrophoresis: Trends and Developments in Pharmaceutical Research
Edited by Suvardhan Kanchi, Salvador Sagrado, Myalowenkosi Sabela, and Krishna Bisetty
Copyright © 2017 Pan Stanford Publishing Pte. Ltd.
ISBN 978-981-4774-12-3 (Hardcover), 978-1-315-22538-8 (eBook)
www.panstanford.com

from the globus pallidum in Parkinson's patients that underwent stereotaxic neurosurgery as part of their treatment. The changes in this amino acid will help us understand the neurochemical circuitry of the basal ganglia in humans so that in the near future, new drugs capable of modifying it will be synthesized to improve the treatment of the disease. In a second study, as part of a metabolomic analysis, the polyamine putrescine was monitored from blood samples of patients with Parkinson's disease. This polyamine seems to play a key role in the appearance of α-synuclein aggregates in the dopamine neuron, and it is thought to be closely involved in the neurodegeneration process. As hypothesized, patients in different stages of the disease had higher concentrations of putrescine when compared to controls. This finding is of great importance for two reasons: (i) it will give us a tool for early detection of Parkinson's disease and (ii) if this polyamine is directly involved in the pathophysiological mechanism of the disease future treatments could appear in an attempt to prevent neurodegeneration.

7.1 Introduction

7.1.1 Parkinson's Disease

Parkinson's disease (PD) is the second most frequent neurodegenerative illness only surpassed by Alzheimer's disease. It affects 1% of the population over 70 years of age [9, 25] and represents a social burden as well as a psychological and economical overload on family (care keepers) and the patient himself/herself [6]. The disease is clinically characterized by motor symptoms such as slowness in movement (bradykinesia), rigidity, resting tremor, postural instability, and also by psychological symptoms such as depression, anhedonia, and disturbances in sleep patterns.

In the late 1950s, it was discovered that PD was caused by the degeneration of dopamine (DA) neurons in the substantia nigra (SNi) and became the first neurological disease linked to an alteration of a specific neurotransmitter [11, 20]. However, the exact etiology of the disease still evades our current knowledge. Genetic factors have been correlated, although less than 5% of PD follows a Mendelian genetic link [38]. Environmental factors have also been implicated (pesticides and herbicides), but no definitive cause has been found [3]. Recent research point to possible epigenetic causes of the disease, suggesting that both a genetic background associated to a specific environment converge in the appearance of the disease [12].

7.1.2 Basal Ganglia Circuitry

Once PD was related to a deficit of DA, researchers started looking at the altered circuitry. The SNi is part of a set of 5-nuclei deep in the brain called basal ganglia (see Fig. 7.1). These nuclei are the SNi, caudate, and putamen (together form the striatum, STR), globus pallidum (GP), and subthalamic nucleus (STh). They are intimately linked and isolated from the rest of the brain. Most of the neural information inputs the basal ganglia through the STR and SNi, and its outputs to the brain are mostly through the internal portion of the GP (GPi) to the thalamus (see Fig. 7.2).

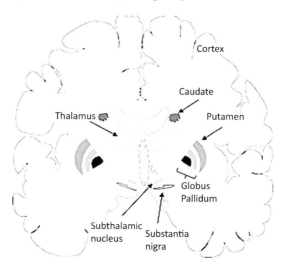

Figure 7.1 Coronal section of the human brain showing the location of the 5 nuclei that constitute the basal ganglia and the thalamus.

The thalamus is another deep-brain nucleus (see Fig. 7.1) and is the main relay station of sensory information to the cerebral cortex. All of the sensations (visual, hearing, tactile, taste, and smell) make a stop in the thalamus before reaching the brain cortex. The interaction of the basal ganglia with the thalamus is the basis of the thalamic filter hypothesis suggesting that the basal ganglia functions as a gatekeeper determining the relevant sensory information that reaches the cortex through the thalamus (Arvid Carlsson Nobel prize 2000).

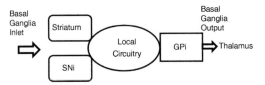

Figure 7.2 Summary of the neutral effect and afferent pathways through the basal ganglia. SNi = substantia nigra; GPi = globus pallidus internal portion.

What do we know of the local circuitry within the basal ganglia? Since PD is caused by a lesion of the DA neuron in the SNi, we will start the circuitry from this nucleus. SNi neurons project to the STR using DA as neurotransmitter. From the STR, two pathways emerge and both utilize the inhibitory neurotransmitter gamma-aminobutyric acid (GABA). One goes directly to the GPi and from here to the thalamus, and the other goes first to the external portion of the GP (GPe). From here, it projects a GABAergic tract to the STh, where it connects to a glutamatergic neuron that finally reaches the GPi. The former is called the direct pathway, and the latter is the indirect pathway (see Fig. 7.3) [5, 8].

In the striatum, DA has both an excitatory and an inhibitory function depending on the receptor it attaches to. Striatal GABAergic projecting neurons in the direct pathway express D1 receptors (excitatory), while the GABAergic neuron that projects in the indirect pathway expresses D2 receptor (inhibitory)(see Fig. 7.3)[5, 33]. Both direct and indirect pathways impinge on the GPi, and from here the GABAergic modulation of the thalamus determining the passage of sensory information to the cortex (thalamic filter hypothesis). In the case of PD, a decrease in DA release in the STR, assuming that the circuitry is correct, will cause an increase in glutamatergic output from the STh to the GPi and a decrease in GABA from the STR to the GPi. This leads to an overactive GABA release in the thalamus from GPi GABA neurons. This has been proven in animal models [34] but not in the ventrolateral thalamus in humans. To test this circuitry, it is imperative to monitor in vivo GABA release in the thalamus in humans. This raises the question on how to collect and analyze GABA in the brain from human patients?

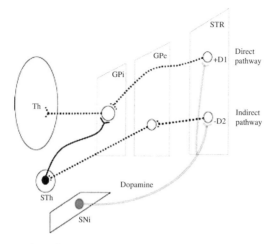

Figure 7.3 Basal ganglia circuitry. SNi = substantia nigra; STR = striatum; GPe = globus pallidum external portion; GPi = globus pallidum internal portion; STh = subthalamic nucleus; Th = thalamus; Gray = dopamine neuron; Dotted = GABA; Solid = glutamate.

7.1.3 In Vivo Monitoring of Brain Molecules

Cerebral microdialysis has become one of the most reliable and widely used techniques for the in vivo monitoring of neurotransmitters in the central nervous system (CNS). It consists of a concentric probe with a double barrel tube with a dialysis membrane at its tip (180 µm diameter) (see Fig. 7.4) [16, 17, 40]. The microdialysis probe is perfused with a saline solution, and neurotransmitters diffuse through the dialysis membrane. This technique is used in both animal and human studies [37]. Initial studies using microdialysis were limited by a poor time resolution. This was due to the lack of analytical techniques sensitive enough for trace compounds in low volume samples. High precision liquid chromatography (HPLC) assays are highly sensitive, but they require large volumes of sample (usually over 10 µL). So most experiments monitored neurotransmitters every 10 min or more. This was overcome in the late 1980s and the beginning of the 1990s when microdialysis was coupled to CZE [18].

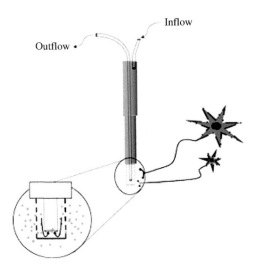

Figure 7.4 Diagram of the concentric microdialysis probe.

7.2 Capillary Zone Electrophoresis

CZE, in the last 25 years, has emerged as a robust analytical technique. Its laser-induced fluorescence detection mode (CZE-LIFD) has permitted sufficient sensitivity to monitor neurotransmitter from microdialysates with very low volumes [18, 19]. This brought down microdialysis time resolution from 10–20 min samples to a few seconds samples [30, 36]. First studies coupling CZE and microdialysis focused on measuring glutamate, the most abundant excitatory neurotransmitter in the CNS, during different behaviors in animals [18, 19, 26, 28, 32]. Samples were derivatized with fluorescein isothiocyanate (FITC) and ran on a carbonate buffer (20 mM). This permitted the measurement of charged amino acid neurotransmitters; however, it was limited for the detection of neutral amino acids. This restriction in analyzing uncharged molecules was surpassed by Terabe et al. by adding a surfactant to the mobile phase to create a pseudostationary phase and separating not only by the electrophoretic mobility of charged molecules but also by solvophobicity due to the partition coefficient of the neutral molecules [35]. This modification of CZE is called micellar electrokinetic chromatography (MEKC) and permitted the measurement of the most abundant inhibitory amino acid neurotransmitter GABA [4, 23, 27]. In our case, we used a small-diameter capillary (25 µM ID) with a high-concentration sodium dodecyl salt as surfactant (SDS; 120 mM) in a 23 mM borate buffer with 1% acetonitrile and were able to separate GABA from other neutral amino acids in microdialysates and cerebrospinal fluid [27] (see Fig. 7.5).

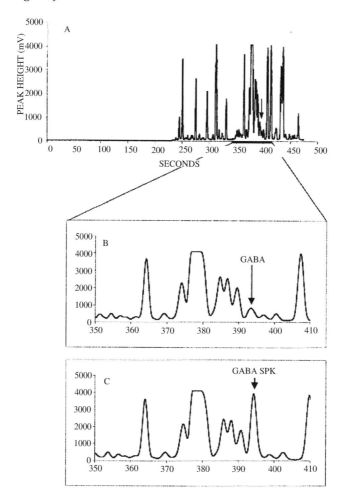

Figure 7.5 (A) Electropherogram from a microdialysis sample taken from thalamus VL nucleus of patient 1. (B) Magnification of electropherogram to show the putative GABA peak. (C) Same sample as (B) spiked with a GABA standard. Reprinted from Ref. [27], Copyright 2012, with permission from Elsevier.

7.2.1 GABA in the Ventrolateral Nucleus of the Thalamus in Two PD Patients

Surgery is necessary in a limited number of PD patients who do not respond well to pharmacological treatment and have a severe disability score, making it impossible for the patient to complete simple daily chores. A left stereotactic Laitinen's pallidotomy (postero-ventrolateral pallidotomy) was performed in a male patient aged 49 (Patient 1), who presented predominant right PD with rigidity and L-dopa-induced dyskinesia. A second patient (Patient 2), female, aged 54, suffering PD with predominant tremor in the right side of her body, underwent left stereotactic Leksell's thalamotomy (thermolesion of the ventral intermediate nucleus -Vim-) (see Fig. 7.6). To confirm the correct placement of the lesioning electrode, high-frequency electrical pulses (HFEP) were delivered (100 Hz) to reversibly inhibit the neural circuit by depolarization block [10]. Patients will immediately improve their clinical condition during HFEP if the electrode is in the correct placement, and in our case, it permitted a functional reversible test before the final lesion.

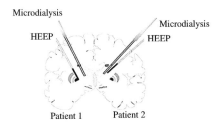

Figure 7.6 Diagram of electrode and microdialysis probe placement in both PD patients. Patient 1 had a pallidotomy (HFEP and lesion in the internal portion of the pallidum) and microdialysis probe in the ventrolateral nucleus of the thalamus, while patient 2 has a thamotomy (HFEP and lesion in the Vim of the thalamus) and microdialysis probe in the VL-thalamus.

In both patients, microdialysis probes were implanted in the thalamus (ventrolateral nucleus). Microdialysis probes were inserted 30 min before sampling to minimize the collection of GABA released by the lesion made during insertion. Probes were perfused with a sterile solution (135 mM NaCl, 3.7 mM KCl, 1.0 mM $MgCl_2$, 1.2 mM $CaCl_2$, and 10 mM $NaHCO_3$, pH 7.4) at a flow rate of 1 µL/min and samples were collected every 10 min. Sampling consisted of two basal samples, one sample during HFEP, three samples post-HFEP, and two final samples following radiofrequency coagulation (see Fig. 7.6).

7.2.2 Sample Preparation

Samples were derivatized with equal volumes of a FITC solution prepared with 1 mg FITC dissolved in 1 mL of acetone and mixed with 1 mL of 20 mM carbonate buffer. Samples reacted overnight and were ran in a CZE apparatus the next day (Meridialysis Inc). Briefly, samples were hydrodynamically injected into the capillary (25 μm ID and 350 μM OD; 40 cm length) by applying a negative pressure (−10 psi) for 1 s.

7.2.3 Running Conditions

The running buffer consisted of a borate buffer (23 mM) with a high concentration of SDS (120 mM) and 1% acetonitrile. Once the sample was injected, 29 kV was applied between the anode and the cathode, and the specific molecules attached to FITC were detected by focusing an argon laser (488 nm) to the capillary with the help of a 60× magnification, 0.8 numerical aperture microscope objective, and a photomultiplier to detect the emitted fluorescence [18].

7.3 Result

In both patients, GABA extracellular levels significantly decreased in the ventrolateral thalamus during HFEP of the GPi in one patient and the thalamus (Vim) in the other patient (see Fig. 7.7). Levels had a tendency to recover to original concentrations once HFEP was stopped to descend again following the thermolesion (see Fig. 7.7).

This result in human microdialysates supports the hypothesis that in PD patients, there is an overactive GABA projection from the GPi to the thalamus and the decrease in GABA levels during HFEP was positively correlated to an improvement of the clinical symptoms. It also confirms the usefulness of CZE as an analytical technique when high temporal resolution and peak resolution are needed to monitor neurotransmitter in small volumes.

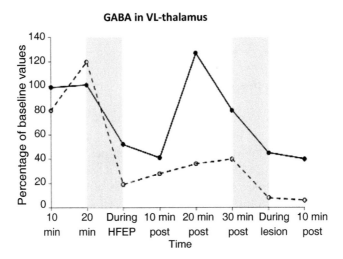

Figure 7.7 Effect of HFEP and thermolesion on GABA levels in the VL-thalamus in PD patients. Patient 1 (solid line) had a pallidotomy (HFEP and lesion in the internal portion of the pallidum), while patient 2 had a thalamotomy (HFEP and lesion in the Vim nucleus of the thalamus).

7.3.1 Polyamine Putrescine and Parkinson's Disease

In this second set of experiments, we showed the usefulness of CZE in the search for biological markers of PD. This is important because to date PD diagnosis is mostly based on clinical findings and discarding other similar diseases [29]. Other techniques such as flurodopa PET scanning and beta-CIT-SPECT are extremely expensive (over $2500 for one analysis) and still are at an experimental level.

Polyamines (PA) such as putrescine, spermidine, and spermine are organic cations involved in the stabilization of nucleic acid and membranes and vital in cell growth and differentiation [22]. In the CNS, these molecules normally interact with neurotransmitter receptors such as glutamate NMDA and AMPA receptors and also with K+ inward rectifying channels [39]. Abnormal levels of PA are suggested to contribute in the pathogenesis of PD. Specifically, they seem to play a key role in the misfolding and aggregation of α-synuclein, a primary protein found in Lewy's bodies, which are histological hallmark of PD [1, 13–15, 21].

We focused on putrescine because it was found to be elevated in cerebrospinal fluid of PD patients [24], and several other studies detected changes in spermine and spermidine but were unable to detect putrescine in blood samples (both in red blood cells (RBCs) or plasma) using HPLC [7, 31]. For this reason, we decided to study levels of putrescine in RBCs and plasma of PD patients and compared them to age- and sex-matched controls using CZE in its MEKC modality.

7.3.2 Blood Sampling

Samples were obtained from patients at the Neurology Unit of the University of Los Andes Hospital, Mérida, Venezuela. Twelve patients had the diagnosis of PD, and 12 patients were age- and sex-matched controls. An automatic lancing device was used, and blood samples were collected into hematocrit tubes and immediately centrifuged for 5 min to separate plasma from RBCs. A volume of 10 µL of RBCs was mixed with the same amount of pure water to rupture the RBC membranes, then vortexed for 30 s and centrifuged for 5 min. Afterward, 10 µL of the supernatant was mixed with 10 µL of acetonitrile to trigger deproteinization, and again the mixture was vortexed for 30 s and centrifuged for 5 min. The last step consisted of mixing 10 µL of the supernatant with the same amount of derivatizing solution (FITC-acetone/20 mM carbonate buffer). Plasma preparation consisted of the same procedure as RBCs except there was no need for the first step (rupturing cell membranes).

7.3.3 Running Conditions

The running buffer consisted of a 40 mM tetrasodium borate solution with 20 mM SDS at pH 9.0. The capillary was 60 cm long and washed hydrodynamically for 6 min with 1 M sodium hydroxide solution, followed by water for 6 min and filled with running buffer for 6 min. After injection, a positive voltage (27 kV) was applied at the anode, and data were collected for 20 min.

Putrescine was well separated from other constituents in the sample (see Fig. 7.8).

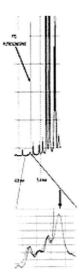

Figure 7.8 FTC-putrescine was revealed by the sample of RBCs with a standard of FTC-putrescine.

Using these conditions, it was possible to measure putrescine in the picomolar range (100 pM) showing a linear relationship at concentrations found in blood samples from PD patients (see Fig. 7.9) (unpublished results).

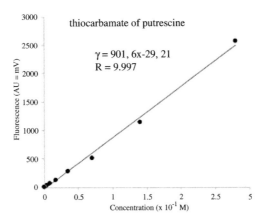

Figure 7.9 A linear relationship between thiocarbamate putrescine concentration and the intensity of fluorescence was found in the range from 5 nM to 2.8 µM.

7.3.4 Putrescine Levels in RBCs and Plasma of PD Patients

Putrescine concentrations in RBCs from PD patients (1.82 ± 0.57 µM, mean ± sem) were significantly higher than control subjects (0.55 ± 0.15 µM, mean ± sem) (see Fig. 7.10). There was no statistically significant difference in the plasma concentrations of putrescine in PD patients versus controls ($P > 0.497$).

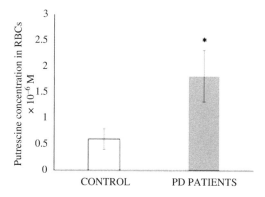

Figure 7.10 Putrescine concentration in RBCs of control patients and PD patients ($N = 12$ per group). Asterisk = $p < 0.05$.

This finding is interesting, showing PD patients have an increase in putrescine concentrations in RBCs, although we cannot propose this increase as a causal effect. At this moment, it does pose the possibility that this polyamine could be used as a biomarker of the disease. Further studies should be done to confirm if there is a connection between the severity of PD disease and the concentrations of putrescine.

Red blood cells function as simple polyamine carriers since they do not have the enzymatic machinery to synthetize polyamines [2]. Ornithine is decarboxylated outside the RBCs (cell metabolism, neurons) to putrescine by ornithine decarboxylase (ODC), and then the RBCs internalize it through its polyamine transporters [22]. A significant increase in putrescine in RBC but not in plasma suggests (a) an increase in RBC transporter activity or (b) an increase in putrescine synthesis in specific organs (e.g., in the central nervous system) and once blood reaches the periphery has already been transported to the interior of the RBCs. The former possibility is hard to prove at this moment since there are no published data on monitoring the polyamine transporter activity. The latter eventuality could hold true since putrescine levels have been found elevated in cerebrospinal fluid of PD patients [24], suggesting that putrescine levels are elevated in CNS tissue.

In conclusion, these experiments show the applicability of CZE in its MEKC modality in measuring biomolecules in PD patients. In the first experiment, the findings help to determine, in part, the anomaly of the brain circuitry involved in the disease. This is already helpful since neurosurgical treatment is based on lesioning areas where there is an overactive system (pallidotomy or thalamotomy) or inhibiting the circuit through high-frequency electrical stimulation (deep-brain stimulation). Our result suggests that the overactive circuit is GABAergic in nature and that by toning it down, there is an immediate clinical improvement. This facilitates new possible strategies in pharmacological treatments of PD patients that in the present are based on modifying the dopamine neurotransmission and opens up the possibility of looking at GABA antagonists that could have a preferential effect in the thalamus.

The second experiment shows the usefulness of CZE in determining polyamines as probable biomarkers of PD. Putrescine was found to be elevated in PD patients, and future studies will try to elucidate if this elevation is specific for this affection and whether it is modified depending on the severity of the disease. Likewise, if putrescine is involved in the pathogenesis of the disease as suggested by others [1, 21], then a possible new era in pharmacological prevention will be opened by modifying enzymes activity in the polyamine metabolism. This has been shown in mice where increasing SAT1 activity (a catabolic enzyme called spermidine/spermine N1 acetyltransferase) decreased α-synuclein aggregation and ameliorated clinical symptoms [21].

References

1. Antony, T., Hoyer, W., Cherny, D., Heim, G., Jovin, T. M., and Subramaniam, V. (2003). Cellular polyamines promote the aggregation of α-synuclein, *J. Biol. Chem.*, **278**, pp. 3235–3240.

2. Assaraf, Y. G., Golenser, J., Spira, D. T., and Bachrach, U. (1984). Polyamine levels and the activity of their biosynthetic enzymes in human erythrocytes infected with the malarial parasite, *Plasmodium falciparum*, *Biochem. J.*, **222**, pp. 815–819.

3. Bellou, V., Belbasis, L., Tzoulaki, I., Evangelou, E., and Ioannidis, J. P. A. (2015). Environmental risk factors and Parkinson's disease: An umbrella review of meta-analyses, *Parkinsonism Related Disorders*, **23**, pp. 1–9.

4. Bergquist, J., Vona, M. J., Stiller, C. O., O'Connor, W. T., Falkenberg, T., and Ekman, R. (1996). Capillary electrophoresis with laser-induced fluorescence detection: A sensitive method for monitoring extracellular concentrations of amino acids in the periaqueductal grey matter, *J. Neurosci. Meth.*, **65**, pp. 33–42.

5. Calabresi, P., Picconi, B., Tozzi, A., Ghiglieri, V., and Di Filippo, M. (2014). Direct and indirect pathways of basal ganglia: A critical reappraisal, *Nature Neurosci.*, **17**, pp. 1022–1030.

6. Casey, G. (2013). Parkinson's disease: A long and difficult journey, *Nurs. New Zealand*, **19**, pp. 20–24.

7. Cipolla, B., Moulinoux, J. P., Quemener, V., Havouis, R., Martin, L. A., Guille, F., and Lobel, B. (1990). Erythrocyte polyamine levels in human prostatic carcinoma, *J. Urology*, **144**, pp. 1164–1166.

8. DeLong, M. R. and Wichmann, T. (2015). Basal ganglia circuits as targets for neuromodulation in Parkinson disease, *JAMA Neurol.*, **72**, pp. 1354–1360.

9. Dorsey, E. R., Constantinescu, R., Thompson, J. P., Biglan, K. M., Holloway, R. G., Kieburtz, K., Marshall, F. J., Ravina, B. M., Schifitto, G., Siderowf, A., and Tanner, C. M. (2007). Projected number of people with Parkinson disease in the most populous nations, 2005 through 2030, *Neurology*, **68**, pp. 384–386.

10. Dostrovsky, J. O., Hutchison, W. D., and Lozano, A. M. (2002). The globus pallidus, deep brain stimulation, and Parkinson's disease, *Neuroscientist*, **8**, pp. 284–290.

11. Fahn, S. (2015). The medical treatment of Parkinson disease from James Parkinson to George Cotzias, *Movement Dis.*, **30**, pp. 4–18.

12. Feng, Y., Jankovic, J., and Wu, Y.-C. (2015). Epigenetic mechanisms in Parkinson's disease, *J. Neurol. Sci.*, **349**, pp. 3–9.

13. Fernandez, C. O., Hoyer, W., Zweckstetter, M., Jares-Erijman, E. A., Subramaniam, V., Griesinger, C., and Jovin, T. M. (2004). NMR of

alpha-synuclein-polyamine complexes elucidates the mechanism and kinetics of induced aggregation, *EMBO J.*, **23**, pp. 2039–2046.

14. Gomes-Trolin, C., Nygren, I., Aquilonius, S.-M., and Askmark, H. (2002). Increased red blood cell polyamines in ALS and Parkinson's disease, *Exp. Neurol.*, **177**, pp. 515–520.

15. Grabenauer, M., Bernstein, S. L., Lee, J. C., Wyttenbach, T., Dupuis, N. F., Gray, H. B., and Bowers, M. T. (2008). Spermine binding to Parkinson's protein alpha-synuclein and its disease-related A30P and A53T mutants, *J. Phys. Chem. B*, **112**, pp. 11147–11154.

16. Hernandez, L., Paez, X., and Hamlin, C. (1983). Neurotransmitters extraction by local intracerebral dialysis in anesthetized rats, *Pharmacol. Biochem. Behav.*, **18**, pp. 159–162.

17. Hernandez, L., Stanley, B. G., and Hoebel, B. G. (1986). A small, removable microdialysis probe, *Life Sci.*, **39**, pp. 2629–2637.

18. Hernandez, L., Joshi, N., Murzi, E., Verdeguer, P., Mifsud, J. C., and Guzman, N. (1993). Colinear laser-induced fluorescence detector for capillary electrophoresis. Analysis of glutamic acid in brain dialysates, *J. Chromatogr. A*, **652**, pp. 399–405.

19. Hernandez, L., Tucci, S., Guzman, N., and Paez, X. (1993). In vivo monitoring of glutamate in the brain by microdialysis and capillary electrophoresis with laser-induced fluorescence detection, *J. Chromatogr. A*, **652**, pp. 393–398.

20. Hornykiewicz, O. (2006). The discovery of dopamine deficiency in the parkinsonian brain, *J. Neural Transm. Supplementum*, **70**, pp. 9–15.

21. Lewandowski, N. M., Ju, S., Verbitsky, M., Ross, B., Geddie, M. L., Rockenstein, E., and Small, S. A. (2010). Polyamine pathway contributes to the pathogenesis of Parkinson disease, *Proc. Nat. Acad. Sci.*, **107**, pp. 16970–16975.

22. Miller-Fleming, L., Olin-Sandoval, V., Campbell, K., and Ralser, M. (2015). Remaining mysteries of molecular biology: The role of polyamines in the cell, *J. Mol. Biol.*, **427**, pp. 3389–3406.

23. Páez, X., Rada, P., and Hernández, L. (2000). Neutral amino acids monitoring in phenylketonuric plasma microdialysates using micellar electrokinetic chromatography and laser-induced fluorescence detection, *J. Chromatogr. B*, **739**, pp. 247–254.

24. Paik, M.-J., Ahn, Y.-H., Lee, P. H., Kang, H., Park, C. B., Choi, S., and Lee, G. (2010). Polyamine patterns in the cerebrospinal fluid of patients with Parkinson's disease and multiple system atrophy, *Clin. Chim. Acta*, **411**, pp. 1532–1535.

25. Pringsheim, T., Jette, N., Frolkis, A., and Steeves, T. D. L. (2014). The prevalence of Parkinson's disease: A systematic review and meta-analysis, *Mov. Dis.*, **29**, pp. 1583–1590.

26. Rada, P., Tucci, S., Murzi, E., and Hernández, L. (1997). Extracellular glutamate increases in the lateral hypothalamus and decreases in the nucleus accumbens during feeding, *Brain Res.*, **768**, pp. 338–340.

27. Rada, P., Tucci, S., Teneud, L., Paez, X., Perez, J., Alba, G., García, Y., Sacchettoni, S., del Corral, J., and Hernandez, L. (1999). Monitoring gamma-aminobutyric acid in human brain and plasma microdialysates using micellar electrokinetic chromatography and laser-induced fluorescence detection, *J. Chromatogr. B*, **735**, pp. 1–10.

28. Rada, P., Moreno, S. A., Tucci, S., Gonzalez, L. E., Harrison, T., Chau, D. T., Hoebel, B. G., and Hernandez, L. (2003). Glutamate release in the nucleus accumbens is involved in behavioral depression during the PORSOLT swim test, *Neurosci.*, **119**, pp. 557–565.

29. Rizzo, G., Martino, D., Arcuti, S., Copetti, M., Fontana, A., and Logroscino, G. (2015). Accuracy of clinical diagnosis of Parkinson's disease: A systematic review and Bayesian meta-analysis, *Mov. Disor.*, **30**, pp. S441.

30. Rossell, S., Gonzalez, L. E., and Hernández, L. (2003). One-second time resolution brain microdialysis in fully awake rats. Protocol for the collection, separation and sorting of nanoliter dialysate volumes, *J. Chromatogr. B*, **784**, pp. 385–393.

31. Saeki, Y., Uehara, N., and Shirakawa, S. (1978). Sensitive fluorimetric method for the determination of putrescine, spermidine and spermine by high-performance liquid chromatography and its application to human blood, *J. Chromatogr.*, **145**, pp. 221–229.

32. Sepulveda, M. J., Hernandez, L., Rada, P., Tucci, S., and Contreras, E. (1998). Effect of precipitated withdrawal on extracellular glutamate and aspartate in the nucleus accumbens of chronically morphine-treated rats: an in vivo microdialysis study, *Pharmacol. Biochem. Behav.*, **60**, pp. 255–262.

33. Shuen, J. A., Chen, M., Gloss, B., and Calakos, N. (2008). Drd1a-tdTomato BAC transgenic mice for simultaneous visualization of medium spiny neurons in the direct and indirect pathways of the basal ganglia, *J. Neurosci.*, **28**, pp. 2681–2685.

34. Stefani, A., Fedele, E., Vitek, J., Pierantozzi, M., Galati, S., Marzetti, F., and Stanzione, P. (2011). The clinical efficacy of L-DOPA and STN-DBS share a common marker: Reduced GABA content in the motor thalamus, *Cell Death Disease*, **2**, pp. e154.

35. Terabe, S. (1992). Selectivity manipulation in micellar electrokinetic chromatography, *J. Pharm. Biomed. Anal.*, **10**, pp. 705–715.

36. Tucci, S., Rada, P., Sepúlveda, M. J., and Hernandez, L. (1997). Glutamate measured by 6-s resolution brain microdialysis: Capillary electrophoretic and laser-induced fluorescence detection application, *J. Chromatogr. B*, **694**, pp. 343–349.

37. Ungerstedt, U. (1991). Microdialysis: Principles and applications for studies in animals and man, *J. Intern. Med.*, **230**, pp. 365–373.

38. Volta, M., Milnerwood, A. J., and Farrer, M. J. (2015). Insights from late-onset familial Parkinsonism on the pathogenesis of idiopathic Parkinson's disease, *Lancet Neurol.*, **14**, pp. 1054–1064.

39. Williams, K. (1997). Modulation and block of ion channels: A new biology of polyamines, *Cellular Signalling*, **9**, pp. 1–13.

40. Zetterström, T., Sharp, T., Marsden, C. A, and Ungerstedt, U. (1983). In vivo measurement of dopamine and its metabolites by intracerebral dialysis: Changes after d-amphetamine, *J. Neurochem.*, **41**, pp. 1769–1773.

Chapter 8

Electrophoretically Mediated Microanalysis for Evaluation of Enantioselective Drug Metabolism

Yolanda Martín-Biosca,[a] Laura Escuder-Gilabert,[a] Salvador Sagrado,[a,b] and María José Medina-Hernández[a]

[a]*Department of Analytical Chemistry, University of Valencia C/Vicente A. Estellés s/n E-46100, Burjassot, Valencia, Spain*
[b]*Instituto Interuniversitario de Investigación de Reconocimiento Molecular y Desarrollo Tecnológico (IDM), Universitat Politècnica de València, Universitat de València, C/ Vicente A. Estellés s/n E-46100, Burjassot, Valencia, Spain*
yolanda.martin@uv.es

8.1 Introduction

The study of xenobiotic metabolism is a key feature for the pharmaceutical and chemical industries since it is a determining factor both in the preclinical development of new drugs and in the toxicity evaluation of chemicals. Metabolic reactions usually make xenobiotics more polar and easier to be excreted and can be classified into two groups attending to the nature of the molecule modification. Phase I includes some reactions that generally lead

---

Capillary Electrophoresis: Trends and Developments in Pharmaceutical Research
Edited by Suvardhan Kanchi, Salvador Sagrado, Myalowenkosi Sabela, and Krishna Bisetty
Copyright © 2017 Pan Stanford Publishing Pte. Ltd.
ISBN 978-981-4774-12-3 (Hardcover), 978-1-315-22538-8 (eBook)
www.panstanford.com

to the introduction or uncovering of key functional groups (e.g., OH, COOH, NH_2, and SH) increasing the molecule polarity, which may facilitate removal from the body. The main enzymatic system responsible for phase I reactions is the cytochrome P450 (CYP), despite other enzymes can also be involved. In phase II metabolism, there is a conjugation between the xenobiotic and some endogenous molecules or groups. It increases the molecule size, favoring its excretion and usually its inactivation. Not all compounds suffer phase I and II processes. Some compounds are directly eliminated after a phase I metabolic reaction or after a phase II conjugation [1, 2].

CYP450 is the main metabolic enzymatic system of mammals, being responsible for the metabolism of a large number of drugs, steroids, carcinogens, and other chemicals. It has been estimated that more than 90% of drug oxidations in humans are mediated by CYPs [3]. There are different isoforms of CYP450 that catalyze the oxidation of different substrates [4]. Human cytochrome P3A4, one of the four most important CYP isozymes in the human liver, is involved in the metabolism of a large number of compounds and its reactions are categorized as oxidation, mainly hydroxylation, *N*-demethylation, and *N*-dealkylation of its substrates [5]. The oxidation of xenobiotic substances by CYP450 is a significant focus of scientists in the areas of toxicology, drug metabolism, and pharmacology. The effects of these oxidations can be manifested in poor drug bioavailability and various acute and chronic toxicities, including adverse drug interactions, cancer susceptibility, and birth defects [6].

Enzymatic metabolism of chiral xenobiotics can show a high degree of stereoselectivity as a consequence of the sterereoselective interaction with optically active macromolecules involved in the process [7]. Stereoselectivity in the enzyme–xenobiotic interaction has been extensively reported and is the subject of a number of reviews [7–9]. There has been reported enantioselective metabolism for the anticoagulant warfarin [10, 11], the β-blocker propranolol [12], the calcium channel antagonist verapamil [13, 14], and the anti-arrhythmic propafenone [15]. Some proton pump inhibitors (omeprazole, lansoprazole, pantoprazole, or tenatoprazole) are also enantioselectively metabolized by different isoforms of CYP [16].

Metabolism studies can be performed in vivo, by administering the drug to the living system and monitoring the free concentration of xenobiotic and its metabolites in body fluids and tissues, or by in vitro assays. In vitro methods are based on the use of liver subcellular fractions as liver microsomes, isolated hepatocytes, liver slices, or individual forms of enzymes, predominantly isoforms of CYP450. In these studies, the components of the enzymatic system are mixed, and after an incubation period, one or more components of the enzymatic reaction are monitored mainly using liquid chromatography. More recently, capillary electrophoresis (CE) has become very useful for enzymatic assays since it is fully automated, requires small sample volumes, provides high peak efficiencies, and is environmentally friendly [17, 18].

8.2 Enzymatic Reactions in CE: Electrophoretically Mediated Microanalysis

For enzymatic studies, CE has been classically employed only as separation system after offline reactions. However, the use of CE not only for the separation but also for performing the reaction inside the capillary allows to decrease notably the volume of reagents employed (from microliters to only few nanoliters in each assay). This is interesting in the study of enzymatic reactions since enzymes and substrates use to be expensive, and to increase the automation of the process since the injection, mixing, reaction, separation, and detection can be done in a unique step [19]. This recent in-capillary reaction procedure is known as electrophoretically mediated microanalysis (EMMA) in the literature.

In the EMMA procedure, an enzyme solution is injected into the capillary and it is allowed to mix with the substrate solution and react before the separation. The most frequent mechanism for mixing the reagents is to apply an electric field to make the substrate and enzyme zones interpenetrate due to their differences in electrophoretic mobility. This is known as EMMA plug–plug mode (Fig. 8.1a). One of the problems of this methodology is the high variability in peak areas of reagents and reaction products due to slight variations in the in-capillary contact period of the enzyme and substrate plugs, which affect the extent of the reaction [20].

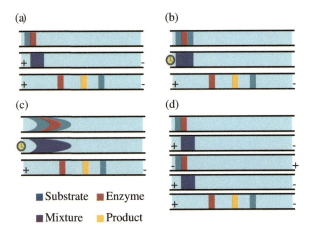

Figure 8.1 Scheme of the plugs injection and mixing in the different EMMA modes: (a) plug–plug, (b) at-inlet, (c) TDLFP, and (d) RPS.

As an alternative to the classical plug–plug mode, the at-inlet EMMA procedure has been proposed. In this modality, the mixing of the enzyme and substrate plugs takes place by simple diffusion, and they react for a given time in the inlet part of the capillary (Fig. 8.1b). This is especially interesting for enzymes that lose activity when exposed to an electric field. Classically, it has been described that longitudinal diffusion is responsible for the mixing of plugs in at-inlet EMMA. However, this is a slow process and cannot be efficient if more than two plugs are involved in the reaction.

Okhonin et al. have described an alternative mode of mixing, which is called transverse diffusion of laminar flow profiles (TDLFP) [21]. In this mode, the plugs are introduced consecutively in the capillary by pressure and, due to the laminar nature of flow inside the capillary, the nondiffused plugs have parabolic profiles with predominantly longitudinal interfaces between them, and transverse diffusion is responsible for the mixing (Fig. 8.1c). They also developed mathematical models to estimate the mixing process and to show the concentration profile of each reagent in the capillary.

In practice, when the plugs in the at-inlet EMMA are small (standard injection plugs of 0.5 psi during 3 or 5 s), there is no difference between the experimental injection sequences shown in the literature for at-inlet EMMA and TDLFP. This fact suggests that transverse diffusion (which is more efficient than the longitudinal one due to the higher contact surface between plugs) can also be responsible for the mixing in the at-inlet EMMA mode.

Other alternative to the classical plug–plug mode, which is called rapid polarity switching (RPS) (Fig. 8.1d), consists in applying a series of positive and negative potentials to the capillary to have some shaking effect that increases the plug mixing [22].

In most of the EMMA modes, the background electrolyte is not the same as the incubation medium in which substrates, enzymes, and cofactors are dissolved, and its pH or components can affect the enzymatic activity. In order to preserve the correct enzymatic activity and provide an adequate medium for the reaction turnover, two plugs of incubation buffer can be injected before and after the reaction plugs. This practice has been called "partial filling technique" (PFT) in the literature [19].

EMMA, in its different modalities, has been employed for some enzymatic reactions to study enzyme kinetics, inhibition of enzymes, and to determine substrates and enzymes in different biological samples [19]. Several applications of EMMA to the evaluation of nonchiral drug metabolism with different CYP450 isoforms or flavin-containing monooxygenases (FMOs) have been reported in the literature [23–31]: dextromethorphan metabolism by CYP3A4 [23]; phenacetin, dextromethorphan, tramadol, and midazolam by different CYP450 isoforms [24]; testosterone and nifedipine by CYP3A4 [25]; diclofenac by CYP2C9 [26, 27]; clozapine metabolism by FMO3 [28]; and also inhibition studies with CYP450 [29].

8.3 Evaluation of Enantioselective Metabolism by EMMA

To evaluate the enantioselective metabolism of racemic compounds, it is necessary to integrate methodologies that allow, first, the interaction of racemic xenobiotics with the enzymatic system and later the determination of the enantiomers of intact xenobiotic or its metabolites in the reaction mixture. There are a few applications that use the EMMA approach for the characterization of the enantioselective metabolism of compounds [32–36]. These methods have been summarized in Table 8.1, showing the most important information. In all these metabolic studies, as well as optimization

Table 8.1 Applications of EMMA in the evaluation of enantioselective metabolism

Substrate (Enzyme)	Mixing mode	Separation electrolyte		Kinetic parameters				Ref
				$K_m\ (K')^a$, μM	V_{max}	$CL_i\ (CL_{max})^e$	n	
Cimetidine (FMO1) (FMO3)	Plug–plug	50 mM phosphate buffer, pH 2.5, 30 mM SBE-β-CD	(+)-CIM (−)-CIM (+)-CIM	4310 4560 4680				[32]
Ketamine (CYP3A4)	Plug–plug	50 mM Tris/phosphoric acid, pH 2.5, 3 % (w/v) HS-γ-CD		Michaelis–Menten model				[33]
			S-KET	122.3	26.67[b]	0.218[f]		
			R-KET	107.5	20.08[b]	0.187[f]		
				Hill model				
			S-KET	245.2[a]	34.67[b]	0.291[e,f]	0.7616	
			R-KET	238.6[a]	26.87[b]	0.256[e,f]	0.7203	
Ketamine (CYP3A4)	TDLFP	50 mM Tris/phosphoric acid, pH 2.5, 3 % (w/v) HS-γ-CD		Michaelis–Menten model				[34]
			S-KET (Rac)	110 ± 15	11.8 ± 0.6^c	0.11 ± 0.02^g		
			S-KET	97 ± 10	17.7 ± 0.6^c	0.18 ± 0.02^g		
			R-KET (Rac)	84 ± 10	8.3 ± 0.3^c	0.10 ± 0.01^g		
			R-KET	68 ± 7	8.5 ± 0.2^c	0.12 ± 0.01^g		

Substrate (Enzyme)	Mixing mode	Separation electrolyte		Kinetic parameters				Ref
				K_m $(K')^a$, µM	V_{max}	CL_i $(CL_{max})^e$	n	
				Hill model				
			S-KET (Rac)	80 ± 8^a	10.1 ± 0.5^c	$0.08 \pm 0.01^{e,g}$	1.37 ± 0.14	
			S-KET	74 ± 6^a	15.7 ± 0.5^c	$0.12 \pm 0.02^{e,g}$	1.31 ± 0.10	
			R-KET (Rac)	66 ± 7^a	7.5 ± 0.3^c	$0.07 \pm 0.01^{e,g}$	1.29 ± 0.13	
			R-KET	63 ± 8^a	8.2 ± 0.4^c	$0.10 \pm 0.01^{e,g}$	1.08 ± 0.11	
Verapamil (CYP3A4)	At-inlet	50 mM sodium phosphate, pH 8.8, 2.5 % (w/v) HS-β-CD	S-VER	51 ± 9	22 ± 2^b 2800 ± 200^d	55 ± 10^h		[35]
			R-VER	47 ± 9	21 ± 2^b 2600 ± 200^d	55 ± 11^h		
Fluoxetine (CYP2D6)	At-inlet	50 mM sodium phosphate, pH 8.8, 1.25 % (w/v) HS-β-CD	S-FLX (Rac)	30 ± 3	28.6 ± 1.2^b	121 ± 13^h		[36]
			R-FLX (Rac)	39 ± 5	34 ± 2^b	111 ± 16^h		
			R-FLX	38 ± 9	33 ± 4^b			

TDLFP, transverse diffusion of laminar flow profiles; SBE-β-CD, sulfobutylether-β-CD; HS-γ-CD, highly sulfated γ-CD; HS-β-CD, highly sulfated β-CD; K_m, Michaelis–Menten constant; V_{max}, maximum formation rate; K', constant of the autoactivation model; CL_i, intrinsic clearance; CL_{max}, maximal clearance; n, Hill coefficient.

[a] values of K' obtained from Hill model. K' is equivalent to K_m when $n = 1$.

V_{max} expressed in: [b] pmol min^{-1} (pmol CYP)$^{-1}$; [c] nmol min^{-1} (mmol CYP)$^{-1}$; [d] pmol min^{-1} (mg CYP)$^{-1}$; [e] values of CL_{max} obtained from Hill model.

CL_i, CL_{max} expressed in: [f] µL min^{-1} (pmol CYP)$^{-1}$; [g] µL min^{-1} (nmol CYP)$^{-1}$; [h] mL min^{-1} (mg CYP)$^{-1}$; Rac means than the enantiomer has been separated from the racemate.

of general parameters of the EMMA method (i.e., reactant concentrations, plug lengths, and injection sequence), due to the complex mixtures of reagents and products, a careful optimization of enantioseparation conditions (i.e., nature and concentration of chiral selector, pH of background electrolyte, temperature, or applied voltage) is necessary.

As can be seen in Table 8.1, in all cases, the addition of an anionic cyclodextrin (CD) to the BGE allowed the separation of the enantiomers of the xenobiotic and/or its metabolite. In four of the five papers reported, a highly sulfated CD was used as chiral selector. Its high density of negative charges makes it an excellent chiral selector for neutral and positively charged compounds. Verapamile/norverapamile [35] and fluoxetine/norfluoxetine [36] enantiomers were separated using highly sulfated β-CD (HS-β-CD), whereas highly sulfated γ-CD (HS-γ-CD) was used to separate ketamine and norketamine enantiomers [33, 34]. Sulfobutylether-β-CD (SBE-β-CD) was chosen to separate the S-oxide metabolites of cimetidine [32]. The partial filling mode was applied in all cases. Mixing of enzyme and substrate/cofactor plugs was achieved either electrophoretically (plug–plug mode) [32, 33] or by longitudinal diffusion (at-inlet mode) [35, 36] and TDLFP [34].

Evaluation of metabolic reactions, which follow the mathematical approaches employed in enzymatic studies, has been conducted by estimation of kinetic parameters. In all the applications reported in Table 8.1. the simplest Michaelis–Menten model (Eq. 8.1) has been used with a satisfactory match with the experimental data:

$$V_0 = \frac{V_{max}[S]}{K_m + [S]} \tag{8.1}$$

The terms V_{max} and K_m are the kinetic parameters of the enzymatic reaction. V_{max} represents the maximum reaction velocity (at infinite substrate concentration), and K_m, the Michaelis–Menten constant, represents the concentration of substrate at which the velocity is half the V_{max}. [S] is the substrate concentration. Metabolic enzymes are usually compared in terms of their K_m for different substrates transformation, which is a measurement of the enzyme–substrate affinity. The estimation of kinetic parameters was carried out by varying the substrate concentration (chiral xenobiotic) keeping enzyme concentration and incubation time constant.

The quantification of the reaction turnover (V_0) could be done considering the metabolite formation rate [32–34] or either the consumption of substrate [35, 36], obtaining in this case information about the overall metabolism. In some applications [33, 34], the Hill equation (Eq. 8.2) has been also used for comparative purposes:

$$V_0 = \frac{V_{max}[S]^n}{K'^n + [S]^n} \tag{8.2}$$

where K' is a constant of the autoactivation model that is equivalent to K_m when $n = 1$, and n is the Hill coefficient. From kinetic parameters, in vitro intrinsic clearance ($CL_i = V_{max}/K_m$) and maximal clearance due to activation (CL_{max}) (Eq. 8.3) can be estimated. The intrinsic clearance, defined as the amount of plasma from which a xenobiotic is removed in a certain time, reflects the capacity of the body (or of a single enzyme) to eliminate a xenobiotic.

$$CL_{max} = \frac{V_{max}}{K'} x \frac{|n-1|}{n} x (|n-1|)^{1/2} \tag{8.3}$$

Principally, nonlinear regression was used to fit experimental data to kinetic models (Michaelis–Menten or Hill equations) [33–36] and estimate their corresponding parameters. K_m and V_{max} can also be calculated from the Lineweaver–Burk plots derived from the Michaelis–Menten plot [32].

8.3.1 Study of Stereoselectivity of Flavin-Containing Monooxygenase Isoforms using Cimetine as Substrate

The first application of EMMA to the evaluation of enantioselective metabolism has been reported by Hai et al. [32], who developed an in-line screening CE method to determine the stereoselectivity of flavin-containing monooxygenase (FMO) isoforms using cimetidine (CIM) as a substrate. CIM, which is mainly metabolized to CIM S-oxide (CSO) (Fig. 8.2) by CYP450 and FMO, is a histamine H2-receptor antagonist used therapeutically in the treatment of peptic ulcer disease and gastric hypersecretory syndromes [37]. In humans, the stereoselectivity of CSO metabolite formation is clinically relevant because there is evidence that the S-oxide is responsible for mental status deterioration observed after chronic administration of CIM

286 | *Electrophoretically Mediated Microanalysis for Evaluation of Enantioselective Drug*

to patients with renal or hepatic dysfunction. Moreover, the CSO metabolite formation is an indicator for the presence of FMO forms in the liver of humans and animals [38].

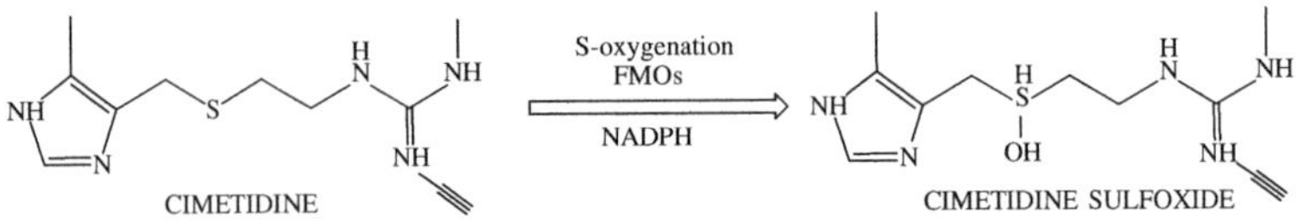

Figure 8.2 The FMO-catalyzed S-oxigenation of CIM to CSO.

The authors [32] investigated the S-oxygenation of CIM using achiral chemical oxidants and (human supersomes) enzymatic metabolism procedures. First, the separation conditions for the substrate CIM and its metabolite CSO were optimized. A screening using several neutral and anionic CDs at low pH (phosphate buffer) was performed. Among all the conditions assayed, only the anionic SBE-β-CD allows the chiral separation of the CSO enantiomers. After optimization of other experimental parameters, the best separation conditions were obtained with a 50 mM phosphate buffer with 30 mM SBE-β-CD (pH 2.5) as a BGE, by application of a 17 kV voltage at normal polarity at a temperature of 25°C. The electrophoretic migration order of CSO was confirmed to be (+) before (−) through the use of single enantiomers obtained by preparative chromatography.

For the EMMA method, the injection and electrophoretic mixing procedure and reaction conditions (i.e., temperature, incubation time) were also optimized. Finally, the in-capillary enzymatic reaction was performed as follows: A plug of reaction buffer (100 mM phosphate, pH 8.3) was first introduced into the capillary at 0.5 psi for 10 s, followed by a plug of CIM and a plug of enzyme dissolved in NADPH solution, and finally a plug of reaction buffer. These last three plugs were hydrodynamically injected at 0.5 psi for 6 s. The enzymatic reaction was initiated by application of a 4 kV voltage for 0.3 min. Later the voltage was turned off for 20 min (in-line incubation), and the capillary inlet was submerged into a full vial of reaction buffer placed on the heated sample tray at 37°C. After incubation, the separation of all the components was performed using the separation conditions previously established.

Using the EMMA method proposed to study the stereospecifity of FMO isoenzymes (FMO1, FMO3, and FMO5), the authors found that the formation of the new chiral center on the CIM sulfur was stereoselective. FMO1 produces more (−)-CSO-enantiomer, while FMO3 generates mainly (+)-CSO-enantiomer. On the other hand, FMO5 shows no activity, probably due to the low substrate affinity of FMO5. Figure 8.3 shows the overlaid electropherograms obtained after in-line screening of FMO metabolism. The kinetic parameters for FMO1 and FMO3 were determined from eight different CIM concentrations ranging from 0.5 to 12 mM and keeping constant the enzyme concentration of 308 nM FMO1 or 340 nM FMO3. Kinetic constants were calculated by adjusting experimental data to the Michaelis–Menten model and obtaining the corresponding Lineweaver–Burk plots (see Table 8.1). The K_m value for human FMO1 using CIM as a substrate, 4.31 and 4.56 mM for formation of (+) and (−)-CSO-enantiomers, respectively, was determined for the first time. On the other hand, the K_m value for human FMO3 was in good agreement with the literature value obtained for the same enzyme calculated from rates of NADPH oxidation ($K_m \sim 4$ mM) [39].

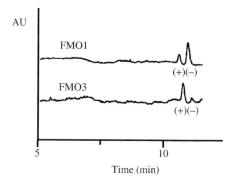

Figure 8.3 The overlaid electropherograms of FMO1 and FMO3 obtained after in-line reactions. See the text for experimental conditions. Reprinted with permission from Ref. [32], Copyright 2009, John Wiley and Sons.

8.3.2 Characterization of the Enantioselective CYP3A4 Catalyzed *N*-demethylation of Ketamine

Ketamine is a chiral phencyclidine derivative, which is used as an anesthetic drug and postoperative analgesic. For (*S*)-ketamine, affinity for the *N*-methyl-D-aspartate receptor was found to be four times higher than for (R)-ketamine and its anesthetic potency two to three times higher than that of the racemic mixture. In vitro and in vivo studies in humans and other animal species have shown that ketamine is metabolized mainly by the hepatic CYP450 enzymes. The stereoselectivity of the CYP3A4-mediated N-demethylation of ketamine was extensively investigated by Thormann et al. using an offline kinetic assay [40] and two different EMMA methodologies [33, 34].

The first EMMA assay proposed by the authors to characterize the enzymatic reaction of ketamine was based on the mixing of reaction plugs by voltage application [33] and the following enantioseparation with HS-γ-CD as chiral selector. Ketamine was incubated in a 50 μm ID bare fused-silica capillary together with human CYP3A4 Supersomes using a 100 mM phosphate buffer (pH 7.4) at 37°C. A plug containing racemic ketamine and the NADPH-regenerating system including all required cofactors for the enzymatic reaction was injected, followed by a plug of the metabolizing enzyme CYP3A4 (500 nM). These two plugs were bracketed by plugs of incubation buffer to ensure proper conditions for the enzymatic reaction. The rest of the capillary was filled with a pH 2.5 running buffer containing 50 mM Tris, phosphoric acid, and 2% w/v of HS-γ-CD. Mixing of reaction plugs was enhanced via application of −10 kV for 10 s. After an incubation of 8 min at 37°C without power application, the capillary was cooled to 25°C within 3 min followed by application of −10 kV for the separation and detection of the formed enantiomers of norketamine. The anionic chiral selector penetrates the reaction mixture interacting with the positively charged enantiomers of ketamine and norketamine and forming negatively charged complex that migrate toward the anode. The EMMA assay was used to determine the kinetic parameters of the CYP3A4-mediated N-demethylation of ketamine (see Table 8.1). For this purpose, experimental data were fitted to the Michaelis–Menten and Hill models using nonlinear regression analysis. The coefficients of determination (R^2) were similar for both models; nevertheless, the F-test confirms that the enzymatic reaction can best be described with the Michaelis–Menten model. As observed in the offline assay [40], kinetic parameters V_{max} and K_m obtained for the formation of (S)-norketamine were higher compared to (R)-norketamine, suggesting a stereoselective N-demethylation of ketamine via CYP3A4. Estimated V_{max} values were comparable to those obtained in the offline study, whereas the determined values for K_m were about twofold larger. Interestingly, the ratio of K_m for (S)-norketamine formation to K_m for (R)-norketamine formation was the same in both the online and the offline studies (1.138 and 1.147, respectively).

Recently, the original EMMA procedure developed for the stereoselectivity evaluation of the CYP3A4-mediated N-demethylation of ketamine [33] was improved using a diffusion-based technique (TDLFP), which allows mixing multiple short plugs (i.e., substrate, enzyme, reaction buffer) inside the nanoliter-scale capillary reactor and without the need of additional optimization of mixing conditions [34]. The authors proposed the new approach as an economical capillary electrophoretic method suitable for online studies of the enantioselective drug metabolism mediated by cytochrome P450 enzymes.

The diffusion-based mixing procedure consists of the alternate hydrodynamic introduction of four plugs of the solution comprising substrate and NADPH and three plugs of the enzyme solution. Due to the friction close to the capillary inner wall, each injected plug has a parabolic profile and penetrates into the proceeding plug in the injection sequence, creating longitudinal interfaces between them as is schematically depicted in Fig. 8.4a. Since even large molecules, such as an enzyme, are able to diffuse across the short distance of a capillary inner diameter in a few seconds, a reaction mixture with close to uniform concentration distributions is rapidly formed by transverse diffusion. In the final setup, each plug of ketamine and NADPH solution was injected for 3 s into the capillary at a pressure of 0.5 psi and each plug of CYP3A4 solution was introduced by the application of a negative pressure of −0.5 psi for 4 s. Plug volumes of 3.58 and 4.77 nL were estimated, respectively. The overall length of the entire seven plug assembly was calculated to be approximately 14.6 mm (about 29 nL or 2.3% of total capillary). The injection of shorter plugs suffered from low repeatability, whereas the introduction of larger plugs caused an overloading of the capillary and loss of separation. The conducted sets of experiments revealed significantly lower repeatability of the injection procedure using only positive or negative pressure compared to the selected final approach that combines both ways. The plug assembly used was assumed to provide a twofold dilution of the plugs.

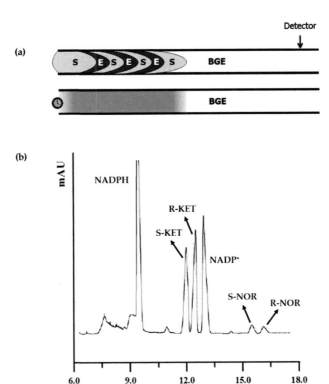

Figure 8.4 (a) Principle of TDLFP. The injection procedure consists of alternate introduction of three plugs of CYP3A4 (E) and four plugs of a solution containing the substrate and NADPH (S) into the capillary previously filled with BGE. Reaction mixture is rapidly formed by transverse diffusion. (b) Electropherogram obtained after the in-capillary reaction (10 min incubation at 37°C) with 200 nM CYP3A4, 400 μM racemic ketamine, and 1 mM NADPH. Separation conditions: 50 μm ID fused-silica capillary (64 and 54 cm total and effective length, respectively); 37°C; 50 mM Tris-phosphate buffer (pH 2.5) containing 3% w/v HS-γ-CD as BGE; −20 kV and a pressure of +0.2 psi. Reprinted with permission from Ref. [34], Copyright 2015, John Wiley and Sons.

Kinetic parameters were estimated from experimental data obtained at an incubation time of 10 min with various amounts of racemic ketamine or its single enantiomers (between 12.5 and 500 μM per enantiomer) and keeping constant a 200 nM concentration of CYP3A4 in the reaction mixture. Data obtained for norketamine formation rate were curve-fitted using nonlinear regression analysis according to the Michaelis–Menten model and the Hill equation (see

Table 8.1). The Hill equation was found to provide better fits for the incubations of both racemic ketamine and its single enantiomers. Estimated K' values were in a good agreement with literature data [34, 40–43], while V_{max} and calculated (CL_i) and (CL_{max}) values were significantly lower compared to literature data. V_{max} is a relative parameter that may differ depending on conditions of the enzyme reaction incubation. In agreement with literature results, the formation rate of (S)-norketamine is significantly higher compared to that of (R)-norketamine, so the N-demethylation of ketamine mediated by CYP3A4 was found to proceed stereoselectively.

It is known that coadministration of a drug that acts as a CYP inhibitor causes changes in the metabolic rates of other compounds, so the study of inhibitory effects of candidate compounds toward CYP3A4 is a routine task in the early stages of the development of a new drug. In order to verify the validity of the methodology proposed in this kind of studies, the new EMMA approach was applied to the determination of the inhibition characteristics in the presence of ketoconazole and dexmedetomidine, two potent inhibitors of CYP3A4 activity. Following the same procedure as in the kinetic study, three series of analyses were carried out with reaction mixtures containing concentrations of ketamine enantiomers equal to 0.5-, one-, and twofold of the previously determined K_m value and different concentrations of inhibitor (0–2.5 µM). The formation rates of (S)- and (R)-norketamine were plotted against the inhibitor concentration and fitted by nonlinear regression to models of competitive, uncompetitive, noncompetitive, and mixed-mode inhibition. Figure 8.5 shows the plots for ketoconazole. The competitive inhibition model was found to be superior for all data sets. The half-maximal inhibitory concentration (IC_{50}) causing the diminishing of the reaction rate by 50% and apparent inhibition constant (K_i') describing the affinity between inhibitor and the enzyme were determined. The calculated IC_{50} and K_i' values were found to be similar for ketoconazole and dexmedetomidin and of the same order of magnitude as data reported in the literature. For both inhibitors, estimated IC_{50} and K_i' values were found to be significantly higher for incubation of racemic ketamine compared to those obtained with incubations of single ketamine enantiomers. These results are in agreement with previous observations that suggested the competition of the enantiomers for the active site of the enzyme.

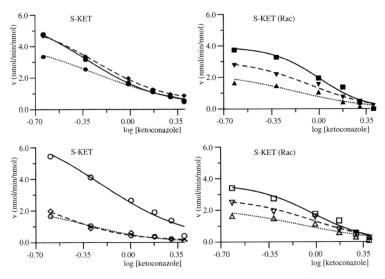

Figure 8.5 Formation rates of (S)-norketamine (S-KET) and (R)-norketamine (R-KET) versus ketoconazole concentration fitted to on-site competition model. Data obtained after incubation with single enantiomers (left part) and with racemic ketamine (right part). Substrate concentration: $0.5 \times K_m$ (dotted lines); K_m (dashed lines); $2 \times K_m$ (solid lines). Reprinted with permission from Ref. [34], Copyright 2015, John Wiley and Sons.

8.3.3 Evaluation of Enantioselective Metabolism of Verapamil and Fluoxetine by the at-Inlet EMMA Mode

Asensi-Bernardi et al. developed two applications of the EMMA methodology to the evaluation of the enantioselective metabolism based on the at-inlet mode [35, 36]. In the first of the papers reported, the enantioselective metabolism of verapamil by CYP3A4 was evaluated. Verapamil, a Ca^{2+}-channel blocker with anti-arrhythmic activity, is a model compound for several metabolic studies. EMMA reaction conditions and the separation and verapamil and norverapamil enantiomers by means of HS-β-CD and the partial filling technique were optimized.

The authors assayed two experimental injection modes: the plug–plug and the at-inlet EMMA modes. Best results were obtained using the at-inlet mode, so a "sandwich" injection with a substrate plug between two enzyme plugs was proposed. Two plugs of incubation buffer were added before and after the "sandwich" plugs in order to give to the enzyme its adequate reaction medium. The scheme of the EMMA configuration used in these studies is shown in Fig. 8.6.

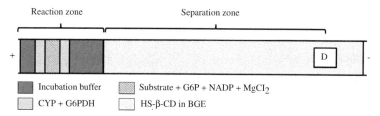

Figure 8.6 Scheme of the reaction and separation zones in the proposed EMMA assay for the evaluation of the enantioselective metabolism of verapil. Reprinted from Ref. [35], Copyright 2013, with permission from Elsevier.

Two different zones can be distinguished in the capillary: the separation zone filled with an HS-β-CD solution in the BGE and the reaction zone, at the inlet part, where all the plugs of the EMMA sequence are injected and left to mix and react during a fixed incubation time. In these conditions, the mixture of reagents is due to both longitudinal and transverse diffusions. After reaction, a voltage is applied for the separation of substrates, products, and enzymes.

The NADPH-regenerating system employed in this work comprises two solutions (A and B) that allow the in situ NADPH formation. Solution A contains G6P, NADP$^+$, and MgCl$_2$ and solution B contains G6PDH. When these solutions are put into contact, the G6PDH oxidizes G6P and forms NADPH, which can act as cofactor for the CYP3A4 reaction. Two configurations were tested for the inclusion of these solutions in the CYP and substrate vials. In the first one, both solutions A and B were included in the substrate vial, to form NADPH before the injection in the CE system. This strategy allows the immediate starting of the reaction due to the fact that the NADPH formed is present when the CYP and verapamil plugs contact in the capillary, but it presents the disadvantage of a possible degradation of NADPH. The other configuration studied consisted of the in-line formation of NADPH by the inclusion of solution B in the CYP vial and solution A in the verapamil vial, which allows always the presence of native NADPH but can retard the beginning of the enzymatic CYP reaction. Both configurations lead to the same metabolite (norverapamil) formation in short incubation times, so the in-line formation of NADPH was selected in order to preserve NADPH from degradation.

With the aim of having a good reaction turnover in the shortest possible incubation time, the in-capillary incubation time and the substrate injection have been optimized. With this purpose, 200 µM verapamil solutions were incubated, changing the incubation times or the substrate plugs. These experiences were carried out in achiral mode, without chiral selector in the separation zone, using separation conditions taken from literature [44]: 50 mM sodium phosphate at pH 8.8 as BGE, capillary temperature 25°C, and separation voltage 20 kV.

The optimum incubation time depends on the enzyme and the concrete conditions of the assay, and it may be taken into account the linear range of each enzyme, since at longer incubation times, they lose activity. Here, incubation times between 1 and 30 min were tested, based on the specifications of the commercial CYP and the previously published papers concerning this topic. The formation of the major product, norverapamil, was selected as response variable. From 1 to 5 min, it increased considerably, but from 5 to 30 min, there was only a slight increase in the amount of norverapamil formed. Also at longer incubation times, there was an important peak broadening due to the diffusion, so 5 min was selected as the best incubation time for further studies.

The length of the substrate plug may be adjusted to maximize the reaction turnover but taking into account that very large plugs will not allow a complete mixing of the substrate with the enzyme. Here, three substrate plugs of 3, 5, and 7 s were tested (with a fixed pressure of 0.5 psi), and it was seen that the area of norverapamil formed was larger with the 7 s injection, so this injection plug was selected for the substrate. Once the EMMA conditions were fixed, the EMMA assay was coupled to the enantioseparation of verapamil and norverapamil in order to measure the enantioselective metabolism of verapamil by CYP3A4. Using the same BGE and conditions, HS-β-CD was tested as chiral selector due to its widely proved good enantioseparation abilities, using the partial filling technique. Solutions of 0.25–2.5% (m/v) HS-β-CD in the BGE were tested. With 0.25% CD, no separation of verapamil and norverapamil enantiomers was achieved. Using 2% HS-β-CD, each pair of enantiomers was separated, but (S)-norverapamil and (R)-verapamil overlapped. A 2.5% CD solution provided good enantioseparation of the four peaks with migration times of 18.2, 19.9, 20.6, and 21.5 min for (S)-verapamil, (S)-norverapamil, (R)-verapamil, and (R)-norverapamil, respectively. In order to shorten these migration times, a little pressure of 0.2 psi was added to the separation voltages and the new migration times were 9.8, 10.5, 10.8, and 11.2 min, respectively. The separation of the four peaks as well as the NADP$^+$ and NADPH can be seen in Fig. 8.7. Using the selected conditions, the total analysis time (including preconditioning, injection, incubation, and separation) was less than 35 min per sample.

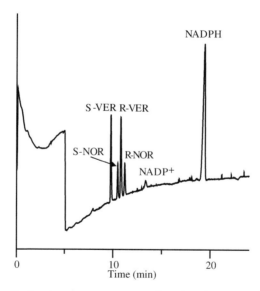

Figure 8.7 Electropherogram corresponding to the separation of the components of the proposed EMMA assay (VER: verapamil; NOR: norverapamil). EMMA conditions: incubation buffer, 100 mM potassium phosphate; pH 7.4; 200 nM CYP3A4; capillary temperature 37°C; and injection sequence as described in Fig. 8.6. Reprinted from Ref. [35], Copyright 2013, with permission from Elsevier.

The developed methodology was applied to estimate the kinetic Michaelis–Menten parameters for the metabolism of verapamil enantiomers by CYP3A4. With this purpose, six solutions containing verapamil in the concentration range 20–200 µM (10–100 µM of each enantiomer) were prepared in duplicate. The experiment was repeated in two different days in order to estimate the uncertainty under intermediate precision conditions. In order to obtain the verapamil peaks in the same diffusion conditions as in the enzymatic assay, calibration curves for quantifying verapamil enantiomers were prepared in both sessions following the proposed methodology. Solution A was not included to avoid the advance of the enzymatic reaction, and corrected peak area was used as response variable. The quantification of the reaction turnover was done considering the consumption of verapamil enantiomers, so these results correspond to the overall metabolism of verapamil by CYP3A4 and not only to the norverapamil formation.

From the nonlinear fitting of experimental data (combining data from both experimental sessions) to the Michaelis–Menten equation, K_m and V_{max} were estimated (see Table 8.1). The enantioselectivity, calculated as the ratio between the V_{max} values of both enantiomers, was 1.08 in favor of the (S)-enantiomer, which is slightly more metabolized than the (R) one. The in vitro intrinsic clearance Cl_i estimated was in agreement with those found in the literature [45].

The at-inlet EMMA methodology developed for the enantioselective enzymatic evaluation of verapamil was also applied by the authors to the study of fluoxetine metabolism by CYP2D6 [36]. For this purpose, the conditions proposed in the previous paper were applied to carry out the EMMA enzymatic reaction. Conditions for the enantioseparation of fluoxetine and its metabolite norfluoxetine were adjusted to achieve adequate chiral resolution. Based on previous experiences, concentrations of HS-β-CD (prepared in the BGE 50 mM sodium phosphate at pH 8.8) in the range 0.1–1.25% (m/v) were assayed and a pressure of 0.4 psi was applied during separation. Figure 8.8 (upper part) shows some of the results obtained using different HS-β-CD concentrations. As can be seen, with 0.1% HS-β-CD, (a) the enantiomers of fluoxetine and norfluoxetine were separated, but they overlapped with the peaks corresponding to the CYP2D6. With 0.15% HS-β-CD, (b) the peak of (R)-fluoxetine overlaps with the NADP$^+$ peak (see Fig. IV.26 for all peaks identification). Increasing the CD concentration to 0.5%, the fluoxetine and norfluoxetine enantiomers migrate after the $NADP^+$/$NADPH$ system, but (S)-norfluoxetine overlaps with the NADPH peak. Finally, with an HS-β-CD concentration of 1.25% (m/v), all the peaks in the reaction mixture were baseline resolved and fluoxetine enantiomers able to be quantified. The electropherogram corresponding to the final separation conditions is shown in Fig. 8.8 (lower part).

Once the separation conditions were adjusted, an experimental design was planned in order to obtain the Michaelis–Menten plots for the enantioselective fluoxetine metabolism by CYP2D6. The EMMA assay was carried out with seven concentration levels of racemic fluoxetine in the range 10–200 µM, each one with two independent replicates. The complete experiment was repeated in two days in order to obtain the data under intermediate precision conditions. All data were used together for estimations. The inter-day precision of the proposed methodology was evaluated. Averaged intermediate precision results for peak areas of calibration curves and samples were 9% and 16%, which can be considered acceptable taking into account the low concentrations used and the intrinsic variability of enzymatic reactions. Velocity rate data were adjusted to the Michaelis–Menten equation for each fluoxetine enantiomer employing a Mardquardt algorithm. From these plots, kinetic parameters for the metabolism of both enantiomers have been estimated (see Table 8.1). Results show a slight enantioselectivity in favor of (R)-fluoxetine, estimated as the ratio between their V_{max} values (V_{max-R}/V_{max-S}) and giving a value of 1.18. The authors found considerable differences between these results and those reported in the literature in a previous study of enantioselective metabolism of fluoxetine by CYP2D6 [46]. These differences can be attributed to the different enzyme sources and experimental methodologies, which strongly affect the estimations.

Finally, in order to check a possible interaction between the enantiomers in their metabolism, an experiment was carried out incubating only a single enantiomer, (R)-fluoxetine. The differences with the estimates obtained using the racemate were very low (1.6% for K_m and 3.9% for V_{max}), so it can be concluded that there is no significant interaction between (R)- and (S)-fluoxetine metabolisms.

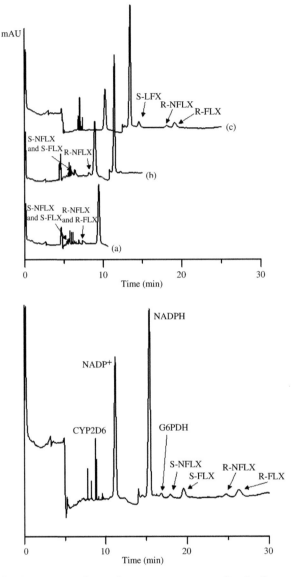

Figure 8.8 Upper part: electropherograms corresponding to the separation of the reaction mixture using different HS-β-CD concentrations: (a) 0.1%, (b) 0.15%, and (c) 0.5%. Experimental conditions: BGE, 50 mM sodium phosphate at pH 8.8, 25°C, 20 kV. Lower part: electropherogram showing the separation of the reaction mixture with 1.25% HS-β-CD. Reprinted with permission from Ref. [36], Copyright 2013, John Wiley and Sons.

8.4 Conclusion

The combination of EMMA for the in-line development of an enzymatic assay and PFT with a cyclodextrin for the chiral separation of substrates and/or metabolites has proved to be a powerful tool for a fast evaluation of in vitro enantioselective metabolism of drugs and suitable for the characterization of a broad range of enzymes. The main advantages of this methodology are its speed of analysis, low consumption of enzymes and other expensive reagents, and high degree of automation, joining the intrinsic advantages of EMMA and PTF in a unique step carried out in the CE system, and making this method an interesting approach for substrate and inhibitor screening in preliminary phases of drug development.

Acknowledgments

The authors acknowledge the Spanish Ministry of Science and Technology (MCYT), the European Regional Development Fund (ERDF) (Project CTQ2015-70904R) for the financial support.

References

1. Gordon Gibson, G. and Skett, P. (2001). *Introduction to Drug Metabolism*, 3rd edition (Cengage Learning, UK).

2. Doménech Berrozpe, J., Martínez Lanao, J., and Plá Delfina, J. M. (1998). *Biofarmacia y Farmacocinética. Volumen II: Biofarmacia* (Ed. Síntesis, Madrid, Spain).

3. Lewis, D. F. V. (2003). P450 structures and oxidative metabolism of xenobiotics, *Pharmacogenomics*, **4**, pp. 387–395.

4. Lewis, D. F. V. (2004). 57 varieties: The human cytochromes P450, *Pharmacogenomics*, **5**, pp. 305–318.

5. Ha, P. T. T., Sluyts, I., Dyck, S. V., Zhang, J., Gilissen, R. A. H. J., Hoogmartens, J., and Schepdael, A. V. (2006). Chiral capillary electrophoretic analysis of verapamil metabolism by cytochrome P450 3A4, *J. Chromatogr. A*, **1120**, pp. 94–101.

6. Li, D., Wang, Y., and Han, K. (2012). Recent density functional theory model calculations of drug metabolism by cytochrome P450, *Coord. Chem. Rev.*, **256**, pp. 1137–1150.

7. Caldwell, J. (1995). Stereochemical determinants of the nature and consequences of drug metabolism, *J. Chromatogr. A*, **694**, pp. 39–48.

8. Brocks, D. R. (2006). Drug disposition in three dimensions: An update on stereoselectivity in pharmacokinetics, *Biopharm. Drug Dispos.*, **27**, pp. 387–406.

9. Campo, V. L., Bernardes, L. S. C., and Carvalho, I. (2009). Stereoselectivity in drug metabolism: Molecular mechanisms and analytical methods, *Curr. Drug Metab.*, **10**, pp. 188–205.

10. Toon, S., Hopkins, K. J., Garstang, F. M., and Rowland, M. (1987). Comparative effects of ranitidine and cimetidine on the pharmacokinetics and pharmacodynamics of warfarin in man, *Eur. J. Clin. Pharmacol.*, **32**, pp. 165–172.

11. Rettie, A. E., Korzekwa, K. R., Kunze, K. L., Lawrence, R. F., Eddy, A. C., Aoyama, T., Gelboin, H. V., Gonzales, F. J., and Trager, W. F. (1988). Hydroxylation of warfarin by human cDNA-expressed cytochrome P-450: A role for P-450 2C9 in the etiology of (S)-warfarin drug interactions, *Chem. Res. Toxicol.*, **28**, pp. 25–39.

12. Walle, T., Webb, J. G., Bagwell, E. E., Walle, U. K., Daniell, H. B., and Gaffney, T. E. (1988). Stereoselective delivery and actions of beta-receptor antagonists, *Biochem. Pharmacol.*, **37**, pp. 115–124.

13. Vogelsang, B., Echizen, H., Schmidt, E., and Eichelbaum, M. (1984). Stereoselective first-pass metabolism of highly cleared drugs: Studies of the bioavailability of L- and D-verapamil examined with a stable isotope technique, *Brit. J. Clin. Pharmaco.*, **18**, pp. 733–740.

14. Thöm, H. A., Sjögren, E., Dickinson, P. A., and Lennernäs, H. (2012). Binding processes determine the stereoselective intestinal and hepatic extraction of verapamil in vivo, *Mol. Pharm.*, **9**, pp. 3034–3045.

15. Afshar, M. and Thormann, W. (2006). Capillary electrophoretic investigation of the enantioselective metabolism of propafenone by human cytochrome P-450 supersomes: Evidence for atypical kinetics by CYP2D6 and CYP3A4, *Electrophoresis*, **27**, pp. 1526–1536.

16. Zhou, Q., Yan, X. F., Pan, W. S., and Zeng, S. (2008). Is the required therapeutic effect always achieved by racemic switch of proton-pump inhibitors? *World J. Gastroentero.*, **14**, pp. 2617–2619.

17. Nehme, H., Nehme, R., Lafite, P., Routier, S., and Morin, P. (2012). New development in in-capillary electrophoresis techniques for kinetic and inhibition study of enzymes, *Anal. Chim. Acta*, **722**, pp. 127–135.

18. Glatz, Z. (2006). Determination of enzymatic activity by capillary electrophoresis, *J. Chromatogr. B*, **841**, pp. 23–37.

19. Hai, X., Yang, B., and Van Schepdael, A. (2012). Recent developments and applications of EMMA in enzymatic and derivatization reactions, *Electrophoresis*, **33**, pp. 211–227.

20. Van Dyck, S., Vissiers, S., Van Schepdael, A., and Hoogmartens, J. (2003). Kinetic study of angiotensin converting enzyme activity by capillary electrophoresis after in-line reaction at the capillary inlet, *J. Chromatogr. A*, **986**, pp. 303–311.

21. Okhonin, V., Liu, X., and Krylov, S. N. (2005). Transverse diffusion of laminar flow profiles to produce capillary nanoreactors, *Anal. Chem.*, **77**, pp. 5925–5929.

22. Sanders, B. D., Slotcavage, R. L., Scheerbaum, D. L., Kochansky, C. J., and Strein, T. G. (2005). Increasing the efficiency of in-capillary electrophoretically mediated microanalysis reactions via rapid polarity switching, *Anal. Chem.*, **77**, pp. 2332–2337.

23. Zeisbergerova, M., Řemínek, R., Mádr, A., Glatz, Z., Hoogmartens, J., and Van Schepdael, A. (2010). On-line drug metabolites generation and their subsequent target analysis by capillary zone electrophoresis with UV-absorption detection, *Electrophoresis*, **31**, pp. 3256–3262.

24. Curcio, R., Nicoli, R., Rudaz, S., and Veuthey, J. L. (2010). Evaluation of an in-capillary approach for performing quantitative cytochrome P450 activity studies, *Anal. Bioanal. Chem.*, **398**, pp. 2163–2171.

25. Zhang, J., Hoogmartens, J., and Van Schepdael, A. (2008). Kinetic study of cytochrome P450 by capillary electrophoretically mediated microanalysis, *Electrophoresis*, **29**, pp. 3694–3700.

26. Řemínek, R. and Glatz, Z. (2010). Study of atypical kinetic behaviour of cytochrome P450 2C9 isoform with diclofenac at low substrate concentrations by sweeping-MEKC combination, *J. Sep. Sci.*, **33**, pp. 3201–3206.

27. Konečný, J., Juřica, J., Tomandl, J., and Glatz, Z. (2007). Study of recombinant cytochrome P450 2C9 activity with diclofenac by MEKC, *Electrophoresis*, **28**, pp. 1229–1234.

28. Hai, X., Konečny, J., Zeisbergerová, M., Adams, E., Hoogmartens, J., and Van Schepdael, A. (2008). Development of electrophoretically mediated microanalysis method for the kinetics study of flavin-containing monooxygenase in a partially filled capillary, *Electrophoresis*, **29**, pp. 3817–3824.

29. Konecny, J., Micikova, I., Reminek, R., and Glatz, Z. (2008). Application of micellar electrokinetic capillary chromatography for evaluation of inhibitory effects on cytochrome P450 reaction, *J. Chromatogr. A*, **1189**, pp. 274–277.

30. Nowak, P., Voźniakiewicz, M., and Kościelkiak, P. (2013). An overview of on-line systems using drug metabolizing enzymes integrated into capillary electrophoresis, *Electrophoresis*, **34**, pp. 2604–2614.

31. Scriba, G. K. E. and Belal, F. (2015). Advances in capillary electrophoresis-based enzyme assays, *Chromatographia*, **78**, pp. 947–970.

32. Hai, X., Adams, E., Hoogmartens, J., and Van Schepdael, A. (2009). Enantioselective in-line and off-line CE methods for the kinetic study on cimetidine and its chiral metabolites with reference to flavin-containing monooxygenase genetic isoforms, *Electrophoresis*, **30**, pp. 1248–1257.

33. Kwan, H. Y. and Thormann, W. (2012). Electrophoretically mediated microanalysis for characterization of the enantioselective CYP3A4 catalyzed N-demethylation of ketamine, *Electrophoresis*, **33**, pp. 3299–3305.

34. Řemínek, R., Glatz, Z., and Thormann, W. (2015). Optimized on-line enantioselective capillary electrophoretic method for kinetic and inhibition studies of drug metabolism mediated by cytochrome P450 enzymes, *Electrophoresis*, **36**, pp. 1349–1357.

35. Asensi-Bernardi, L., Martín-Biosca, Y., Escuder-Gilabert, L., Sagrado, S., and Medina-Hernández, M. J. (2013). In-line capillary electrophoretic evaluation of the enantioselective metabolism of verapamil by cytochrome P3A4, *J. Chromatogr. A*, **1298**, pp. 139–145.

36. Asensi-Bernardi, L., Martín-Biosca, Y., Escuder-Gilabert, L., Sagrado, S., and Medina-Hernández, M. J. (2013). Fast evaluation of enantioselective drug metabolism by electrophoretically mediated microanalysis: Application to fluoxetine metabolism by CYP2D6, *Electrophoresis*, **34**, pp. 3214–3220.

37. Lipsy, R. J., Fennerty, B., and Fagan, T. C., (1990). Clinical review of histamine2 receptor antagonists, *Arch. Intern. Med.*, **150**, pp. 745–751.

38. Schentag, J. J., Cerra, F. B., Calleri, G. M., Leising, M. E., French. M. A., and Bernhard, H. (1981). Age, disease, and cimetidine disposition in healthy subjects and chronically ill patients, *Clin. Pharmacol. Ther.*, **29**, pp. 737–743.

39. Overby, L. H., Carver, G. C., and Philpot, R. M. (1997). Quantitation and kinetic properties of hepatic microsomal and recombinant flavin-containing monooxygenases 3 and 5 from humans, *Chem. Biol. Interact.*, **106**, pp. 29–45.

40. Kwan, H. Y. and Thormann, W. (2011). Enantioselective capillary electrophoresis for the assessment of CYP3A4-mediated ketamine

demethylation and inhibition in vitro, *Electrophoresis*, **32**, pp. 2738–2745.

41. Portmann, S., Kwan, H. Y., Theurillat, R., Schmitz, A., Mevissen, M., and Thormann, W. (2010). Enantioselective capillary electrophoresis for identification and characterization of human cytochrome P450 enzymes which metabolize ketamine and norketamine in vitro, *J. Chromatogr. A*, **1217**, pp. 7942–7948.

42. Schmitz, A., Thormann, W., Moessner, L., Theurillat, R., Helmja, K., and Mevissen, M. (2010). Enantioselective CE analysis of hepatic ketamine metabolism in different species in vitro, *Electrophoresis*, **31**, pp. 1506–1516.

43. Kharasch, E. D. and Labroo, R. (1992). Metabolism of ketamine stereoisomers by human liver microsomes, *Anesthesiology*, **77**, pp. 1201–1207.

44. Zhang, J., Ha, P. T. T., Lou, Y., Hoogmartens, J., and Van Schepdael, A. (2005). Kinetic study of CYP3A4 activity on verapamil by capillary electrophoresis, *J. Pharm. Biomed. Anal.*, **39**, pp. 612–617.

45. Kroemer, H. K., Echizen, H., Heidemann, H., and Eichelbaum, M. (1992). Predictability of the in vivo metabolism of verapamil from in vitro data: Contribution of individual metabolic pathways and stereoselective aspects, *J. Pharmacol. Exp. Ther.*, **260**, pp. 1052–1057.

46. Margolis, J. M., O'Donell, J. P., Mankowski, D. C., Ekins, S., and Scott Obach, R. (2000). (R)-, (S)-, and racemic fluoxetine N-demethylation by human cytochrome P450 enzymes, *Drug Metab. Dispos.*, **28**, pp. 1187–1191.

Chapter 9

Capillary Electrophoresis for the Quality Control of Intact Therapeutic Monoclonal Antibodies

Anne-Lise Marie,[a] Grégory Rouby,[b] Emmanuel Jaccoulet,[a] Claire Smadja,[a] Nguyet Thuy Tran,[a] and Myriam Taverna[a]

[a]*Institut Galien Paris Sud, UMR8612, Protein and Nanotechnology in Analytical Science (PNAS), CNRS, Univ. Paris Sud, Université Paris Saclay, 5 rue Jean-Baptiste Clément, 92290 Châtenay-Malabry, France*
[b]*LFB, 91940 Les Ulis, France*
myriam.taverna@u-psud.fr

9.1 Introduction

Since the commercialization of the first therapeutic monoclonal antibody (mAb) product in 1986, Orthoclone OKT3, this class of biopharmaceuticals has grown significantly, and many new mAbs have been regularly approved in the United States or Europe. In 1990, the first chimeric mAbs were approved, followed by the approval of humanized and then fully human mAbs. More recently, bispecific antibodies, antibody–drug conjugates, and crystallizable

---

Capillary Electrophoresis: Trends and Developments in Pharmaceutical Research
Edited by Suvardhan Kanchi, Salvador Sagrado, Myalowenkosi Sabela, and Krishna Bisetty
Copyright © 2017 Pan Stanford Publishing Pte. Ltd.
ISBN 978-981-4774-12-3 (Hardcover), 978-1-315-22538-8 (eBook)
www.panstanford.com

fragment (Fc)-fusion proteins containing the antibody constant region fused to another protein have appeared. The commercial clinical pipeline of antibody therapeutics is now totaling over 470 molecules [1]. MAbs are employed for the treatment of a variety of severe diseases, including cancers, multiple sclerosis, ankylosing spondylitis, Crohn's disease as well as chronic plaque psoriasis, asthma, or rheumatoid arthritis [2]. MAbs are glycoproteins that belong to the immunoglobulin (Ig) superfamily, which can be divided into five isotypes: IgA, IgD, IgE, IgG, and IgM. Only IgGs (also called recombinant mAbs) are produced for therapeutic purposes through genetic engineering [3].

Manufacturing processes for therapeutic mAbs have evolved since the first licensed mAb product, which was produced in the ascites of mice [4, 5]. Early murine mAbs were derived from hybridoma cell lines, using diverse production technologies. Later, mAbs that contained a combination of rodent- and human-derived sequences, resulting in chimeric or humanized mAbs, have appeared. Currently, most therapeutic human mAbs in clinical trials are obtained from either immunization of transgenic mice expressing human antibody genes or phage–display recombinants. The rapid growth in product demand for mAbs triggered parallel efforts to increase production capacity through construction of large bulk manufacturing plants as well as improvements in cell culture processes to raise product titers.

MAbs are microheterogeneous proteins, mainly due to their post-translational modifications (PTMs), but also due to their susceptibility to chemical and physical degradations. The cell type used during the upstream process has an impact on the glycosylation pattern of a mAb, but of less importance than those related to the production process itself. Such complex molecules with a variety of functional groups are susceptible to instability through different degradation pathways that could happen as a result of exposure to different environmental changes and stresses during their multistep production, handling, shipping, storage, and even just before administration to patients (for instance after their compounding performed at the hospital) [6]. The product may contain multiple natural variants (charge variants, isoforms, glycoforms) eventually associated to many possible product-related impurities. Moreover, because mAbs exhibit great molecular complexity, they are quite sensitive to changes in the manufacturing process. This is the reason why quality control (QC) of therapeutic mAbs is challenging, requiring many different complementary techniques to assess their purity, identity, lot-to-lot consistency, and dosage. Beck et al. published in 2013 a comprehensive review on the different methods available to characterize mAbs [7].

Capillary electrophoresis (CE) has already proved to be a useful technique for this purpose. With its multiple separation modes relying on either the isoelectric point (pI), or molecular mass, or charge-to-mass ratio of antibodies, CE brings the selectivity required to check for the presence of abnormal charge variants, aggregates, fragments, mAbs not fully or properly glycosylated, forms with incomplete formation of disulfide bridges, etc. Being quantitative, this technique can also be useful for the determination of product concentration. In addition, identity control can be made based on the physicochemical properties of the mAb.

The QC of mAbs can be performed on the intact molecules or after their fragmentation using mostly proteases (e.g., IdeS, papain, pepsin), glycosidases, or after chemical reduction of disulfide bonds with dithiothreitol (DTT). In this review, only the applications dealing with the QC of intact mAbs will be considered.

The first part of this review gives a description of what are impurities and drug-related substances for therapeutic proteins and in particular for mAbs. We explain how these compounds are formed during the manufacturing and/or storage of the drug, and their potential impact on the biological activity of the biopharmaceutical. In the second part, the sources of natural heterogeneity of mAbs (post-translational modifications) are briefly introduced, while the specific domain of QC performed after mAb compounding at the hospital is presented. Finally, the third part is the state-of-the-art of CE-based methods (mostly CZE, CGE, and CIEF) applicable for (i) in-process controls, (ii) controls of drug-related substances, and (iii) identity control after mAb compounding. We discuss the specificity of each CE mode, their respective advantages and drawbacks, and their possible evolution toward miniaturization or coupling with mass spectrometry.

9.2 Impurities and Drug-Related Substances

9.2.1 Process-Related Impurities

The International Conference on Harmonisation (ICH) of technical requirements for registration of pharmaceuticals for human use classifies in the ICH guideline Q6B [8] the impurities of recombinant therapeutic proteins (biopharmaceuticals that also include mAbs) in two major categories, either process or product-related substances. Different issues are associated with the characterization and quantification of these related substances, whatever their origin. The required methods for purity assessment need to be sensitive enough to detect impurities present at trace levels, but also very selective to afford the resolution between structurally close co-variants, product impurities, and eventual excipients. All mAb variants can be detected by multiple techniques, major ones are high-performance liquid chromatography (HPLC), sodium dodecyl sulfate-polyacrylamide gel electrophoresis (SDS-PAGE), and capillary electrophoresis (CE). Due to its high peak capacity, CE is very well suited for the profiling of drugs, in particular for detecting product-related impurities of mAbs. Considering the amount of active protein required for therapeutic dose (generally quite small), the intrinsic and natural heterogeneity of mAbs (see Section 9.3.1), and the presence of various excipients or process-related impurities, the impurity profiling of mAbs remains challenging.

Product-related impurities will be described in details in Section 9.2.2. They are actually isoforms of the desired product. As in the case of physicochemical characterization, scientific and technical approaches should be employed to fully and exhaustively document these impurities. The amount of degradation products must be monitored during all manufacturing steps and/or storage of the drug. ICH defines three categories of degradation products: "truncated forms," produced by chemical or protease cleavage of peptide bonds, "other modified forms" corresponding to deamidated, isomerized, misfolded, and oxidized forms, and finally "aggregates" that include dimers and also higher oligomers.

Process-related impurities encompass those that are derived from the manufacturing process. Commonly, the manufacturing process for mAb production in cell culture starts with the cell amplification in bioreactor, followed with a clarification step (part of the upstream process purification). Then the most popular technique to purify more than 99% of mAbs from cell culture supernatants is Protein A affinity chromatography, which is followed by polishing steps to remove contaminants and process impurities (part of the downstream process purification). After the formulation step, the last critical one is called "fill and finish." This step consists of several finished operations such as mixing the active substance into its final form (i.e., liquid or lyophilized powder), and filling and sealing within final containers (prefilled syringes or other delivery systems) in order to release the drug product batches.

Process-related impurities are classified into three major categories [8]: cell substrate–derived, cell culture–derived, and downstream-derived impurities. Cell substrate–derived impurities include proteins derived from the host organism (host cell genomic, vector, or total DNA). Cell culture–derived impurities entail a wide variety of compounds from inducers, antibiotics (e.g., neomycin, tobramycin, kanamycin, ampicillin, penicillin, tetracycline, gentamicin) used to control bacterial contamination or for plasmid selection, serum (mostly extracted from bovines), and other medium components as growth promoter/expresser, insulin, redox reagents, antioxidants. Downstream-derived impurities include enzymes, chemical and biochemical processing reagents (e.g., detergents, oxidizing and reducing agents), inorganic salts (e.g., heavy metals, arsenic, non-metallic ions), solvents, carriers, chromatographic ligands (e.g., monoclonal antibodies or resin ligands), process additives (e.g., polyethylene glycol), and other leachables like protein A. Another type of impurities called contaminants include all adventitiously introduced materials not intended to be part of the manufacturing process, such as chemical and biochemical materials (e.g., microbial proteases), and/or microbial species. ICH Harmonized Tripartite Guidelines [9] force the pharmaceutical companies to "strictly avoid" contaminants that can expand to any step in the process even during fill and finish stages.

In this chapter, only the QC of the mAb itself (identity, quantification) and its product-related impurities will be considered.

9.2.2 Product-Related Substances

During their biosynthesis, extraction, purification, formulation, and shelf-life, mAbs can undergo a wide variety of physicochemical modifications, including oxidation, deamidation, fragmentation, glycation, unfolding/misfolding, aggregation, and formation of disulfide bond isoforms [10, 11, 6]. These modifications are the result of enzymatic or non-enzymatic processes, and some of them provide mAb charge heterogeneity, with acidic and basic species. Deamidation, glycation, N-terminal glutamine cyclization, and fragmentation contribute to the formation of acidic variants, while C-terminal lysine and C-terminal proline amidation induce the formation of basic species [10, 11, 7]. All these degradations, leading to the so-called product-related substances, may impact the quality, safety, and efficacy of the antibody-based therapeutics. This section summarizes the different mAb degradations, with a brief description of their mechanism, and their potential therapeutic impact.

9.2.2.1 Chemical degradations

Chemical modifications correspond to the change or loss of specific functional groups or the breakage of covalent bonds. They encompass oxidation, deamidation, isomerization, racemization, pyroglutamate formation, fragmentation, glycation, disulfide bond modification, and covalent oligomerization.

Protein oxidation is a covalent modification of amino acids induced by reactive oxygen species [12]. This chemical degradation is catalyzed by metals and light. Methionine (Met) is one of the most susceptible residues to oxidation [13]. Oxidation of Met to sulfoxide increases the molecular mass by 16 Da and makes the side chain of Met more polar. Long-term storage, incubation at elevated temperatures or with oxidizing reagents are factors promoting Met oxidation in mAbs. Wei et al. [14] reported that exposure of two humanized mAbs to ultraviolet, light, or tert-butyl hydroperoxide led primarily to oxidation of Met101, Met255, and Met431 of the heavy chain (HC), these methionine residues being more exposed in the 3D structure. In addition to Met, tryptophan (Trp) is another amino acid prone to oxidation in mAbs. Interestingly, the oxidized Trp residues of humanized mAbs that have been reported so far are all located in the complementarity determining regions (CDRs), which may reflect the fact that they are the only solvent-exposed Trp residues [11]. Cysteine (Cys) residues, when they are not involved in disulfide bonds, are also likely to be oxidized. For example, Kroon et al. reported oxidation of an unpaired Cys residue of a murine mAb after storage at 2–8°C for three years [15]. However, in general, the presence of extra Cys residues is avoided for the development of recombinant mAbs to prevent instability caused by the highly reactive free sulfhydryl (SH). Nowadays, it is well recognized that oxidation of therapeutic proteins may limit their clinical efficacy or stability.

Deamidation is another frequent chemical modification of mAbs [12, 16, 17]. Deamidation is a spontaneous non-enzymatic process, which involves either asparagine (Asn) or glutamine (Gln) residues. During deamidation, an Asn residue is converted into aspartic (Asp) or isoaspartic (isoAsp) acid (Fig. 9.1), whereas a Gln residue is converted into glutamic acid. The reaction produces an intra-molecular cyclic succinimide (Asu) intermediate, which is not stable although its presence has been detected for several mAbs [18–21]. Deamidation increases the mass by 1 Da. More importantly, it introduces one additional negative charge to the antibody and generates acidic species. The rates of deamidation depend on the primary sequence of the protein, but also on its secondary and tertiary structures. It is now recognized that Asn followed by glycine is the most susceptible site for deamidation [11, 12]. Buffer composition, ionic strength, pH, and temperature can also affect the rates of deamidation [11]. In 2006, Wakankar et al. reported formulation strategies that could improve the stability of mAbs to deamidation [22]. Depending on the site of action, deamidation can have functional implications. For example, the formation of a stable succinimide intermediate at Asn55 in the CDR2 region of IgG1 HC led to a 70% drop in the drug potency [18.]

The mechanism of isomerization is similar to that of deamidation since both reactions proceed through the formation of an Asu intermediate. Once it is formed, it can open to produce either Asp or isoAsp residues (Fig. 9.1). Asp itself can also cyclize to form the same Asu species, thereby allowing conversion from Asp to isoAsp. This reaction, called Asp-isoAsp interconversion, is more commonly referred to as Asp isomerization [12]. As in deamidation, the formation of the Asu intermediate is the rate-limiting step in the isomeri-

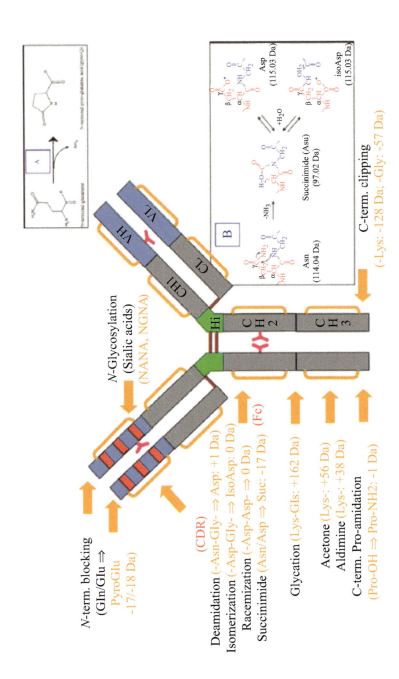

Figure 9.1 IgG charge microvariants found after post-translational modifications (PTMs) or degradations. Inset (A): Pyroglutamate formation. Inset (B): Deamidation. Reprinted with permission from Ref. [7]. Copyright 2013, American Chemical Society.

Impurities and Drug-Related Substances | **313**

zation reaction, and the formulation conditions highly influence the reaction rate. For instance, it has been reported that addition of magnesium salts to humanized mAb formulations resulted in an improved stability of the antibody to Asp isomerization [22].

Racemization is the transformation of L-Asp (or L-isoAsp) into D-Asp (or D-isoAsp) [23]. It is a chemical modification related to deamidation, since it follows the same chemical pathway with the formation of an Asu intermediate. Racemization may be induced by high temperatures or pH stresses [24]. For example, racemization of aspartate residue in a thermally stressed antibody has been reported [25].

The *N*-terminal residue of human IgG heavy chains is commonly glutamine (Gln) or glutamate (Glu). It has been reported that most of the *N*-terminal Gln of human IgG HC is cyclized into pyroglutamate (pyroGlu) [11]. Dick et al. reported that 90% of the HC *N*-terminal Gln of a recombinant mAb was cyclized to pyroGlu in the bioreactor at day 15, while the other 10% was cyclized during purification, formulation, and sample handling [26]. Formation of *N*-terminal pyroGlu from Gln or Glu decreases the molecular mass by 17 or 18 Da, respectively (Fig. 9.1). More significantly, cyclization of Gln to pyroGlu causes a loss of the *N*-terminal primary amine, which is positively charged at physiological pH and, therefore, results in antibodies with more acidic species [11]. Given its position far from the CDR surface, the presence of one or two HC *N*-terminal pyroGlu does not seem to affect the antibody potency [10].

Fragmentation is a very general term that usually corresponds to disruption of a covalent bond in a protein as a result of either spontaneous or enzymatic reaction. Here is considered non-enzymatic fragmentation, which is a function of the primary sequence, the flexibility of the local structure, the solvent conditions (pH, temperature), and the presence of metals or radicals [27]. Fragmentation of mAbs can be generated during protein production in the cell culture or during the purification process and can accrue during storage. It was shown that most of the backbone fragmentation events in mAbs occurred at one of the following residues: Asp, Gly, Ser, Thr, Cys, or Asn, the side chains of these residues (with the exception of Gly) facilitating peptide bond cleavage via specific mechanisms like β-elimination [11, 27] (Fig. 9.2). However, other mechanisms may also contribute to fragmentation like free-radical-induced hydrolysis or direct hydrolysis. For example, the hinge region of mAbs may easily undergo fragmentation by direct hydrolysis [28], or β-elimination [29]. The effect of fragmentation on the function of a mAb depends on whether cleavage sites are observed in the variable or constant regions, and on the mechanism of action of the molecule [27]. Finally, fragmentation may have an impact on the quality of mAb products by altering the aggregation rates and leading to inactive molecules.

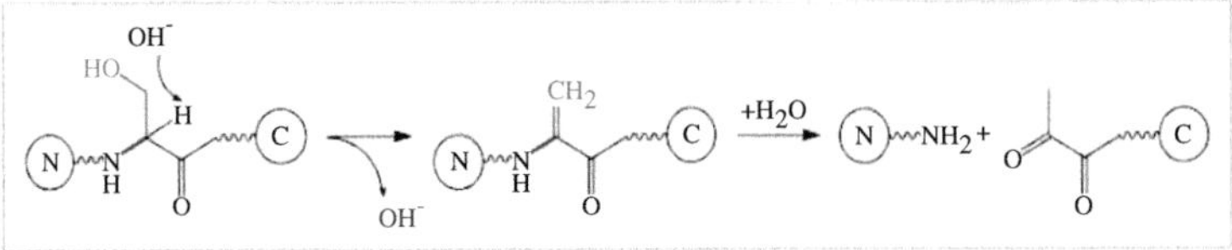

Figure 9.2 β-elimination at Ser residue. Reprinted with permission from Ref. [27], Copyright 2011, Taylor & Francis.

Unlike glycosylation, glycation is a non-enzymatic reaction that initially involves the addition of a reducing sugar, or its derivatives, to amine groups of proteins. This leads to the formation of a Schiff base, which can subsequently undergo rearrangement into irreversible conjugates, called advanced glycation end products (AGEs) [12, 30]. The formation of AGEs affects the structure and functionality of proteins. Glycation often occurs at low level during the production or storage of mAbs, and could be due to glucose addition during cell culture [31] or to the presence of sucrose or other sugars used as excipients during formulation [32, 33]. Short-term incubation of antibodies in dextrose infusion bags before patient administration could also lead to glycated products [32]. Glycation can occur non-specifically on lysine residues of both HC and light chain (LC) [34], which leads to a loss of one positive charge, making the antibody more acidic.

IgG molecules are composed of two HC and two LC. Each LC is connected to each HC by one disulfide bond, and each HC is connected to the other HC by 2–11 disulfide bonds [35] (see Section 9.3.1). In addition, IgGs have 12 domains, each of which contains one intra-chain disulfide bond. Disulfide bond formation is important for the assembly and structural integrity of antibodies. Heterogeneity related to disulfide bonds can be introduced at different stages [11, 36]. For example, incomplete formation of intra-chain disulfide bonds provides a source of free sulfhydryl, which can trigger disulfide bond scrambling, especially under denaturing conditions. The formation of free SH groups can also derive from β-elimination, a reaction normally accelerated under basic conditions [35]. A single incomplete disulfide bond results in two free SH groups and increases the mass by 2 Da. The presence of free SH may affect the structure, stability, and also the biological function of the antibody. For example, it has been reported that incomplete formation of the disulfide bond in the HC variable domain of a recombinant mAb resulted in a significant decrease in potency [35].

Covalent oligomerization arises from the formation of a chemical bond between two or more monomers. Disulfide bond formation resulting from previously unpaired thiols is a common mechanism for covalent oligomerization [37]. Andya et al. reported the formation of antibody dimers and trimers that were covalently linked by intermolecular disulfides following the storage and reconstitution of freeze-dried recombinant mAb formulations [38]. They showed that the addition of carbohydrate excipients such as sucrose and trehalose to the mAb formulation stabilized the protein structure and reduced its oligomerization during storage. Oxidation of tyrosine may also result in covalent dimerization through the formation of dityrosine [37]. For example, dityrosine formation is likely to be involved in the covalent dimerization of Palivizumab, an IgG1 mAb [39]. Aggregates and oligomers are undesirable because they may reduce the therapeutic activity, but also induce an immunogenic reactivity [37].

9.2.2.2 Biochemical degradations

C-terminal lysine clipping of the heavy chain is one of the most common modifications of recombinant mAbs. The HC C-terminal lysine residue is encoded in the gene sequence, but this residue can be completely or partially removed by the action of carboxypeptidases during the production cell culture process [10, 11]. This enzymatic activity is modulated by metal cofactors such as zinc and copper, the former facilitating carboxypeptidase activity, the latter being rather inhibitory [40]. The removal of one C-terminal lysine (Lys) decreases the molecular mass by 128 Da, and the mAb charge by one unit (Fig. 9.1). Incomplete removal of the C-terminal Lys results in a mixed population of antibodies with zero, one, or two C-terminal Lys. Experiments have shown that C-terminal Lys variants do not impact the in vitro potency, effector function, or pharmacokinetics of mAbs [10]. However, it is important to monitor these variants to ensure the consistency of the development and manufacturing process.

Besides C-terminal lysine variants, minor C-terminal isoforms also exist. For example, in some IgG1 antibodies, C-terminal amidation of proline residues may happen after removal of the HC C-terminal dipeptide Gly-Lys. The significance of amidation and whether this modification occurs prior to or following secretion from the Chinese hamster ovary (CHO) cells have not been established. However, given its position at the end of the protein chain, it may not constitute a critical quality attribute of mAbs [10].

9.2.2.3 Physical degradations

Proteins, because of their polymeric nature and their ability to form superstructures (e.g., secondary, tertiary, and quaternary), can undergo structural changes without any change in their chemical composition [12]. These modifications include denaturation, unfolding, aggregation, precipitation, misfolding, and conformational transitions. These physical instabilities, which alter the native three-dimensional structure of a mAb, may impact the biological activity of the therapeutic protein.

Denaturation corresponds to the loss of the globular or three-dimensional structure of the protein. This globular structure is referred to as the native state [12]. The denaturation phenomenon induces the unfolding of the protein, which may lead to its aggregation, and even to its precipitation. A variety of factors may cause denaturation (heat, light exposure, lyophilization, freezing/thawing, organic solvents, salts). In particular, elevated temperatures and high concentrations of chaotropes are common stresses that induce protein unfolding [24, 41]. The unfolding of an IgG is a complex process, in which the denatured state is obtained from the native protein through several, at least partly independent, intermediate states [42]. Moreover, depending on the type of denaturing stress, the denaturation process follows different paths: the antigen-binding fragment (Fab) is more sensitive to heat treatment, whereas the crystallizable fragment (Fc) is more sensitive to the lowering of the pH. The main consequence of protein denaturation is a loss of therapeutic activity.

MAbs like all other proteins are prone to aggregation. Aggregates are formed mainly because of intermolecular interactions of hydrophobic regions as a result of partial or transient unfolding of the proteins [11, 43, 44]. During cell cultivation and downstream processing, mAbs encounter various stresses such as high temperature, agitation, extreme pH, high concentrations of chaotropes, ultra/diafiltration, and exposure to air or metal surfaces [37, 45, 46]. These stresses result in the formation of aggregates, the size of which varies from soluble submicron range to visible precipitates. Purification processes can normally remove most of the aggregates. However, aggregates can be further generated during formulation, filling, and storage [11, 37]. Nicoud et al. [47] studied the role of some excipients on mAb aggregation and showed that for one IgG1, NaCl accelerated the aggregation kinetics, while sorbitol delayed it by specifically inhibiting protein unfolding. Only very low levels of aggregates are considered acceptable in mAb-based products because aggregates can trigger an immunogenic response [48, 49], or cause adverse effects during administration [37].

The correct folding of a protein after biosynthesis is important because it greatly determines its activity [50]. However, the production at industrial scale of a monoclonal antibody may saturate protein-folding machinery, which could result in a higher probability of producing partially or incorrectly folded molecules [11].

Under stress conditions, mAbs can undergo conformational changes affecting their higher-order structures [6]. For example, it has been shown that thermal and acidic stresses resulted in changes in the secondary structure of a mAb due to the possible conformational transitions and isoform inter-conversions [51, 52], and that oxidative stress exaggerated conformational changes of a deglycosylated fully humanized IgG1 mAb [53].

9.3 Identity and Heterogeneity of mAbs

9.3.1 Source of Natural Heterogeneity of Proteins: Post-Translational Modifications

The vast majority of marketed mAbs belongs to the subclass IgG1 and to some extent IgG2 and IgG4. The different heavy chains (c1, c2, c3, and c4) divide the IgGs into their subclasses 1–4. During their production, mAbs can undergo chemical modifications that are considered part of the degradations detailed in Section 9.2.2. Glycation of recombinant antibodies can also happen during cell cultures [54, 55], but these glycated mAbs can be considered as degradation products. In contrast, post-translational modifications (PTMs) are a common and natural event during protein production in mammalian cells. For mAbs, these PTMs mainly include glycosylation, disulfide bonding, and folding.

Like most extracellular glycoproteins, therapeutic proteins and specifically mAbs undergo glycosylation in the endoplasmic reticulum and Golgi network of cells. Glycosylation is crucial for the structural stability and biological activity of the molecule. The N-glycosylation on the antibody can influence stability, function, and pharmacokinetics. For instance, deglycosylation renders mAbs more prone to unfolding and aggregation [56]. Protein glycosylation is O-linked on serine (Ser) and threonine (Thr) residues, or N-linked on asparagine (Asn) residues. IgG antibodies are typically glycosylated on one Asn residue on the Fc region. MAbs have one conserved N-linked glycosylation at the Fc part and at position Asn297. Approximately 20% contain a second N-linked glycosylation site in their variable region. Both sites are located on the HC. Glycosylation pattern is strongly related to the cell line used for mAb production. CHO (hamster), Sp2/0 (mouse), and NS0 (mouse) cell lines are the most common in commercial production, and the glycosylation patterns of their products differ noticeably [57, 58].

Glycosylation of biopharmaceuticals shows a high grade of heterogeneity. Figure 9.3 shows a schematic representation of glycans frequently found for mAbs. Different structures exist with fucose residue added on N-acetylglucosamine, or without core-fucosylation. A variation of mannose residues is also observed [36]. MAbs are usually free of *N*-glycans with more than two antennae, and furthermore the sialic acid content is low compared with other glycoproteins [36]. This can be explained by the fact that the glycosylation site in the CH2 domain at Asn297 is buried in the protein structure. Typically, antibodies contain a high percentage of complex bi-antennary glycans with core-fucosylation [59]. Murine myeloma cell lines can include immunogenic and non-human α-1,3-galactose linkages, and CHO cells are incapable of adding sialic acid in human-type α-2,6-linkages [60]. In addition, controlled expression of glycosyltransferases within a cell line can lead to products with fairly specific glycoform profiles [61–63].

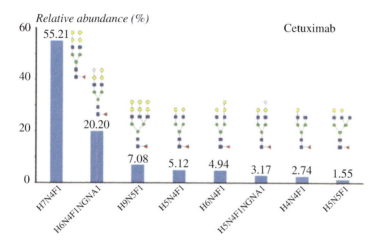

Figure 9.3 Glycoform semiquantitative analysis results regarding Fd domain glycosylation of cetuximab (green circle/yellow circle, hexose (mannose/galactose); blue square, *N*-acetylhexosamine; red triangle, fucose; diamond, sialic acid). Reprinted with permission from Ref. [64], Copyright 2014, Taylor & Francis.

Heavy and light chains are connected by disulfide bridges giving the antibody its Y-shaped structure. In addition, two intra-chain disulfide bonds stabilize each LC, and four stabilize each HC. Typically in IgG1 molecules, all cysteine residues are linked through disulfide bonds, which further stabilize the folding: 16 disulfide bridges per IgG1, and 7 per chain. The number of disulfide bonds in the hinge region is an intrinsic feature of each of the four human

IgG subclasses: IgG1 and IgG4 hinges both have two, IgG2 has four, and IgG3 has 11 inter-chain disulfide bonds [35] (Fig. 9.4). These bonds are incredibly stable in IgG1 and IgG3 subclasses. IgG2 and IgG4 hinge disulfide bonds, by contrast, are significantly more labile, which can trigger disulfide bond scrambling (see Section 9.2.2.1) in IgG2 antibodies and/or the formation of diabodies (dimeric antibody fragments), or single antibodies with two different CDR specificities in the case of IgG4 [36].

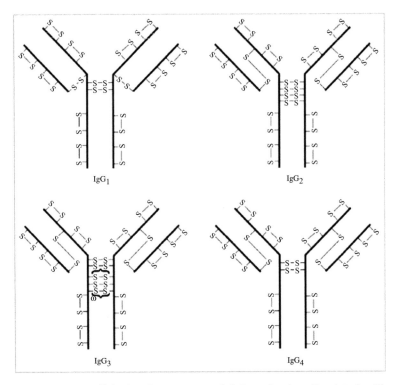

Figure 9.4 Disulfide bond structures of IgG molecules. Reprinted with permission from Ref. [35], Copyright 2012, Taylor & Francis.

9.3.2 Identity Control Issues after mAb Compounding

In hospital, anticancer drugs are compounded before their administration to patients, and some of them are mAbs. Marketed therapeutic mAbs are available under high concentrated solutions or lyophilized powders. Prior to the infusion to patient, therapeutic mAbs, as well as other anticancer drugs, have to be compounded into sterile bag (i.e., diluted) with a compatible diluent such as 0.9% sterile sodium chloride or 5% dextrose solutions. This process of pharmaceutical compounding is operated by specialized pharmaceutical technicians. However, considering (i) an increasing number of available therapies and administered mAbs, (ii) an increasing number of patients and prescriptions, and (iii) the need for high-throughput, medication errors (nature of administered mAb drug and dosage) may occur. The implementation of QC of compounded drugs before administration to patient is, therefore, mandatory to ensure the production safety and to limit these medication errors. However, it is worth noting that in hospital pharmacy, the QC of therapeutic mAbs already approved by the Food and Drug Administration and European Medicine Agency does not require in-depth characterization, i.e., micro-heterogeneity determination or degradation profile. Indeed, these controls have been already performed to release the batches. The QC at the hospital needs to answer specific questions, to be as rapid and simple as possible, and must provide identification and quantitation at the same time.

The main challenge is the mAb identification in the infusion bag with an analytical method allowing unambiguous discrimination between different possible therapeutic mAbs. This could be obtained by exploiting mAb physicochemical properties such as isoelectric point (pI) and global net charge. The advantage of these parameters is their dependence on the amino acid residue content, mAb spatial conformation, and solvation. Thus, although mAbs can exhibit high similarities, the nature and position of the charged lateral chains of amino acids (that can be exposed or buried) on the variable fragment (Fv) may provide a source of discrimination between mAbs. MAb pI can be assessed through CE [65]. In particular, imaged capillary isoelectric focusing (iCIEF) has the potential to provide rapid mAb identification. Indeed, this method allows separation of heterogeneous mAbs based on the different pI of the variants yielding to specific electrophoregram for a specific mAb. Another approach relies on the global net charge of the mAbs, which is expected to change with pH media. Li et al. [66] have been able to estimate the effective charge of bevacizumab and ranibizumab in

PBS buffer pH 7.4 from their intrinsic electrophoretic mobilities. In this work, the Henry equation, which takes into account the influence of the surrounding ions related to the electrophoretic conditions, was employed to determine the effective charge. The results obtained showed that the two mAbs exhibited similar pI (~8.8) but a significant difference of their effective charge and hydrodynamic radius. These results point out the relevance of CE for mAb separation and discrimination. However, the high concentration of saline solutions or dextrose employed for mAb compounding can be detrimental for the electrophoretic analysis. To circumvent this problem, dilution of the sample can be performed.

9.4 Quality Control of mAbs at Different Stages of Their Production

9.4.1 In-process Controls

Biotechnology production processes involve a multiplicity of host cells, most of the time mammalian ones but also yeasts or bacteria in research departments. The fermentation medium is composed from many different feeder components that have to be removed by extraction or clarification and several purification steps in order to obtain, with the best yield, the quality expected for the therapeutic protein. Only a few articles describe the use of CE for in-process control, which are mostly data indicating the purity level or quantity of the selected product in real time as presented hereafter.

In 1999, Hunt et al. [67] demonstrated the capability of capillary gel electrophoresis (CGE) for detecting a manufacturing inconsistency during recombinant protein production (see Section 9.4.2.2). The selectivity of the QC method was demonstrated with a sample contaminated with bacteria during the upstream process (Fig. 9.5). The presence of impurities was highlighted in the contaminated sample by comparison with the reference sample in both reduced and non-reduced conditions.

A typical technology for detecting mAb during its production can refer to the work published by Ohashi et al. [68]. This group compared Agilent 2100 Bioanalyzer to SDS-PAGE. The Bioanalyzer method (see Section 9.4.2.2) allows determination of IgG concentration in cell culture supernatants with high reproducibility and is commonly used as a QC to detect common degradations and aggregation patterns of antibodies.

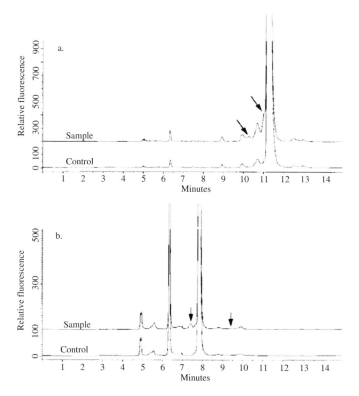

Figure 9.5 CE-SDS separations of (a) non-reduced and (b) reduced preparations of recombinant mAb control sample labeled with 5-TAMRA.SE, and a sample contaminated with microbes during cell culture fermentation. Arrows indicate the appearance of new peaks in the infected sample preparation. Reprinted with permission from Ref. [67], Copyright 1999, American Chemical Society.

A few years later, Sunday et al. [69] described a CE-SDS assay for recombinant mAbs (IgG4). The CGE mode to monitor the product purity used a linear polyacrylamide (LPA)-coated capillary (100 μm × 30 cm) to control the electroosmotic flow (EOF), and a replaceable sieving matrix from Beckman (proprietary entangled polymer) with UV detection (214 nm). The method was based on the Beckman protocol and was qualified with respect to precision, repeatability,

324 | *Capillary Electrophoresis for the Quality Control of Intact Therapeutic Monoclonal*

linearity (from 1 to 5 mg/mL), and limit of detection (LOD) and quantification (LOQ) estimation. CGE could monitor the product purity and showed the relation between the progressive elimination of impurities and the increase in purified mAb sample content. Using an alternative purification process, the removal of impurities was not observed (due to a preliminary affinity chromatographic step), and the control was more based on the monitoring of the entire mAbs and half mAbs (inter-hinge cleavage).

9.4.2 Control of Drug-Related Substances

9.4.2.1 Capillary zone electrophoresis

Capillary zone electrophoresis (CZE), due to its great simplicity of implementation, and the possibility to use mild conditions favorable to the preservation of the protein conformation, is very attractive to analyze intact mAbs. As in CZE the separation of analytes is based on their charge-to-size ratio, this mode is particularly well suited to detect degradations that affect the mAb charge like deamidation or C-terminal clipping, or its mass only like oxidation. Another advantage of CZE arises from the possibility to couple it with mass spectrometry (MS) in order to clearly characterize the separated peaks. A major issue in CZE is the adsorption of proteins onto the capillary wall, which degrades the separation performance. However, adsorption can be minimized through different strategies, the most efficient being the coating of the capillary. This can be accomplished via permanent static coatings or dynamic ones. Although CZE is not yet much employed in the pharmaceutical industries compared to CGE or CIEF for the quality control of therapeutic antibodies, in the last few years several groups have developed CZE-based methods to analyze, for example, charge variants of intact mAbs.

In 2011, He et al. [70] developed a CZE method based on their previous work [71] for the rapid analysis of charge heterogeneity of mAbs. The separation was carried out in a short capillary (total and effective lengths of 30 and 10 cm, respectively), which was dynamically coated with triethylenetetramine (TETA). The background electrolyte (BGE) consisted of 400 mM ε-aminocaproic acid (EACA)-acetic acid, 0.05% (w/v) hydroxypropylmethylcellulose (HPMC), and 2 mM TETA, pH 5.7. In these conditions, a selective and reproducible separation of mAb charge variants was achieved under very high electric field strength (1000 V/cm). The developed CZE method was applied to the separation of charge variants of multiple mAbs with pI in the range 7.0–9.5. Compared with other existing methods for charge variant analysis, this method offers several advantages. In particular, it enables a shorter run time since the separation is achieved in 2–5 min, whereas iCIEF, the other separation mode, requires 6–9 min to obtain a similar profile. One year later, Shi et al. [72] developed a CZE method with a higher resolution than that reported by He et al. [70]. In this case, a dynamic coating composed of TETA but also polyethylene oxide (PEO) was employed. The optimized CZE running buffer consisted of 20 mM acetate–acetic acid, pH 6.0, with the co-addition of 0.3% PEO and 2 mM TETA (Fig. 9.6). By analyzing a mAb1 with this CZE method, five charge variants were separated with relative standard deviations (RSDs) less than 0.6% and 3.2% for migration times and corrected peak areas, respectively.

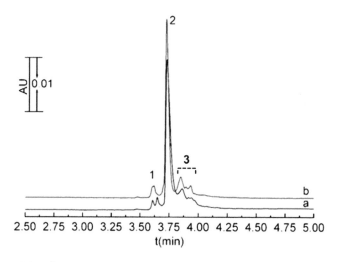

Figure 9.6 Charge variant separations conducted by using: (a) the method developed by Shi et al., CZE running buffer: 20 mM acetic-acetate, pH 6.0, 2 mM TETA, 0.3% PEO; (b) the reference method developed by He et al., CZE running buffer: 400 mM EACA, pH 5.7, 2 mM TETA, 0.05% HPMC. (1) Basic peaks, (2) main peak, and (3) acidic peaks. Capillary effective length: 20 cm; voltage: +30 kV; injection: 0.5 psi for 10 s; test sample: mAb1. Adapted from Ref. [72], Copyright 2012, with permission from Elsevier.

Espinosa de la Garza et al. developed, in 2013, two CZE methods to analyze the charge heterogeneity of diverse mAbs, including their biosimilars [73]. A neutral coated capillary with a hydrophilic polyacrylamide-based surface was used to prevent nonspecific and undesirable adsorption of proteins, but also to diminish the EOF and improve the resolution during the separation. UV detection was fixed at 214 nm to maximize the sensitivity for low-concentration variants. About 0.05% (w/v) HPMC was systematically added in the BGE. However, depending on the mAb, the best resolution was obtained with a BGE containing 200 mM EACA-acetic acid, and 30 mM lithium acetate, pH 4.8 (for rituximab, trastuzumab, and ranibizumab), or 150 mM EACA-acetic acid, and 20 mM lithium acetate, pH 5.5 (for infliximab and bevacizumab). In each case, lithium acetate base was used to level the BGE buffering capacity. Both CZE methods allowed detecting and quantifying the acidic and basic isoforms of the different mAbs. The RSDs on migration time and isoform content were less than 1% and 2%, respectively, whatever the analyzed antibody.

The same year, Gassner et al. [74] investigated the optimal CZE conditions associated with good isoform resolution, minimized adsorption, and acceptable repeatability for the analysis of intact mAbs. Various static capillary coatings were compared and evaluated. It was shown that for a positive coating (e.g., polybrene (PB)-dextran sulfate (DS)-PB layers, polyethylene imine (PEI), or UltraTrol^{TM} HR) and at pH < pI, the first criterion is that the EOF electrophoretic mobility (μEOF) and the mAb effective mobility have to be relatively close to each other to obtain the best isoform resolution. In contrast, for neutral coatings (e.g., hydroxypropylcellulose (HPC), polyvinyl alcohol (PVA), or UltraTrol^{TM} LN), adsorption was the main concern. Indeed, if the coating does not completely cover the surface charges, μEOF will be greater than zero and electrostatic interactions with the residual silanols will tend to decrease protein recovery and resolution through irreversible adsorption. The importance of the BGE pH was also demonstrated since the resolution of mAb isoforms is driven by their charge difference. The nature and concentration of the BGE were also studied, and an ampholyte such as EACA was found to decrease adsorption compared with ammonium acetate.

Very recently, an international cross-company study was performed to assess if CZE possesses all the required features (according to ICH Q2) for charge heterogeneity profiling of mAbs [75]. The analyses were carried out in bare fused-silica capillaries, previously flushed with the separation buffer composed of 400 mM EACA, 2 mM TETA, and 0.05% (w/v) HPMC, pH 5.7. It was shown that CZE was applicable for mAbs across a broad pI range between 7.4 and 9.5. The coefficient of correlation was above 0.99, demonstrating the good linearity of the method. Precision by repeatability was around 1%, and accuracy by recovery around 100%. The LOD was 1% and 0.3% for an initial sample concentration of 1 and 3.5 mg/mL, respectively, whereas the LOQ was 3.3% for 1 mg/mL and 1% for 3.5 mg/mL. Compared to ion-exchange chromatography and isoelectric focusing (IEF), CZE demonstrated a better resolution for the investigated mAb.

In parallel to conventional CZE, several groups developed methods based on microchip zone electrophoresis (MZE) for high-throughput analysis of mAbs. For example, Han et al. developed, in 2011, a MZE method for the analysis of mAb charge heterogeneity [76]. The method involved derivatization of protein molecules with Cy5 N-hydroxysuccinimide ester, which did not change the protein charge profile and enabled fluorescence detection on a commercial microchip instrument. The sample preparation could be performed in 96-well microtiter plates within 1 h, and each sample analysis took only 80 s (compared to 6 min with conventional CZE) (Fig. 9.7). Charge profiles similar to those obtained by conventional CZE were found for all the antibodies tested (pI in the range 7.5–9.2). The separation efficiency corresponded to 1.2×10^4 theoretical plates (1.0 μm plate height). In 2014, Wheeler et al. reported a new MZE method for the high-throughput screening of charge variants of mAbs in the pI range of 7–10 [77]. The mAb variants were also fluorescently labeled without altering the overall charge, and the separations were achieved in less than 90 s (i.e., 8–90 times faster than the conventional methods). Compared to the method of Han's group [76], no reconditioning of the microchannel surface was necessary between each sample separation. It was shown that this method enabled to monitor C-terminal lysine modifications, and thereby changes in product consistency resulting from these modifications, but also to quantify the extent of deamidation, making the method applicable routinely for drug stability studies.

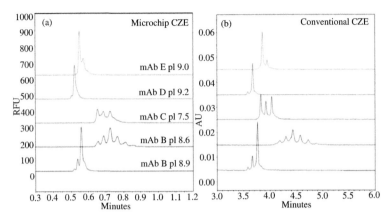

Figure 9.7 Comparison of microchip CZE (a) and conventional CZE (b) for five marketed antibody therapeutics. The main peak pI values determined by iCIEF are indicated. For mAb C, the pI values of the three major peaks from left to right are 7.81, 7.63, and 7.46, respectively. Reprinted with permission from Ref. [76], Copyright 2011, American Chemical Society.

In these last few years, several group tried to couple CZE with MS for a clear characterization of mAb isoforms. The coupling of CZE with MS is not easy due to the constraining conditions in terms of BGE composition and salt concentration. Indeed, CZE separation of intact proteins is highly affected by the pH and ionic strength of the BGE, and this represents a major drawback especially for the electrospray ionization (ESI)-MS detection, which requires volatile buffers. Moreover, some kinds of coatings, like the dynamic ones, are not compatible with MS. However, in 2014, Leize-Wagner's team succeeded in analyzing an intact model humanized mAb by developing a CZE-matrix-assisted laser desorption/ionization (MALDI)-MS approach, MALDI-MS presenting a greater tolerance to the presence of salts compared to ESI-MS [78]. For CE, a neutral HPC-coated capillary was used, and the BGE consisted of 400 mM EACA-acetic acid, 0.05% (w/v) HPMC, pH 5.7. In spite of the lack of resolution of the MALDI-MS detection, this CZE-MS method enabled, for the first time, the detection of intact mAb charge variants.

Very recently, Redman et al. developed an integrated microfluidic CE-ESI-MS device for the rapid separation and identification of intact mAb charge variants [79]. The need for dynamic coating and zwitterionic BGE additives was eliminated by utilizing surface chemistry within the device channels to control analyte adsorption and EOF while maintaining separation efficiency. This CE-MS coupling enabled to separate three C-terminal lysine variants of Infliximab using a BGE composed of 10% 2-propanol/0.2% acetic acid, pH 3.17. In addition to these lysine variants, minor acidic and basic species were detected, some of which could result from double deamidation. The general applicability of the method was demonstrated by analyzing one additional mAb: an IgG2. This latter proved to have similar modifications to Infliximab with lower relative abundances of the lysine variants.

9.4.2.2   Capillary gel electrophoresis

SDS-PAGE is a technique frequently used to analyze the purity of recombinant proteins and allows, at the same time, determination of molecular weight (MW) in comparison with standard. This method has been successfully transposed into the capillary format. Approaches to perform SDS-PAGE in capillaries have been well described in several reviews [80–84]. In CGE, separation takes place inside a capillary filled with a gel that acts as a molecular sieve. Proteins with similar charge-to-mass ratios due to their complexation with SDS (SDS binds globular protein at a ratio of 1.4 g of SDS by g of protein) are separated according to their molecular size and shape. The parameters to control and optimize for SDS-CGE separation are the nature of the gel, its concentration, the electric field, and the capillary coating. Indeed, capillary coating is required to increase the reproducibility of the separation with SDS gel or polymer network sieve and to eliminate the EOF.

The term "gel" in CGE is somewhat ambiguous. In the European Pharmacopeia, gels are classified regarding their characteristics [85]. Two types of gels are used in CE: permanently coated gels and dynamically coated ones. "Permanently coated gels" refer to covalently crosslinked polyacrylamide/bis-acrylamide, which are prepared inside the capillary by polymerization of the monomers. They are bound to the silanols of the fused-silica wall and cannot be removed. This preparation of crosslinked polyacrylamide in capillaries requires extreme care and can be used for antibody analysis under reducing or non-reducing conditions (e.g., analysis of an intact antibody). Polyacrylamide gels are difficult to polymerize homogeneously inside a capillary tube [86]. This approach is less and less employed for the QC of proteins due to several issues (reproducibility, cross-contamination, stability of the covalent bonds, time-consuming preparation of the capillary, and interferences at low UV).

The second sort of gel is called "dynamically coated gel," but authors generally use the term CGE-SDS and a replaceable matrix. In this case, hydrophilic and hydrosoluble polymers, such as LPA (at a concentration from 1% to 6%) that replaced crosslinked polyacrylamide, cellulose derivatives like hydroxyalkylcellulose (6–15%), dextran, PVA or agarose (0.05–1.2%) [87, 88], are dissolved in the BGE at a given concentration depending on the kind of polymer network required. The polymer solution creates a physical gel of an entangled polymer in situ through which proteins are separated following the Ogston model [89]. It is possible to increase the porosity of the gels by using polymers of higher molecular mass (at a fixed polymer concentration), e.g., dextran matrix exists within a broad molecular mass range from 1270 to 2000 kDa, or by decreasing the polymer concentration (for a given polymer molecular mass). Considering the easiness of preparation and performance of the separation of SDS-protein complex, the reasonable polymer size is around 100,000 Da [90]. This approach with replaceable sieving matrix can lead to better separation reproducibility, especially when the gel is replaced before each run.

Polysaccharide matrices for protein separation are excellent sieving matrices and have the advantage to be UV transparent compared to LPA. For example, Hu et al. [90] used slightly branched HPC (2%) in BGE (20 mM TRIS/20 mM Tricine and 0.1% SDS at pH 8) for standard protein separation. Finally, cellulose-derived separation media such as hydroxyethylcellulose or HPMC have a lower viscosity. Recently, Zaifang et al. [91] provided an overview of the sieving matrices used for CGE of proteins. In QC departments, for mAb separation, commercial kits from Beckman, Agilent, and BioRad that contain specific polymeric solutions as sieving matrices are widespread. In 2010, Zhang et al. [92] demonstrated the capability of their validated standard procedure for mAb purity control. In this procedure, bare fused-silica capillaries (30.2 cm × 50 μm) were employed, separation was performed at −15 kV, and detection was done at 220 nm. The BGE was provided by Beckman, and gel buffer consisted of 0.2% SDS and a proprietary polymeric formulation at pH 8. The authors optimized the sample preparation parameters, including sample buffer pH, alkylation and reduction conditions. They concluded that to preserve the tested mAb from fragmentation, a citrate phosphate buffer at pH 6.5 for sample preparation, instead of Beckman original one, was the optimal condition. Using another commercial kit (BioRad CE-SDS run buffer), Guo et al. [93] demonstrated that the method could resolve, under non-reduced conditions, structural isoforms of IgG2 mAbs with different arrangements of disulfide bonds.

Laser-induced fluorescence (LIF) detectors can be used in CGE to improve detection sensitivities. In this case, proteins are covalently labeled with fluorescent dyes, such as naphthalene-2,3-dicarboxaldehyde with a post-column fluorescence derivatization [94], 3-(2-furoyl)quinoline-2-carboxaldehyde (FQ) [92, 95], or 5-carboxytetramethylrhodamine, succinimidyl ester (5-TAMRA.SE) [67]. Nevertheless, sample preparations need to be robust and reproducible, especially the pH conditions that are known to have an impact on the derivatization rate of fluorescent labeling (pH impacts the acid–base properties of the target sites). Balland's group from Amgen [95] described a fluorescence derivatization method for protein analysis combined with the commercial CGE-SDS kit from Beckman. MAbs were derivatized with FQ in the presence of potassium cyanide (KCN). The use of KCN may be an issue for laboratories (safety), but it allows minimizing sample preparation artifacts such as mAb fragmentation or aggregation and improves detection sensitivity of labeled mAbs with LOD as low as 10 ng/mL.

Genentech [67, 96] was one of the first pharmaceutical companies to develop and validate a CGE-based method for the analysis of mAb. BioRad CE-SDS kit was used with a fused-silica capillary (24 cm × 50 µm). Analyses were performed on BioRad Biofocus 3000 CE system. They used LIF detection (argon ion laser with an excitation wavelength of 488 nm and an emission band pass filter of 560 nm) by derivatizing the investigated mAb with 5-TAMRA.SE, allowing a separation based on size of product-related substances and impurities (host cell proteins or process-related products). The protein Tag with fluorescent detection significantly improved the sensitivity, making possible the detection of minor species undetected with UV detection (half-antibody and HC under reducing conditions). Accuracy and precision of the assay were determined to monitor consistency and purity in compliance with acceptance criteria defined for batch release. One of the critical steps of the method was the labeling as some labeling agents may trigger mAb aggregation [67]. They demonstrated the capability of the method to discriminate impurities that are non-product-related from product-related ones. This method was validated according to ICH Q2 guidelines and is currently used as batch release method. The importance of sample preparation was discussed again by Lee et al. in 2000 [97], demonstrating that the heat treatment performed during the preparation of SDS-antibody complexes led to an increase in the tested mAb fragmentation.

CE-SDS gel has been applied to characterize glycan occupancy and number of glycosylation sites on mAbs [98, 99]. Glycans cannot bind SDS, thus the charge-to-mass ratio of glycosylated proteins is lower than that of the non-glycosylated counterparts, leading to a smaller migration time. The glycosylated and non-glycosylated forms of one mAb could be well separated using CE-SDS gel separation buffer from Beckman kit, and by analyzing the mAb after disulfide bond reduction. The electropherogram of Fig. 9.8 shows the separation of two main peaks corresponding to LC (MW ~25 kDa) and HC (MW ~50 kDa), a minor peak corresponding to non-glycosylated HC (NGHC), and an additional peak at p95 representing less than 1% of product-related impurities.

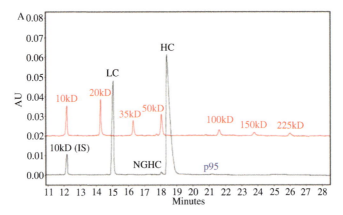

Figure 9.8 (A) CGE-SDS electropherogram of reduced final product mAb (black trace) with purity of 99% containing LC, HC, NGHC, and p95. The apparent MWs were calculated using the MW markers showed in the red trace. Reprinted with permission from Ref. [98], Copyright 2008, John Wiley and Sons.

The CE-SDS gel method is also suitable for formulation buffer screening and stability studies, which are compulsory for biopharmaceutical products. As shown in Fig. 9.9, mAb stability is better in the formulation buffer A than in the formulation buffer B, and protein clipping degradations are observed when mAb samples are incubated at elevated temperatures.

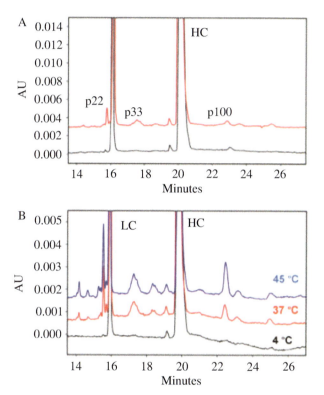

Figure 9.9 Reduced CE-SDS gel electropherograms. (A) For formulation screening (black trace, buffer A; red trace, buffer B). The samples were placed at 45°C for 1 week. (B) For accelerated temperature stress stability. The samples were incubated at different temperatures for 4 weeks. Reprinted with permission from Ref. [98], Copyright 2008, John Wiley and Sons.

Kaschak et al. reported a study of characterization of one glycated IgG1 [34]. As in the work of Hunt, they used a pre-column labeling to achieve LIF detection. They presented a study of glycation monitoring of both LC and HC. Glycated LC profile shows one shoulder on the tailing of the non-glycated LC peak. After purification, they identified with MS the sites implied by this glycation.

Pharmaceutical companies are looking for technologies that allow rapid and efficient screening at the best cost. For instance, Qiagen proposed a multicapillary CE method with LED-induced fluorescent detection for purity assays [100]. The developed method offered the possibility to perform high-throughput CGE analysis (24 samples per hour) (QIAxcel electrophoresis instrument). A 12-channel capillary cartridge was employed (containing bare fused-silica capillaries), the BGE consisted of 60 mM TRIS borate and 0.1% SDS at pH 8.5, and the separation was carried out at 6 kV with Beckman SDS gel buffer. MAbs were labeled with a covalent fluorophore (pyrylium-type dye bound to primary amino groups). The major issue of the labeling was the reducing conditions required. The ideal DTT concentration was established in a range of 5–10 mM. The method enabled to detect different fragmentation products of IgG and also to resolve non-glycosylated HC from glycosylated ones, allowing a reproducible relative quantitation for purity assessment.

Another type of CGE-based method relying this time on microfluidic technology was developed by Vasilyeva et al. [101] to support quantitative analysis of recombinant mAb samples. CE principles were transferred to a chip format that integrates all separation, staining, virtual destaining, and detection steps. Study was performed on Agilent 2100 Bioanalyzer. The detection system used a fluorescent dye (Agilent Technologies No. 5065-4430 proprietary, with excitation/emission at 650/680 nm, respectively), which is able to non-covalently bind protein-SDS complexes. The Protein 200 LabChip kit and automated standard operating procedure (Agilent Technologies) was used for sample preparation. LIF detector measured the labeled proteins after electrophoretic separation in a polymer solution. Bousse et al. [102] optimized the polymer solution containing polydimethylacrylamide at 3.25% in a TRIS-Tricine buffer at pH 7.6 (120 mM Tricine, 42 mM TRIS), 0.25% SDS (8.7 mM final concentration), and 4 µM of the dye used. The chips developed in collaboration with Caliper Technologies Corporation were made from "soda lime" glass; the size of each chip was 17.5 mm square. The method under non-reducing conditions proved to be quantitative and was used to check the presence of half-antibody fragments in IgG4 samples (Fig. 9.10).

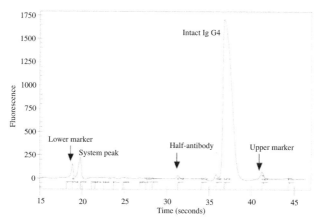

Figure 9.10 GelChip-CE analysis of an IgG4 sample. Four mL of an IgG4 sample diluted to 5 mg/mL was combined with 2 mL of the sample buffer containing 15 mM *N*-ethylmaleimide (NEM). The sample was heated in boiling water for 3 min and loaded onto the primed chip. Reprinted with permission from Ref. [101], Copyright 2004, John Wiley and Sons.

Agilent 2100 Bioanalyzer provided a fast standardized method with excellent reproducibility (RSD on purity content < 2%) and could be used to study the stability of antibodies under stress conditions (kit Protein 200 plus assay). The most interesting aspect of the Bioanalyzer technology is definitively the high-throughput analysis (run duration less than 1 min), which can be used in the downstream process of antibody production, and to compare the stability of mAbs stored in different formulations as demonstrated by Hamel's group [68].

In another work [103], microchip electrophoresis (LabChip® GXII, Perkin Elmer) showed impressive precision and accuracy results on marketed antibodies. Profiles were obtained in only a few minutes with a separation buffer from Caliper Life Sciences, and an LIF detection using anhydrophobic fluorescent dye (excitation/emission, 635/700 nm, respectively). The tricky part of the optimization was the selection of conditions able to evaluate the proportion of non-glycosylated HC, to identify glycan types, and to screen the presence of dimers [104, 105]. The high reproducibility and resolution are an advantage of this technology for the analysis of stressed products, or for stability studies required in pharmaceutical development.

9.4.2.3 Capillary isoelectric focusing

During the last two decades, capillary isoelectric focusing (CIEF) has proven its ability to combine the resolving power of isoelectric focusing (IEF) on gel format with advantages of CE quantitation and high throughput. As the separation is based on pIs, CIEF has emerged as an alternative for the characterization of charge heterogeneity of biomolecules and also as an identity assessment and, therefore, for QC. However, lack of robustness and reproducibility of the method has prevented the implementation of CIEF in routine analysis. In the past few years, some works focusing on method development and validation according to the ICH guideline Q2R1 were performed to facilitate regulatory and industrial acceptance of CIEF as a suitable and reliable technology to assess the purity and identity of therapeutic mAbs in biopharmaceutical industries [106].

In CIEF, the capillary is filled with a mixture containing the sample proteins, ampholytes, pI markers, and eventual additives. A pH gradient is formed by carrier ampholytes under electric field; then the proteins migrate and focus into discrete zones at the pH corresponding to their pI. There are two CIEF modes. In the two-step CIEF, the focused zones are shifted to the detector by hydrodynamic or chemical mobilization, the latter leading generally to a higher resolution. In the one-step CIEF, focusing and mobilization occur simultaneously using a low EOF, which is maintained in uncoated capillaries. This mode provides faster analyses. Protein adsorption and stable capillary coating to fully eliminate EOF are the two major obstacles remaining for CIEF. Although many capillary coating protocols have been published and may sometimes provide exceptional results, commercially coated capillaries are preferred for robustness and reproducibility of a CIEF method. One important issue is related to the stability of the capillary coating considering that one end of the capillary is immersed into an anolyte solution (acidic solution), while the other into a catholyte, one which is mainly composed of a diluted base (e.g., NaOH). Excellent reviews related to CIEF technologies provide technological aspects (mode, material, coating, additives, sample loading, pI markers, focusing time, pH gradients, detection method) as well as fundamental synopsis of microchip IEF (mIEF) [107–109]. Koshel and Wirth [110] reviewed technological progress from gel IEF to CIEF to packed capillaries with immobilized gradients for CIEF. More recently, Righetti et al. [111]

proposed a tutorial paper to guide the reader through the history of CIEF, then through the main steps of the process, from sample preparation to analysis of proteins and peptides, while commenting on the constraints and caveats of the technique.

One interesting paper concerns the work of Mack et al. [112] who present the first systematic and thorough study in CIEF, aimed at defining and optimizing all experimental parameters critical to method reproducibility and robustness. A PVA-coated capillary was selected as it remains stable even in extremely basic conditions. Addition of urea (3 M) into carrier ampholytes improved the protein solubility. Sacrificial ampholytes (1.7 mM iminodiacetic acid and 40 mM arginine) were introduced into the CIEF mixture for trapping a stable focused ampholyte train (4<pH<10) in the desired part of the capillary, and for eliminating sample loss caused by the bidirectional isotachophoresis inherent to IEF during focusing. Taking into account these parameters, the two-step CIEF method was then developed using chemical mobilization (catholyte replaced by a 350 mM acetic acid solution) combined with the use of anolyte (200 mM) and catholyte (300 mM) at high concentrations. Intermediate precision studies (RSD < 0.1% for the estimated pI values, and RSD < 3% for isoform proportion) performed with three mAbs demonstrated the excellent performance of the method that met the rigorous demands of therapeutic mAb QC. Since then, numerous groups adapted this standard separation to establish a high-resolution, robust, and reproducible CIEF method for the evaluation of mAb charge heterogeneity.

Maeda et al. [113] reported a powerful tool for reproducible evaluation of charge variants of both native mAb and new type mAb pharmaceutical more complex, gemtuzumab ozogamicin, which is a recombinant humanized IgG4 mAb conjugated with a cytotoxic antitumor antibiotic. This time, a neutral or DB-1 capillary and HPMC in the CIEF solution to avoid protein adsorption were used. Excellent repeatability and intermediate precision in estimated pI values of charge variants were obtained with RSDs no more than 0.06% and 0.95%, respectively. RSDs of charge variant compositions were less than 5%. However, authors highly recommended selecting appropriate commercially available pI markers based on the target pI ranges of charge variants in case of formation of nonlinear pH gradient during CIEF. For accurate determination of pI values, two appropriate pI markers that do not interfere with the peaks of the charge variants were added to the sample mixture as reference (Fig. 9.11).

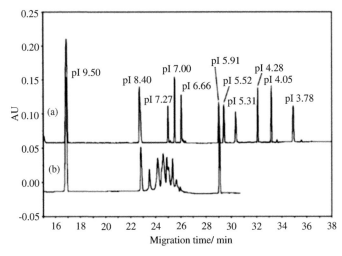

Figure 9.11 CIEF analysis of IgG4 for determination of pI values. (a) Eleven pI markers used for the calibration curve, and (b) IgG4 with two pI markers 8.40 and 5.91 which migrate before and after IgG4 peaks, and pI marker 9.50 as reference. Reprinted from Ref. [113], Copyright 2010, with permission from Elsevier.

Lin et al. [114] optimized the method published by Mack [112] to largely improve the resolution of the heterogenous peaks of mAb by using a mixture of 5% Pharmalyte 8-10.5 and 1% Pharmalyte 3-10. As the acidic forms of some mAbs have been demonstrated to reduce the binding ability to their antigens and hence lower potency [17, 20], this method was applied to select among four clones the one that possessed the least amount of acidic charge variants, and thus exhibiting a profile similar to the reference mAb of interest. It was used in a biopharmaceutical company to provide charge heterogeneity information in clone screening step.

Later, other groups [115, 116] adjusted several parameters of Mack's method but this time using a fluorocarbon (FC)-coated capillary. Suba et al. [116] thoroughly optimized all CIEF parameters to achieve the maximum resolution combined with short migration time (<18 min) and validated the established method according to the ICH guideline Q2R1. This method was demonstrated highly specific to mAbs exhibiting pI range between 7.0 and 9.0, as illustrated by the profiles obtained for three different commercially available therapeutic mAbs (Fig. 9.12).

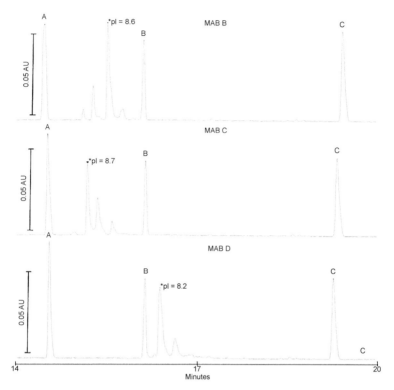

Figure 9.12 CIEF electrophoregrams of three different commercially available therapeutic mAbs (MAB B, MAB C, MAB D). pIs are given for the main peaks (*). A, B, and C are the synthetic peptide markers 9.0, 8.4, and 7.0. Reprinted from Ref. [116], Copyright 2015, with permission from Elsevier.

Recently, Bonn et al. [117] modified several parameters of the established Beckman CIEF kit to improve the robustness and reproducibility of the method. Implementation of a capillary revival solution (formamide-containing solution from Beckman) in the wash step, in combination with a commercially available PVA-coated (instead of polyacrylamide) capillary, inhibited capillary degradation and increased the reproducibility. This method optimization provided capillary robustness and performance longevity, achieving as much as a 20-fold increase in the number of consecutive runs (from 4–5 to more than 100 injections per capillary) before capillary degradation. Implementation of the system suitability, including number of isoforms, pI value, peak area-to-height ratio, and resolution, demonstrated the ability of the method to define predictable negative trends correlating with poor capillary performance. These key parameters calculated for two similar clinical immunodiagnostic mAbs, which have six isoforms, allowed to unambiguously identify them since differentiating factors were obtained in the percentage areas of the first isoform and the area-to-height ratio of the fourth isoform, while most isofoms differed in pI only by 0.03 units.

A relatively different CIEF, called iCIEF, first demonstrated by Wu and Pawliszyn in 1992 [118], uses a whole-column imaging detection and does not require the mobilization step that often causes either focused band broadening (hydrodynamic mobilization) or distortion of pH gradient (chemical mobilization) [119, 120]. In comparison to standard CIEF, iCIEF improves migration time and peak area reproducibilities, resolution, and affords lower analysis times (10–20 min instead of 60 min). This technology was developed by Convergent Bioscience, which marketed in 2000 the iCE280 IEF Analyzer (now commercialized by ProteinSimple), using a glass cartridge containing a transparent and short (5 cm) capillary, which is coated inside with FC (Fig. 9.13). This commercial instrument was reported to increase by at least twofold the analytical high throughput [121]. It has become a standard tool to evaluate charge heterogeneity of various recombinant proteins in particular therapeutic mAbs and has been routinely used in the QC environment in the last 10 years [122–124]. Besides being used for identity and charge isoform stability, iCIEF was also useful for monitoring a mAb and its binding to target antigen [125]. Analysis of a mixture of a mAb and its antigen at different concentrations showed a pI shift from 9.15 to 8.71 for the main peak of the mAb but also for all the acidic peaks (pI 8.9–9.1), together with the main peak. Very recently, Dada et al. [126] demonstrated the possibility to characterize acidic and basic variants of IgG1 therapeutic mAbs analyzed by iCIEF using a preparative immobilized pH gradient IEF (IPG-IEF) fractionation in combination with CE-SDS and MS. Deamidation, sialylation, glycation, and fragmentation were identified as the main modifications contributing to acidic variants of the mAb, while C-terminal lysine, C-terminal proline amidation, and uncyclized

N-terminal glutamine were the major species contributing to the basic variants.

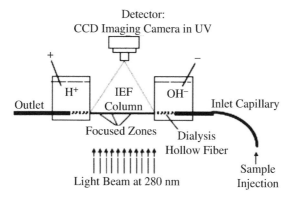

Figure 9.13 Diagram of iCE280 Analyzer. Reprinted with permission from Ref. [124], Copyright 2009, John Wiley and Sons.

Salas-Solano's group, a member of an international team including as much as 11 biopharmaceutical companies, evaluated the precision and robustness of CIEF [127] and iCIEF [128] methods to assess the charge heterogeneity of recombinant mAbs. CIEF method was based on the one optimized by Mack et al. [112]. Each laboratory used a ProteinSimple-assembled iCIEF cartridge with FC-coated capillary and the iCE280 Analyzer. Different analysts from each laboratory participated to these intercompany study using different ampholyte lots and multiple instruments. The apparent pI and the relative distribution of charge isoforms (peak area percentage) of recombinant mAbs were determined by each laboratory, showing very good precision between laboratories with RSD less than 0.8% and 11%, respectively, for iCIEF method; and less than 0.5% and 4.4%, respectively, for CIEF method. Surprisingly, these results showed the superiority of conventional CIEF in terms of precision of isoform distribution.

mIEF is another kind of iCIEF except that it is performed in a real microchip electrophoresis system MCE-2010 from Shimadzu, originally developed for high-throughput zone electrophoresis of DNA. Its features being unfavorable for IEF (cross channels, only one part of the separation channel imaged), only a few papers have used this apparatus for IEF of therapeutic proteins. Vlčková et al.

[129] reported for the first time IEF analysis of three therapeutic proteins, in particular a bevacizumab, using a quartz microchip coated with linear polyacrylamide. However, it was necessary to adjust the pH gradient by using a mixture of broad and narrow pH range ampholytes and adding N,N,N',N'-tetramethylethylenediamine (TEMED) to ensure the focusing of all selected compounds within the visible (imaged) part of the separation channel. This optimization of mIEF method is thus more complicated and time consuming compared to the specialized instrument iCE280. Recently, Kinoshita et al. [130] customized the MCE-2010 system for routine evaluation of QC of mAb pharmaceuticals using a non-coated quartz chip with short and straight channel. Charge variants could be resolved within 380 s with profiles well correlated to conventional CIEF ones. High reproducibility in terms of measured position (RSD of 0.81–1.28%) and calculated pI values (RSD < 0.28%) was obtained and found comparable to those obtained by iCIEF (0.2% or less) [123] and superior to the CIEF ones [113]. However, manual washing of the chip is still required, and further improvements in functionalization and automation will be required for high-throughput IEF analyses in the pharmaceutical industries.

CIEF has been coupled with various detection techniques, including UV, LIF, and MS. Although online coupling of CIEF with ESI-MS provides high separation efficiency and rich identity information, this direct hyphenation is challenging. CIEF-ESI-MS suffers from electrospray instabilities and ion suppression due to high ampholyte concentrations and the presence of non-volatile acids and bases in anolytes and catholytes, resulting in technical difficulties. To circumvent this problem, different strategies have been pursued and are deeply summarized in a recent review [131]. In 2012, Dovichi's group [132] reported a rapid CIEF-ESI-MS/MS method for the detection of host cell protein impurities in a recombinant humanized mAb product. A commercial polyacrylamide-coated capillary and ampholyte concentration as low as 0.4% were used. Three kinds of sheath flow buffers serving for both chemical mobilization and electrospray ionization were investigated: 50% methanol and different acids (0.05% acetic acid, 0.05 and 1% formic acid). High concentrations of formic acid were required for the mobilization step resulting in a narrow separation window, which was not desirable as it limits the number of tandem mass spectra accumulated during the separation.

9.4.3 Control of Compounded mAb before Patient Administration

The CE approaches reported previously for the analysis of intact mAbs after their compounding include CZE, CIEF, and imaged CIEF. Only a few illustrations can be found in the literature.

The CZE mechanism takes into account the whole physical and chemical properties of the mAbs. Indeed, mAb solvation is influenced by both charged residues and glycosylation. Accordingly, the hydrodynamic radius (Rh) of different mAbs may vary slightly as well as the global charge, leading thereby to potential differences in their electrophoretic mobility. Kwong Glover et al. [133] reported the separation by CZE of a mixture of pertuzumab and trastuzumab after their compounding using a neutral coated capillary. The two mAbs only differ by a few amino acids, and their theoretical pIs are quite close (8.48 and 8.39 for pertuzumab and trastuzumab, respectively). Their separation was obtained with a BGE containing 40 mM EACA-acetic acid at pH 4.5. The selected buffer pH led to a significant difference in the ionization state of the two mAbs, allowing separation of their main peaks despite their similar formulation. Therefore, we can assume that providing a high reproducibility of migration times, mAb identification by this CZE method is possible. Recently, Jaccoulet et al. developed a method allowing mAb identity assessment by demonstrating the separation of a complex mixture containing two humanized (bevacizumab, trastuzumab) and two chimeric (cetuximab and rituximab) mAbs using an internal standard and the estimation of relative migration times (tm) [134]. The separation relied on a counter EOF mode. The combination of a polybrene cationic coating and a separation buffer containing 75 mM sodium phosphate at pH 3.0, 0.15 mM perchloric acid, and 0.01% polysorbate 80 allowed a good resolution between the four mAbs. Perchloric acid was added as an ion-pairing agent to enhance the separation, while polysorbate improved the repeatability by preventing residual adsorption of excipients or mAbs. The electrophoregram of the four mAb mixture shows four distinct peaks corresponding to each investigated anticancer mAbs frequently administered at the Georges Pompidou Hospital in Paris, in less than 12 min (Fig. 9.14). By using an internal standard, very precise and accurate relative tm were determined for each mAb. The repeatability and the intermediate precision were less than 0.32% and 1.3%, respectively.

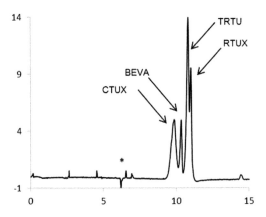

Figure 9.14 CZE separation of a mixture of cetuximab (CTUX), bevacizumab (BEVA), trastuzumab (TRTU), and rituximab (RTUX). BGE: 75 mM sodium phosphate at pH 3.0, 0.15 mM perchloric acid, and 0.01% polysorbate 80. Capillary: 50 μm × 60.5 cm. Coating: 0.2% PB. Applied voltage: −20 kV. *Internal standard: glutamine. Reprinted with permission from Ref. [134], Copyright 2015, John Wiley and Sons.

Another approach relies on CIEF or iCIEF modes to separate charge variants of each mAb and, therefore, to obtain specific mAb profile to help discriminating them from each other. However, mobilization of the focused zone is the critical step in CIEF, limiting the repeatability of the method. Furthermore, the analysis time is usually over 40 min, which is not compatible with the high-throughput analysis required at the hospital. On the contrary, iCIEF has the potential to fulfill the requirements for hospital QC of the mAbs by avoiding the time-consuming and reproducibility issues linked to the mobilization step.

In 2012, Salas-Solano et al. demonstrated the robustness of iCIEF for routine analysis of a recombinant mAb throughout different laboratories [128]. The ampholyte solution and the carrier ampholytes pH 5–8 including 1% methyl cellulose solution and 8 M urea were identical for all the laboratories. The two pI markers were 5.8 and 7.6. In this experiment, the focusing time was reduced to 4.5 min by applying a voltage of 3 kV. The same iCIEF profile of the recombinant mAb was obtained in the different laboratories showing seven charge isoforms of the mAb. The apparent pIs of therapeutic mAbs were determined by each laboratory with an RSD value lower than 0.8%. Quantitatively, the RSD of the relative peak areas across the laboratories were 8% and 11% for the smallest acidic and basic isoforms, respectively, and 1.5% for the main peak. These results showed a very short focusing time demonstrating thereby the potential reliability of iCIEF and its interest for hospital implementation. Kwong Glover et al. reported the use of iCIEF using a FC-coated capillary for the separation of a mixture of pertuzumab and trastuzumab treated with carboxypeptidase enzyme [133]. The mAb separation was obtained with ampholyte solutions containing methyl cellulose (0.35%), a mixture of pharmalyte pH 3–10 (0.47%) and pharmalyte pH 8–10.5 (2.66%) as carrier ampholytes, and pI markers (0.2%). The focusing step was performed by applying 1 kV for 1 min followed by 3 kV for 10 min. As the salt concentration was high after the sample compounding (150 mM of sodium chloride), a dilution of the samples in water was necessary to prevent the disturbance of the pH gradient. Talebi et al. obtained iCIEF profile of three mAbs with an identical FC-coated capillary [135]. The ampholyte solution consisted in methyl cellulose (1%), sucrose (20%), urea (5 M), and pharmalyte pH 3–10. Prefocusing and focusing times were 1.5 kV (1 min) and 3 kV (5 min), respectively. The iCIEF profile showed the acidic variants, the basic variants, and the main peak for the three mAbs with a very good resolution (Fig. 9.15). The profiles obtained for the three corresponding mAbs were significantly different, making their identification possible based on this heterogeneity evaluation.

Both CZE and iCIEF approaches demonstrated their usefulness for the routine QC of compounded mAbs. CZE has been recently demonstrated as a powerful tool for identification of very close compounded mAbs. Although iCIEF is widely used for quantitation purpose of mAb charge variants, precise identification could be possible providing high reproducibilies of charge heterogeneity distributions of mAbs. In that way, iCIEF could be a promising tool for routine QC of mAbs before patient administration.

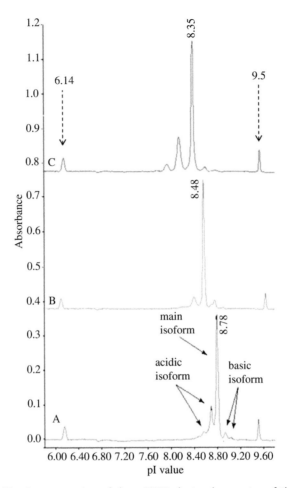

Figure 9.15 Representation of three iCIEF electrophoregrams of three IgG2 mAbs (A: mAb 1, B: mAb 2, C: mAb 3). Conditions: fused-silica capillary (50 mm, 100 μm i.d.) coated with fluorocarbon. iCIEF buffer solution: pI markers (dotted arrows 6.14 and 9.5), carrier ampholytes pH 3–10, 1% methylcellulose, 5 M urea, 20% sucrose, and mAb sample. Injection volume: 35 μL. Prefocusing: 1.5 kV (1 min) and focusing: 3 kV (5 min). Detection: 280 nm. Reprinted from Ref. [135], Copyright 2013, with permission from Elsevier.

9.5 Conclusion

This chapter shows that CE technology using absorbance or fluorescence-based detection methods has been applied for a long time in biopharmaceutical industries for the characterization of mAb purity and heterogeneity. The application of CE techniques for biopharmaceutical analysis is continuously growing. The method has been already implemented in Pharmacopoeias for therapeutic proteins but not yet for mAbs. CE with its different modes (in particular CZE, CGE, and CIEF) and its high-resolving power is now a well-established technique for the determination of drug-related substances of mAbs. It successfully contributes to the assessment of their identity from the drug substance or drug product, but also from the compounded pharmaceutical preparations at the hospital (infusion bags).

CIEF and CGE appear to be the most employed modes for the QC of mAbs. These two modes are fully complementary and address different types of questions for the QC. CIEF is probably the most powerful mode, but at the same time method development and optimization can be quite long. CIEF can resolve structurally close charge variants (sialylated, deamidated forms, C-terminal lysine variants), but this mode suffers from several drawbacks. One is related to the disturbance of the pH gradient caused by the presence of salts in the sample, precluding the use of CIEF for in-process controls. The second drawback is associated to the low sensitivity of UV detection due to the presence of UV-absorbing ampholytes in the BGE. iCIEF technique also showed good applications for the development of mAbs, reducing significantly the analysis time, while improving the reproducibility of the method. iCIEF is very attractive for QC but needs dedicated instrumentation. CGE is, in contrast, more routinely employed, although it is mostly used for separation of size variants (HC, LC, fragments) based on mass differences. This mode can also detect, with a certain level of success, aggregates and not fully glycosylated mAbs or those lacking one occupancy site. However, CGE cannot be employed to detect minor modifications of mAbs such as oxidation, deamidation, or racemization. The current evolution of CGE-SDS toward miniaturization, with commercialized instrumentation that brings speed and possibility of automation, renders this mode very attractive for the pharmaceutical companies despite its quite low resolution power. CZE can be considered the simplest method, which combines the charge and size of the mAb as the separation criterion. Although very performing applications have been demonstrated for mAbs with this mode, up to now only seldom applications have been reported in the field of QC. Nevertheless, caution has to be taken with CZE to reduce the possible adsorption of mAbs onto the capillary wall. Coating strategies are often necessary to achieve reproducibility, high resolution, accuracy of quantitation, and to improve the recoveries. Capillary coatings, be it dynamic or permanent ones, add a supplementary parameter that needs to be investigated, especially regarding their chemical and physical stability over time and rinses. In parallel to conventional CZE, several groups developed methods based on microchip zone electrophoresis for high-throughput profiling of mAb charge heterogeneity. These methods, using commercial microchip instrumentation, are relatively more complicated notably because they require the labeling of the proteins with a fluorescent dye. Yet they enable to dramatically decrease the analysis time (approximately 10 times shorter), even if the resolution of the CZE profiles is slightly lower than that obtained with conventional CZE. In addition, buffer exchange/desalting steps can be performed in microtiter plates, rendering these microchip-based methods applicable to crude cell culture samples.

Because of the significant developments and interest in CE-MS over the last few years, and since CGE and CIEF are hardly compatible with MS, CZE is being considered a valuable alternative that can be easily coupled with MS using ESI or MALDI interfaces. CE-MS analyses not only provide useful information on the glycosylation patterns of mAbs, but also enable the detection of intact mAb charge variants (e.g., lysine variants or deamidated forms).

In general, identity, heterogeneity, impurity content, and activity of each new batch of mAbs should be thoroughly investigated before release [3]. Due to the structural complexity of mAbs and their tendency to get easily degraded under certain stresses, this can only be achieved through the use of multiple analytical methods with a wide range of separation mechanisms, including liquid chromatographic ones (reversed-phase liquid chromatography, size-exclusion chromatography, ion-exchange chromatography), and electrophoretic ones (SDS-PAGE, CE). Finally, CE modes, and in particular CIEF, CGE, and CZE taken together, can serve to provide complementary insights into the QC of mAbs by assessing not only their identity through molecular mass (CGE) or pI (CIEF) estimations, but also by detecting different kinds of drug-related substances from fragments (CGE) to chemical degradations (CZE and CIEF) through abnormal glycosylation profiles (CIEF). CE techniques have, in some cases, complemented HPLC for better coverage of related substances. For instance, CZE showed several times advantages over a LC method in the quantitation and stability testing of biopharmaceuticals, mainly with regard to analysis time and sample and buffer consumption [136].

The major challenges for further implementation of CE in the pharmaceutical industry at the moment are probably the knowledge transfer and an increased implementation and ease of use of CE-MS instruments [136].

References

1. Reichert, J. M. (2016). Antibodies to watch in 2016, *mAbs*, **8**, pp. 197–204.

2. Ecker, D. M., Jones, S. D., and Levine, H. L. (2015). The therapeutic monoclonal antibody market, *mAbs*, **7**, pp. 9–14.

3. Fekete, S., Gassner, A. L., Rudaz, S., Schappler, J., and Guillarme, D. (2013). Analytical strategies for the characterization of therapeutic monoclonal antibodies, *Trends Anal. Chem.*, **42**, pp. 74–83.

4. Dübel, S. (2007). *Handbook of Therapeutic Antibodies* 1st eds. Becker, H., Volume III, Chapter 9 "Muronomab-CD3 (Orthoclone OKT3)", (Wiley-VCH, Weinheim), pp. 905–940.

5. Kelley, B. (2009). Industrialization of mAb production technology: The bioprocessing industry at a crossroads, *mAbs*, **1**, pp. 443–452.

6. Tamizi, E. and Jouyban, A. (2016). Forced degradation studies of biopharmaceuticals: Selection of stress conditions, *Eur. J. Pharm. Biopharm.*, **98**, pp. 26–46.

7. Beck, A., Wagner-Rousset, E., Ayoub, D., Van Dorsselaer, A., and Sanglier-Cianférani, S. (2013). Characterization of therapeutic antibodies and related products, *Anal. Chem.*, **85**, pp. 715–736.

8. ICH Q6B Specifications: Test Procedures and Acceptance Criteria for Biotechnological/Biological Products, 1999, issued as CPMP/ICH/365/96.

9. ICH Q5E Comparability of Biotechnological/Biological Products Subject to Changes in their Manufacturing Process, 2004, issued as CPMP/ICH/5721/03.

10. Brorson, K. and Jia, A. Y. (2014). Therapeutic monoclonal antibodies and consistent ends: Terminal heterogeneity, detection, and impact on quality, *Curr. Opin. Biotechnol.*, **30**, pp. 140–146.

11. Liu, H., Gaza-Bulseco, G., Faldu, D., Chumsae, C., and Sun, J. (2008). Heterogeneity of monoclonal antibodies, *J. Pharm. Sci.*, **97**, pp. 2426–2447.

12. Manning, M. C., Chou, D. K., Murphy, B. M., Payne, R. W., and Katayama, D. S. (2010). Stability of protein pharmaceuticals: An update, *Pharm. Res.*, **27**, pp. 544–575.

13. Levine, R. L., Moskovitz, J., and Stadtman, E. R. (2000). Oxidation of methionine in proteins: Roles in antioxidant defense and cellular regulation, *Life*, **50**, pp. 301–307.

14. Wei, Z., Feng, J., Lin, H. Y., Mullapudi, S., Bishop, E., Tous, G. I., Casas-Finet, J., Hakki, F., Strouse, R., and Schenerman, M. A. (2007). Identification of a single tryptophan residue as critical for binding activity in a humanized monoclonal antibody against respiratory syncytial virus, *Anal. Chem.*, **79**, pp. 2797–2805.

15. Kroon, D. J., Baldwin-Ferro, A., and Lalan, P. (1992). Identification of sites of degradation in a therapeutic monoclonal antibody by peptide mapping, *Pharm. Res.*, **9**, pp. 1386–1393.

16. Zhang, Y. T., Hu, J., Pace, A. L., Wong, R., Wang, Y. J., and Kao, Y. H. (2014). Characterization of asparagine 330 deamidation in an Fc-fragment of IgG1 using cation exchange chromatography and peptide mapping, *J. Chromatogr. B*, **965**, pp. 65–71.

17. Vlasak, J., Bussat, M. C., Wang, S., Wagner-Rousset, E., Schaefer, M., Klinguer-Hamour, C., Kirchmeier, M., Corvaïa, N., Ionescu, R., and Beck, A. (2009). Identification and characterization of asparagine deamidation in the light chain CDR1 of a humanized IgG1 antibody, *Anal. Biochem.*, **392**, pp. 145–154.

18. Yan, B., Steen, S., Hambly, D., Valliere-Douglass, J., Vanden Bos, T., Smallwood, S., Yates, Z., Arroll, T., Han, Y., Gadgil, H., Latypov, R. F., Wallace, A., Lim, A., Kleemann, G. R., Wang, W., and Balland, A. (2009). Succinimide formation at Asn 55 in the complementarity determining region of a recombinant monoclonal antibody IgG1 heavy chain, *J. Pharm. Sci.*, **98**, pp. 3509–3521.

19. Chelius, D., Rehder, D. S., and Bondarenko, P. V. (2005). Identification and characterization of deamidation sites in the conserved regions of human immunoglobulin gamma antibodies, *Anal. Chem.*, **77**, pp. 6004–6011.

20. Harris, R. J., Kabakoff, B., Macchi, F. D., Shen, F. J., Kwong, M., Andya, J. D., Shire, S. J., Bjork, N., Totpal, K., and Chen, A. B. (2001). Identification of multiple sources of charge heterogeneity in a recombinant antibody, *J. Chromatogr. B*, **752**, pp. 233–245.

21. Kwong, M. Y. and Harris, R. J. (1994). Identification of succinimide sites in proteins by N-terminal sequence analysis after alkaline hydroxylamine cleavage, *Protein Sci.*, **3**, pp. 147–149.

22. Wakankar, A. A. and Borchardt, R. T. (2006). Formulation considerations for proteins susceptible to asparagine deamidation and aspartate isomerization, *J. Pharm. Sci.*, **95**, pp. 2321–2336.

23. Reubsaet, J. L. E., Beijnen, J. H., Bult, A., Van Maanen, R. J., Marchal, J. A. D., and Underberg, W. J. M. (1998). Analytical techniques used to study the degradation of proteins and peptides: Chemical instability, *J. Pharm. Biomed. Anal.*, **17**, pp. 955–978.

24. Manning, M. C., Patel, K., and Borchardt, R. T. (1989). Stability of protein pharmaceuticals, *Pharm. Res.*, **6**, pp. 903–918.

25. Zhang, J., Yip, H., and Katta, V. (2011). Identification of isomerization and racemization of aspartate in the Asp-Asp motifs of a therapeutic protein, *Anal. Biochem.*, **410**, pp. 234–243.

26. Dick Jr., L. W., Kim, C., Qiu, D., and Cheng, K. C. (2007). Determination of the origin of the N-terminal pyro-glutamate variation in monoclonal antibodies using model peptides, *Biotechnol. Bioeng.*, **97**, pp. 544–553.

27. Vlasak, J. and Ionescu, R. (2011). Fragmentation of monoclonal antibodies, *mAbs*, **3**, pp. 253–263.

28. Cordoba, A. J., Shyong, B. J., Breen, D., and Harris, R. J. (2005). Non-enzymatic hinge region fragmentation of antibodies in solution, *J. Chromatogr. B*, **818**, pp. 115–121.

29. Gaza-Bulseco, G. and Liu, H. (2008). Fragmentation of a recombinant monoclonal antibody at various pH, *Pharm. Res.*, **25**, pp. 1881–1890.

30. Vistoli, G., De Maddis, D., Cipak, A., Zarkovic, N., Carini, M., and Aldini, G. (2013). Advanced glycoxidation and lipoxidation end products (AGEs and ALEs): An overview of their mechanisms of formation, *Free Radic. Res.*, **47**, pp. 3–27.

31. Yuk, I. H., Zhang, B., Yang, Y., Dutina, G., Leach, K. D., Vijayasankaran, N., Shen, A. Y., Andersen, D. C., Snedecor, B. R., and Joly, J. C. (2011). Controlling glycation of recombinant antibody in fed-batch cell cultures, *Biotechnol. Bioeng.*, **108**, pp. 2600–2610.

32. Fischer, S., Hoernschemeyer, J., and Mahler, H. C. (2008). Glycation during storage and administration of monoclonal antibody formulations, *Eur. J. Pharm. Biopharm.*, **70**, pp. 42–50.

33. Gadgil, H. S., Bondarenko, P. V., Pipes, G., Rehder, D., McAuley, A., Perico, N., Dillon, T., Ricci, M., and Treuheit, M. (2007). The LC/MS analysis of glycation of IgG molecules in sucrose containing formulations, *J. Pharm. Sci.*, **96**, pp. 2607–2621.

34. Kaschak, T., Boyd, D., and Yan, B. (2011). Characterization of glycation in an IgG1 by capillary electrophoresis sodium dodecyl sulfate and mass spectrometry, *Anal. Biochem.*, **417**, pp. 256–263.

35. Liu, H. and May, K. (2012). Disulfide bond structures of IgG molecules: Structural variations, chemical modifications and possible impacts to stability and biological function, *mAbs*, **4**, pp. 17–23.

36. Hmiel, L. K., Brorson, K. A., and Boyne II, M. T. (2015). Post-translational structural modifications of immunoglobulin G and their effect on biological activity, *Anal. Bioanal. Chem.*, **407**, pp. 79–94.

37. Cromwell, M. E. M., Hilario, E., and Jacobson, F. (2006). Protein aggregation and bioprocessing, *AAPS J.*, **8**, pp. 572–579.

38. Andya, J. D., Hsu, C. C., and Shire, S. J. (2003). Mechanisms of aggregate formation and carbohydrate excipient stabilization of lyophilized humanized monoclonal antibody formulations, *AAPS PharmSci.*, **5**, pp. 1–11.

39. Iwura, T., Fukuda, J., Yamazaki, K., Kanamaru, S., and Arisaka, F. (2014). Intermolecular interactions and conformation of antibody dimers present in IgG1 biopharmaceuticals, *J. Biochem.*, **155**, pp. 63–71.

40. Luo, J., Zhang, J., Ren, D., Tsai, W. L., Li, F., Amanullah, A., and Hudson, T. (2012). Probing of C-terminal lysine variation in a recombinant monoclonal antibody production using Chinese hamster ovary cells with chemically defined media, *Biotechnol. Bioeng.*, **109**, pp. 2306–2315.

41. Guerini Rocco, A., Mollica, L., Ricchiuto, P., Baptista, A. M., Gianazza, E., and Eberini, I. (2008). Characterization of the protein unfolding processes induced by urea and temperature, *Biophys. J.*, **94**, pp. 2241–2251.

42. Vermeer, A. W. P. and Norde, W. (2000). The thermal stability of immunoglobulin: Unfolding and aggregation of a multi-domain protein, *Biophys. J.*, **78**, pp. 394–404.

43. Wang, W. (2005). Protein aggregation and its inhibition in biopharmaceutics, *Int. J. Pharm.*, **289**, pp. 1–30.

44. Brader, M. L., Estey, T., Bai, S., Alston, R. W., Lucas, K. K., Lantz, S., Landsman, P., and Maloney, K. M. (2015). Examination of thermal unfolding and aggregation profiles of a series of developable

therapeutic monoclonal antibodies, *Mol. Pharmaceutics*, **12**, pp. 1005–1017.

45. Vázquez-Rey, M. and Lang, D. A. (2011). Aggregates in monoclonal antibody manufacturing processes, *Biotechnol. Bioeng.*, **108**, pp. 1494–1508.

46. Arosio, P., Barolo, G., Müller-Späth, T., Wu, H., and Morbidelli, M. (2011). Aggregation stability of a monoclonal antibody during downstream processing, *Pharm. Res.*, **28**, pp. 1884–1894.

47. Nicoud, L., Sozo, M., Arosio, P., Yates, A., Norrant, E., and Morbidelli, M. (2014). Role of cosolutes in the aggregation kinetics of monoclonal antibodies, *J. Phys. Chem. B*, **118**, pp. 11921–11930.

48. Ratanji, K. D., Derrick, J. P., Dearman, R. J., and Kimber, I. (2014). Immunogenicity of therapeutic proteins: Influence of aggregation, *J. Immuno.*, **11**, pp. 99–109.

49. Rosenberg, A. S. (2006). Effects of protein aggregates: An immunologic perspective, *AAPS J.*, **8**, pp. 501–507.

50. Crommelin, D. J. A. (2008). *Pharmaceutical Biotechnology: Fundamentals and Applications*, 3rd Ed., eds. Crommelin, D. J. A., Sindelar, R. D., and Meibohm, B. (Informa Healthcare, New York).

51. Kats, M., Richberg, P. C., and Hughes, D. E. (1995). Conformational diversity and conformational transitions of a monoclonal antibody monitored by circular dichroism and capillary electrophoresis, *Anal. Chem.*, **67**, pp. 2943–2948.

52. Kats, M., Richberg, P. C., and Hughes, D. E. (1997). pH-dependent isoform transitions of a monoclonal antibody monitored by micellar electrokinetic capillary chromatography, *Anal. Chem.*, **69**, pp. 338–343.

53. Liu, H., Gaza-Bulseco, G., Xiang, T., and Chumsae, C. (2008). Structural effect of deglycosylation and methionine oxidation on a recombinant monoclonal antibody, *Mol. Immunol.*, **45**, pp. 701–708.

54. Brady, L. J., Martinez, T., and Balland, A. (2007). Characterization of nonenzymatic glycation on a monoclonal antibody, *Anal. Chem.*, **79**, pp. 9403–9413.

55. Quan, C., Alcala, E., Petkovska, I., Matthews, D., Canova-Davis, E., Taticek, R., and Stacey, M. (2008). A study in glycation of a therapeutic recombinant humanized monoclonal antibody: Where it is, how it got there, and how it affects charge-based behavior, *Anal. Biochem.*, **373**, pp. 179–191.

56. Gester, A. (2013). High-throughput chromatography screenings for modulating charge-related isoform patterns, *BioProcess. Int.*, **11**, pp. 46–53.

57. Hossler, P., Khattak, S. F., and Li, Z. J. (2009). Optimal and consistent protein glycosylation in mammalian cell culture, *Glycobiology*, **19**, pp. 936–949.

58. Read, E. K., Park, J. T., and Brorson, K. A. (2011). Industry and regulatory experience of the glycosylation of monoclonal antibodies, *Biotechnol. Appl. Biochem.*, **58**, pp. 213–219.

59. Jefferis, R. (2009). Recombinant antibody therapeutics: The impact of glycosylation on mechanisms of action, *Trends Pharmacol. Sci.*, **30**, pp. 356–362.

60. Costa, A. R., Rodrigues, M. E., Henriques, M., Oliveira, R., and Azeredo, J. (2014). Glycosylation: Impact, control and improvement during therapeutic protein production, *Crit. Rev. Biotechnol.*, **34**, pp. 281–299.

61. Wright, A. and Morrison, S. L. (1998). Effect of C2-associated carbohydrate structure on Ig effector function: Studies with chimeric mouse human IgG1 antibodies in glycosylation mutants of Chinese hamster ovary cells, *J. Immunol.*, **160**, pp. 3393–3402.

62. Shields, R. L., Lai, J., Keck, R., O'Connell, L. Y., Hong, K., Meng, Y. G., Weikert, S. H., and Presta, L. G. (2002). Lack of fucose on human IgG1 N-linked oligosaccharide improves binding to human Fcγ RIII and antibody-dependent cellular toxicity, *J. Biol. Chem.*, **277**, pp. 26733–26740.

63. Shinkawa, T., Nakamura, K., Yamane, N., Shoji-Hosaka, E., Kanda, Y., Sakurada, M., Uchida, K., Anazawa, H., Satoh, M., Yamasaki, M., Hanai, N., and Shitara, K. (2003). The absence of fucose but not the presence of galactose or bisecting N-acetylglucosamine of human IgG1 complex-type oligosaccharides shows the critical role of enhancing antibody-dependent cellular cytotoxicity, *J. Biol. Chem.*, **278**, pp. 3466–3473.

64. Gahoual, R., Biacchi, M., Chicher, J., Kuhn, L., Hammann, P., Beck, A., Leize-Wagner, E., and François, Y. N. (2014). Monoclonal antibodies biosimilarity assessment using transient isotachophoresis capillary zone electrophoresis-tandem mass spectrometry, *mAbs*, **6**, pp. 1464–1473.

65. Lehermayr, C. (2011). Assessment of net charge and protein-protein interactions of different monoclonal antibodies, *J. Pharm. Sci.*, **100**, pp. 2551–2562.

66. Li, S. K. (2011). Effective electrophoretic mobilities and charges of anti-VEGF proteins determined by capillary zone electrophoresis, *J. Pharm. Biomed. Anal.*, **55**, pp. 603–607.

67. Hunt, G. and Nashabeh, W. (1999). Capillary electrophoresis sodium dodecyl sulfate non gel sieving analysis of a therapeutic recombinant monoclonal antibody: A biotechnology perspective, *Anal. Chem.*, **71**, pp. 2390–2397.

68. Ohashi, R., Otero, J. M., Chwistek, A., and Hamel, J. F. P. (2002). Determination of monoclonal antibody production in cell culture using novel microfluidic and traditional assays, *Electrophoresis*, **23**, pp. 3623–3629.

69. Sunday, B. R., Sydor, W., Guariglia, L. M., Obara, J., and Mengisen, R. J. (2003). Process and product monitoring of recombinant DNA-derived biopharmaceuticals with high-performance capillary electrophoresis, *J. Capill. Electrophor. Microchip. Technol.*, **8**, pp. 87–99.

70. He, Y., Isele, C., Hou, W., and Ruesch, M. (2011). Rapid analysis of charge variants of monoclonal antibodies with capillary zone electrophoresis in dynamically coated fused-silica capillary, *J. Sep. Sci.*, **34**, pp. 548–555.

71. He, Y., Lacher, N. A., Hou, W., Wang, Q., Isele, C., Starkey, J., and Ruesch, M. (2010). Analysis of identity, charge variants, and disulfide isomers of monoclonal antibodies with capillary zone electrophoresis in an uncoated capillary column, *Anal. Chem.*, **82**, pp. 3222–3230.

72. Shi, Y., Li, Z., Qiao, Y., and Lin, J. (2012). Development and validation of a rapid capillary zone electrophoresis method for determining charge variants of mAb, *J. Chromatogr. B*, **906**, pp. 63–68.

73. Espinosa de la Garza, C. E., Perdomo-Abúndez, F. C., Padilla-Calderón, J., Uribe-Wiechers, J. M., Pérez, N. O., Flores-Ortiz, L. F., and Medina-Rivero, E. (2013). Analysis of recombinant monoclonal antibodies by capillary zone electrophoresis, *Electrophoresis*, **34**, pp. 1133–1140.

74. Gassner, A. L., Rudaz, S., and Schappler, J. (2013). Static coatings for the analysis of intact monoclonal antibody drugs by capillary zone electrophoresis, *Electrophoresis*, **34**, pp. 2718–2724.

75. Moritz, B., Schnaible, V., Kiessig, S., Heyne, A., Wild, M., Finkler, C., Christians, S., Mueller, K., Zhang, L., Furuya, K., Hassel, M., Hamm, M., Rustandi, R., He, Y., Salas-Solano, O., Whitmore, C., Park, S. A., Hansen, D., Santos, M., and Lies, M. (2015). Evaluation of capillary zone electrophoresis for charge heterogeneity testing of monoclonal antibodies, *J. Chromatogr. B*, **983–984**, pp. 101–110.

76. Han, H., Livingston, E., and Chen, X. (2011). High throughput profiling of charge heterogeneity in antibodies by microchip electrophoresis, *Anal. Chem.*, **83**, pp. 8184–8191.

77. Wheeler, T. D., Sun, J. L., Pleiner, S., Geier, H., Dobberthien, P., Studts, J., Singh, R., and Fathollahi, B. (2014). Microchip zone electrophoresis for high-throughput analysis of monoclonal antibody charge variants, *Anal. Chem.*, **86**, pp. 5416–5424.

78. Biacchi, M., Bhajun, R., Saïd, N., Beck, A., François, Y. N., and Leize-Wagner, E. (2014). Analysis of monoclonal antibody by a novel CE-UV/MALDI-MS interface, *Electrophoresis*, **35**, pp. 2986–2995.

79. Redman, E. A., Batz, N. G., Mellors, J. S., and Ramsey, J. M. (2015). Integrated microfluidic capillary electrophoresis-electrospray ionization devices with online MS detection for the separation and characterization of intact monoclonal antibody variants, *Anal. Chem.*, **87**, pp. 2264–2272.

80. Karger, B. L., Cohen, A. S., and Guttman, A. (1989). High-performance capillary electrophoresis in the biological sciences, *J. Chromatogr. B*, **492**, pp. 585–614.

81. Wu, D. and Regnier, F. E. (1992). Sodium dodecyl sulfate-capillary gel electrophoresis of proteins using non-cross-linked polyacrylamide, *J. Chromatogr. A*, **608**, pp. 346–356.

82. Guttman, A. and Nolan, J. (1994). Comparison of the separation of proteins by sodium dodecyl sulfate-slab gel electrophoresis and capillary sodium dodecyl sulfate-gel electrophoresis, *Anal. Biochem.*, **221**, pp. 285–294.

83. Guttman, A., Shieh, P., Lindahl, J., and Cooke, N. (1994). Capillary sodium dodecyl sulfate gel electrophoresis of proteins II. On the Ferguson method in polyethylene oxide gels, *J. Chromatogr. A*, **676**, pp. 227–231.

84. Guttman, A. (1996). Capillary sodium dodecyl sulfate gel electrophoresis of proteins, *Electrophoresis*, **17**, pp. 1333–1341.

85. European Pharmacopoeia 8.8 Chapter 2.2.47 Electrophoresis 2016.

86. Tsuji, K. (1991). High-performance capillary electrophoresis of proteins: Sodium dodecyl sulphate-polyacrylamide gel-filled capillary column for the determination of recombinant biotechnology-derived proteins, *J. Chromatogr. A*, **550**, pp. 823–830.

87. Ganzler, K., Greve, K. S., Cohen, A. S., Karger, B. L., Guttman, A., and Cooke, N. C. (1992). High-performance capillary electrophoresis of SDS-protein complexes using UV-transparent polymer networks, *Anal. Chem.*, **64**, pp. 2665–2671.

88. Lausch, R., Scheper, T., Reif, O. W., Schlösser, J., Fleischer, J., and Freitag, R. (1993). Rapid capillary gel electrophoresis of proteins, *J. Chromatogr. A*, **654**, pp. 190–195.

89. Boileau, J. and Slater, G. W. (2001). An exactly solvable Ogston model of gel electrophoresis. VI. Towards a theory for macromolecules, *Electrophoresis*, **22**, pp. 673–683.

90. Hu, S., Zhang, Z., Cook, L. M., Carpenter, J., and Dovichi, N. J. (2000). Separation of proteins by sodium dodecylsulfate capillary electrophoresis in hydroxypropylcellulose sieving matrix with laser-induced fluorescent detection, *J. Chromatogr. A*, **894**, pp. 291–296.

91. Zaifang, Z., Joann, J. L., and Shaorong, L. (2011). Protein separation by capillary gel electrophoresis: A review, *Anal. Chim. Act.*, **709**, pp. 21–31.

92. Zhang, J., Burman, S., Gunturi, S., and Foley, J. F. (2010). Method development and validation of capillary sodium dodecyl sulfate gel electrophoresis for the characterization of a monoclonal antibody, *J. Pharm. Bio. Anal.*, **53**, pp. 1236–1243.

93. Guo, A., Han, M., Martinez, T., Ketchem, R. R., Novich, S., Jochheim, C., and Balland, A. (2008). Electrophoretic evidence for the presence of structural isoforms specific for the IgG2 isotype, *Electrophoresis*, **29**, pp. 2550–2556.

94. Ye, M., Hu, S., Quigley, Wes, W. C., and Dovichi, N. J. (2004). Post-column fluorescence derivatization of proteins and peptides in capillary electrophoresis with a sheath flow reactor and 488 nm argon ion laser excitation, *J. Chromatogr. A*, **1022**, pp. 201–206.

95. Michels, D. A., Brady, L. J., Guo, A., and Balland, A. (2007). Fluorescent derivatization method of proteins for characterization by capillary electrophoresis-sodium dodecyl sulfate with laser-induced fluorescent detection, *Anal. Chem.*, **79**, pp. 5963–5971.

96. Hunt, G., Moorhouse, K. G., and Chen, A. B. (1996). Capillary isoelectric focusing and sodium dodecyl sulfate capillary gel electrophoresis of recombinant humanized monoclonal antibody HER2, *J. Chromatogr. A*, **744**, pp. 295–301.

97. Lee, H. G. (2000). High-performance sodium dodecyl sulfate-capillary gel electrophoresis of antibodies and antibody fragments, *J. Immun. Met.*, **234**, pp. 71–84.

98. Rustandi, R. R., Washabaugh, M. W., and Wang, Y. (2008). Applications of CE SDS gel in development of biopharmaceutical antibody-based products, *Electrophoresis*, **29**, pp. 3612–3620.

99. Rustandi, R. R., Anderson, C., and Hamm, M. (2013). Application of capillary electrophoresis in glycoprotein analysis, *Met. Mol. Biol.*, **988**, pp. 181–197.

100. Szekrenyes, A., Roth, U., Kerékgyártó, M., Székely, A., Kurucz, I., Kowalewski, K., and Guttman, A. (2013). High-throughput analysis of therapeutic and diagnostic monoclonal antibodies by multicapillary SDS gel electrophoresis in conjunction with covalent fluorescent labeling, *Anal. Bioanal. Chem.*, **404**, pp. 1485–1494.

101. Vasilyeva, E., Woodard, J, Taylor, F. R., Kretschmer, M., Fajardo, H., Lyubarskaya, Y., Kobayashi, K, Dingley, A., and Mhatre, R. (2004). Development of a chip-based capillary gel electrophoresis method for quantification of a half-antibody in immunoglobulin G4 samples, *Electrophoresis*, **25**, pp. 3890–3896.

102. Bousse, L., Mouradian, S., Minalla, A., Yee, H., Williams, K., and Dubrow, R. (2001). Protein sizing on a microchip, *Anal. Chem.*, **73**, pp. 1207–1212.

103. Yagi, Y., Kakehl, K., Hayakawa, T., and Suzuki, S. (2014). Application of microchip electrophoresis sodium dodecyl sulfate for the evaluation of change of degradation species of therapeutic antibodies in stability testing, *Anal. Sci.*, **30**, pp. 483–488.

104. Chen, X., Tang, K., Lee, M., and Flynn, G. C. (2008). Microchip assays for screening monoclonal antibody product quality, *Electrophoresis*, **29**, pp. 4993–5002.

105. Chen, X. and Flynn, G. C. (2009). A high throughput dimer screening assay for monoclonal antibodies using chemical cross-linking and microchip electrophoresis, *J. Chromatogr. B*, **877**, pp. 3012–3020.

106. Apostol, I., Miller, K. J., Ratto, J., and Kelner, D. N. (2009). Comparison of different approaches for evaluation of the detection and quantitation limits of a purity method: A case study using a capillary isoelectrofocusing method for a monoclonal antibody, *Anal. Biochem.*, **385**, pp. 101–106.

107. Silvertand, L. H. H., Toraño, J. S., van Bennekom, W. P., and de Jong, G. J. (2008). Recent developments in capillary isoelectric focusing, *J. Chromatogr. A*, **1204**, pp. 157–170.

108. Shimura, K. (2009). Recent advances in IEF in capillary tubes and microchips, *Electrophoresis*, **30**, pp. 11–28.

109. Sommer, G. J. and Hatch, A. V. (2009). IEF in microfluidic devices, *Electrophoresis*, **30**, pp. 742–757.

110. Koshel, B. M. and Wirth, M. J. (2012). Trajectory of isoelectric focusing from gels to capillaries to immobilized gradients in capillaries, *Proteomics*, **12**, pp. 2918–2926.

111. Righetti, P. G., Sebastiano, R., and Citterio, A. (2013). Capillary electrophoresis and isoelectric focusing in peptide and protein analysis, *Proteomics*, **13**, pp. 325–340.

112. Mack, S., Cruzado-Park, I., Chapman, J., Ratnayake, C., and Vigh, G. (2009). A systematic study in CIEF: Defining and optimizing experimental parameters critical to method reproducibility and robustness, *Electrophoresis*, **30**, pp. 4049–4058.

113. Maeda, E., Urakami, K., Shimura, K., Kinoshita, M., and Kakehi, K. (2010). Charge heterogeneity of a therapeutic monoclonal antibody conjugated with a cytotoxic antitumor antibiotic, calicheamicin, *J. Chromatogr. A*, **1217**, pp. 7164–7171.

114. Lin, J., Tan, Q., and Wang, S. (2011). A high-resolution capillary isoelectric focusing method for the determination of therapeutic recombinant monoclonal antibody, *J. Sep. Sci.*, **34**, pp. 1696–1702.

115. Cao, J., Sun, W., Gong, F., and Liu, W. (2014). Charge profiling and stability testing of biosimilar by capillary isoelectric focusing, *Electrophoresis*, **35**, pp. 1461–1468.

116. Suba, D., Urbányi, Z., and Salgó A. (2015). Capillary isoelectric focusing method development and validation for investigation of recombinant therapeutic monoclonal antibody, *J. Pharm. Biomed. Anal.*, **114**, pp. 53–61.

117. Bonn, R., Rampal, S., Rae, T., and Fishpaugh, J. (2013). CIEF method optimization: Development of robust and reproducible protein reagent characterization in the clinical immunodiagnostic industry: CE and CEC, *Electrophoresis*, **34**, pp. 825–832.

118. Wu, J. and Pawliszyn, J. (1992). Capillary isoelectric focusing with a universal concentration gradient imaging system using a charge-coupled photodiode array, *Anal. Chem.*, **64**, pp. 2934–2941.

119. Goodridge, L., Goodridge, C., Wu, J., Griffiths, M., and Pawliszyn, J. (2004). Isoelectric point determination of norovirus virus-like particles by capillary isoelectric focusing with whole column imaging detection, *Anal. Chem.*, **76**, pp. 48–52.

120. Wu, J. and Pawliszyn, J. (1994). Application of capillary isoelectric focusing with absorption imaging detection to the analysis of proteins, *J. Chromatogr. B*, **657**, pp. 327–334.

121. Li, N., Kessler, K., Bass, L., and Zheng, D. (2007). Evaluation of the iCE280 Analyzer as a potential high-throughput tool for formulation development, *J. Pharm. Biomed. Anal.*, **43**, pp. 963–972.

122. Janini, G., Saptharishi, N., Waselus, M., and Soman, G. (2002). Element of a validation method for MU-B3 monoclonal antibody using an imaging capillary isoelectric focusing system, *Electrophoresis*, **23**, pp. 1605–1611.

123. Sosic, Z., Houde, D., Blum, A., Carlage, T., and Lyubarskaya, Y. (2008). Application of imaging capillary IEF for characterization and quantitative analysis of recombinant protein charge heterogeneity, *Electrophoresis*, **29**, pp. 4368–4376.

124. He, X. Z., Que, A. H., and Mo, J. J. (2009). Analysis of charge heterogeneities in mAbs using imaged CE, *Electrophoresis*, **30**, pp. 714–722.

125. Anderson, C. L., Wang, Y., and Rustandi, R. R. (2012). Applications of imaged capillary isoelectric focusing technique in development of biopharmaceutical glycoprotein-based products: CE and CEC, *Electrophoresis*, **33**, pp. 1538–1544.

126. Dada, O. O., Jaya, N., Valliere-Douglass, J., and Salas-Solano, O. (2015). Characterization of acidic and basic variants of IgG1 therapeutic monoclonal antibodies based on non-denaturing IEF fractionation, *Electrophoresis*, **20**, pp. 2695–2702.

127. Salas-Solano, O., Babu, K., Park, S. A. S., Zhang, X., and Zhang, L. (2011). Intercompany study to evaluate the robustness of capillary isoelectric focusing technology for the analysis of monoclonal antibodies, *Chromatographia*, **73**, pp. 1137–1144.

128. Salas-Solano, O., Kennel, B., Park, S. S., Roby, K., Sosic, Z., Boumajny, B., Free, S., Reed-Bogan, A., Michels, D., McElroy, W., Bonasia, P., Hong, M., He, X., Ruesch, M., Moffatt, F., Kiessig, S., and Nunnally, B. (2012). Robustness of iCIEF methodology for the analysis of monoclonal antibodies: An interlaboratory study, *J. Sep. Sci.*, **35**, pp. 3124–3129.

129. Vlčková, M., Kalman, F., and Schwarz, M. A. (2008). Pharmaceutical applications of isoelectric focusing on microchip with imaged UV detection, *J. Chromatogr. A*, **1181**, pp. 145–152.

130. Kinoshita, M., Nakatsuji, Y., Suzuki, S., Hayakawa, T., and Kakehi, K. (2013). Quality assurance of monoclonal antibody pharmaceuticals based on their charge variants using microchip isoelectric focusing method, *J. Chromatogr. A*, **1309**, pp. 76–83.

131. Hühner, J., Lämmerhofer, M., and Neusüß, C. (2015). Capillary isoelectric focusing-mass spectrometry: Coupling strategies and applications, *Electrophoresis*, **36**, pp. 2670–2686.

132. Zhu, G., Sun, L., Wojcik, R., Kernaghan, D., McGivney, J. B., and Dovichi, N. J. (2012). A rapid cIEF-ESI-MS/MS method for host cell protein

analysis of a recombinant human monoclonal antibody, *Talanta*, **98**, pp. 253–256.

133. Kwong Glover, Z. W., Gennaro, L., Yadav, S., Demeule, B., Wong, P. Y., and Sreedhara, A. (2013). Compatibility and stability of pertuzumab and trastuzumab admixtures in i.v. infusion bags for coadministration, *J. Pharm. Sci.*, **102**, pp. 794–812.

134. Jaccoulet, E., Smadja, C., Prognon, P., and Taverna, M. (2015). Capillary electrophoresis for rapid identification of monoclonal antibodies for routine application in hospital, *Electrophoresis*, **36**, pp. 2050–2056.

135. Talebi, M., Nordborg, A., Gaspar, A., Lacher, N. A., Wang, Q., and He, X. Z. (2013). Charge heterogeneity profiling of monoclonal antibodies using low ionic strength ion-exchange chromatography and well-controlled pH gradients on monolithic columns, *J. Chromatogr. A*, **1317**, pp. 148–154.

136. El Deeb, S., Watzig, H., Abdelhady, D., Albishri, H. M., Sänger-Van de Griend, C., and Scriba, G. K. E. (2014). Recent advances in capillary electrophoretic migration techniques for pharmaceutical analysis, *Electrophoresis*, **35**, pp. 170–189.

Chapter 10

Molecular Simulation of Chiral Selector–Enantiomer Interactions through Docking: Antimalarial Drugs as Case Study

Myalowenkosi Sabela, Suvardhan Kanchi, Deepali Sharma, and Krishna Bisetty

Department of Chemistry, Steve Biko Campus, Durban University of Technology, P.O. Box 1334, Durban 4000, South Africa

myalosabela@gmail.com

10.1 Introduction

Chiral recognition has been among the most attractive fields of research, particularly to the chromatography family due to a large number of concerns in the fields, which include life science, food science, drug discovery, and environmental science. In particular, different stereoisomers of drugs attribute to different physiological responses; hence it is critical to use pure molecules to be able to improve and control the therapeutic effects. In capillary electrophoresis (CE) methods, chiral selectors are used to improve

---

Capillary Electrophoresis: Trends and Developments in Pharmaceutical Research

Edited by Suvardhan Kanchi, Salvador Sagrado, Myalowenkosi Sabela, and Krishna Bisetty

Copyright © 2017 Pan Stanford Publishing Pte. Ltd.

ISBN 978-981-4774-12-3 (Hardcover), 978-1-315-22538-8 (eBook)

www.panstanford.com

the separation of geometric isomers. The major requirement for enantioseparation in CE is believed to be the complexation between the enantiomeric analyte and the chiral selector, which could be a psuedo-stationary phase. The mass-to-charge ratio governs the movement of the free analyte, the selector, and the analyte–selector complex toward the detector. It is for this reason that the theoretical basis for the mechanism of separations involving chiral selectors added to the background electrolyte in CE methods is well documented in the literature. A number of authors have reported on the simulated enantioseparation: iodiconazole [1], fenoprofen [2], phenylazetidin derivatives [3], bupivacaine and propranolol [4], 1-aminoindan, 1-(1-naphthyl)ethylamine, 1,2,3,4-tetrahydro-1-naphthylamine [5], baclofen [6], (+)-catechin and (–)-epicatechin [7], ethyl-3-hydroxybutyrate [8] using cyclidextrin derivatives as chiral selectors. Unfortunately, all these reports do not consider the influence of the capillary surface of the enantioseparation. Instead, there are a few reports that only consider stationary phase surface interactions. For example, Refferty et al. reported the effects of stationary phase and solute chain length by carrying out Monte Carlo simulations of dimethyl triacontyl (C30), dimethyl octadecyl (C18), dimethyl octyl (C8), and trimethyl (C1) silane grafted, and bare silica stationary phases in contact with a water/methanol mobile [9]. Alkane solutes were unretained at the bare silica surface, while alcohol solutes were only slightly enriched at the silica surface due to hydrogen bonding with surface silanols and surface bound solvent. With regard to solute size, it appeared that the retention mechanism was not affected by the chain length of the solute. While Lipa et al. investigated through molecular dynamics, the chromatographic models that represent both monomeric and polymeric stationary phases with alkylsilane surface coverages and bonding chemistries [10], temperature, and chain length [11]. The results were satisfactory, and they exemplify the significance of simulation for better understanding of chromatographic separation mechanism and elucidate the molecular-level structural features that control shape-selective separations.

The most commonly used software in such reports are Gold version 3.0 [1, 12], Gaussian [4, 5, 7, 13–15], Discovery Studio Client 2.5 [1], AutoDock 3.0 [6, 16], GROMACS [3, 17], Insight II/Discover program [2, 18], and AMBER [15].

Cyclodextrins (CDs) are cyclic oligosaccharides consisting of six (α-CDs), seven (β-CDs), and eight (γ-CDs) glucopyranose units with a truncated cone providing a hydrophobic cavity. The glucose units are linked by α-1,4-glycosidic linkage. The potential use of natural CDs and their synthetic derivatives has been extensively explored as can be seen in other sections of this book. Modified CDs can exhibit different properties in relation to the native ones, which can easily be used for improving the selectivity of enantiomeric separation. Besides their separation role, they have also been used to improve certain properties of the drugs, such as solubility, stability, and/ or bioavailability. One of the most useful applications of CDs is dosage form design to enhance the solubility of poorly water-soluble drugs by complex formation [19]. With the extensive application of cyclodextrins in improving drug solubility and oral absorption, there is an urgent need for a quantitative structure–activity relationship (QSAR) models to predict the binding constants of the drug molecules [20]. It is also interesting to note that due to the differences in chemical structure (i.e., shape and size) and chemical properties of interest, the binding behavior of drug molecules with cyclodextirns should differ. In this work, molecular simulations were carried out in the presence of a chiral selector imitating separation conditions from the literature as much as possible. In general, separation of the drugs is carried out for various reasons, which include purity, understanding metabolites by racemic compounds [21], competitive binding, etc. Among the several chiral selectors, cyclodextrins (CDs) and their derivatives are the most widely used as running buffer additives in CE, and it is for that reason, we choose β-CD as our model chiral selector. The selected carboxylmethyl-beta-cyclodextrin (CM-β-CD) is also know to facilitate therapy [22], as anticancer drug carrier [23]. The most important descriptor in both models is the calculated log P, indicating that drugs lipophilicity in relation to the strength of complexation [24].

In this work, simulation was performed with reference to the experimental report by Nemeth et al., which was based on enantioseparation of antimalaria drug derivatives using CM-β-CD [25]. Our interest on these molecules was triggered by the fact that Malaria is one of the most important parasitic diseases affecting and killing millions of people throughout the world. Malaria remains a significant infectious disease with even artemisinin-based therapies now facing resistance in the field. Development of new therapies is urgently needed, either by finding new compounds with unique modes of action, or by reversing resistance toward known drugs with "chemosensitizers" or "chemoreversal" agents (CRA) [26]. The synthetic antimalarial drugs (shown in Fig. 10.1) include chloroquine (CLQ) and primaquine (PRQ), which are all administered as racemates [26–28]. Hence, understanding chiral resolution of such derivatives is imperative. For instance, erythro-mefloquine (MFQ) molecule has two centers of asymmetry; only the erythro pair of enantiomers is administered as a drug. Their structures consist of two or three condensed aromatic rings and aliphatic or alicyclic side chains containing center(s) of chirality and amine group(s). The 8-aminoquinolines, PRQ, CLQ, quinolone methanol, and MFQ are weakly basic compounds due to their amino functional groups, as can be seen in Fig. 10.2.

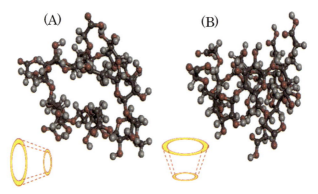

Figure 10.1 Structure of highly carboxylmethyl-beta-cyclodextrin (CM-β-CD). (A) Top view and (B) side view.

As it is well acknowledged that a deeper understanding of biological processes requires a multidisciplinary approach employing the tools of biology, chemistry, and physics. Such understanding involves study of bioactive molecules and their functions, which include quantification and purification. Density functional theory (DFT) finds increasing use in applications related to biological systems as advancements in methodology, and implementations have reached a point where predicted properties of reasonable to high quality can be obtained. DFT studies have the ability to complement experimental investigations, or even venture experimentally unexplored territory.

Introduction | 367

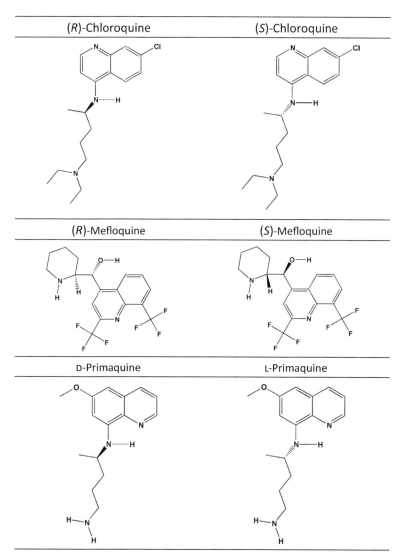

Figure 10.2 Enantiomeric structures of antimalarial drugs chloroquine, mefloquine, and primaquine used for docking simulation.

In the present work, we use a wide range of properties that can be calculated with DFT, such as geometries and energies, to understand the mechanism of chiral separation. A wide range of spectroscopic parameters are nowadays accessible with DFT, including quantities related to infrared and optical spectra and X-ray absorption as well as all the magnetic properties connected with electron paramagnetic resonance spectroscopy, except relaxation times. Such information is beneficial in understanding differences in inclusion complexes of cyclodextrin.

In general, molecular modeling is carried out to further elaborate the complexation modes by looking at interactions between the target and small chemical ligands. Moreover, it can also be used for the selection of the best chiral selector prior to experimental analysis. Although this approach does not simultaneously consider ligand-chiral selector (complex formation) and ligand-CD complex surface capillary adsorption, we have more reason to believe that it is close to experimental mechanism of interaction that results in separation.

10.2 Procedure

10.2.1 Experimental Protocol

The author used the following chemicals as supplied: (±) Chloroquine diphosphate (Sigma), mefloquine hydrochloride (Sigma, racemic mixture of the (−)-(11S,2_R)- and (+)-(11R,2_S)-erythro enantiomers), (±)primaquine diphosphate (Aldrich). Drug sample stock solutions (1 mg/ml) were prepared in methanol and were further diluted threefold with double distilled water prior to measurements. CE was performed with an Agilent Capillary Electrophoresis 3DCE system with bare fused silica capillary connected to online DAD UV–vis detector set for UV absorption detection at 220 nm. Sodium phosphate buffer (50 mM) at pH 2.5 was applied as background electrolyte (BGE), while the capillary was thermostated at 20°C. Samples were hydrodynamically injected and then separated in the presence of 20 mM concentration CMβCD.

10.2.2 Simulation Protocol

10.2.2.1 Ligand and receptor preparation

The starting structures of the selected CM-β-CD and antimalarial drugs shown in Figs. 10.1 and 10.2 were downloaded from

PubChem database. The ligands to be used for docking calculations to cyclodextrin required 3D coordinates and particular ionization and tautomerization states. This is a crucial step in receptor–ligand interactions and other areas, so it is important to correctly prepare the ligands. Different protonation states, isomers, and tautomers typically have different 3D geometries and binding characteristics. Ligand preparation is critical in a sense that it allows us to (i) standardize charges for common groups, (ii) add hydrogens, (iii) enumerate ionization states, (iv) ionize functional groups, and (v) remove any duplicates of the ligands. The maximum number of tautomers was set to 10.

10.2.2.2 Minimization

We performed minimization on a series of ligand poses using CHARMm forcefield. The selected ligands were prepared with an ionization pH-based method, which was set in a range of 2 to 4 with reference to the experimental literature reports as a guide [25]. Acid ionization was mainly for carboxylate, while base ionization could only be for any 1°, 2°, or 3° amine. Optimization was terminated when either the maximum steps, gradient, or energy tolerance has been met.

CM-β-CD was prepared using the receptor option. The smart minimizer algorithm was used for minimization with 2000 max steps and without gradient energy change. This protocol was also an explicit solvent model with dielectric constant set to 1. The results retrieved after minimization include (i) initial potential energy (kcal/mol) of the molecule before minimization, (ii) final RMS gradient (kcal/mol$\times$Å) of the minimized molecule, and (iii) final energy (kcal/mol) of the minimized molecule given as CHARMm energy. In general, the differences between density functionals are usually small for structural parameters, making the choice of functionals not critical for the success of a geometric optimization. However, after structural optimization, additional molecular properties were calculated, as shown in Table 10.1.

The stable conformations were obtained with a minimum energy with the heat of formation of (R)-chloroquine, (S)-chloroquine, (R)-mefloquine, (S)-mefloquine, D-primaquine, and L-primaquine (kJ/mol), respectively. The minimization steps were performed with the normal full potential [29].

Table 10.1 Molecular properties calculated from the optimized structures with ionization pH of 3

Enantiomers		CHARMm Energy	Log P	Molecular Weight	Molecular Volume
Chloroquine	(R)	86.51	3.197	321.89	277.82
	(S)	84.36	3.197	321.89	278.51
Mefloquine	(R)	70.91	3.071	379.32	266.51
	(S)	70.85	3.071	379.32	264.79
Primaquine	D	−1.84	1.999	259.347	222.6
	L	223.85	0.953	262.37	227.75

10.2.2.3 Molecular properties

The molecular properties of interest calculated after optimization include ALogP, minimized energy, molecular volume, HOMO energy ($Dmol^3$), LUMO energy ($Dmol^3$), band gap energy ($Dmol^3$), and binding energy ($Dmol^3$). Geometries predicted by DFT tend to be quite reliable, and the optimized structures usually agree closely with X-ray diffraction (XRD) or extended X-ray absorption fine structure (EXAFS) data [30].

10.2.3.4 HOMO–LUMO calculations

The electrons transferred from the highest occupied molecular orbitals (HOMO) in a molecule are related to its ability to undergo redox reactions. During the reduction process, the electrons move to the lowest unoccupied molecular orbital (LUMO). Therefore, using DFT calculations, the density of the electrons in the molecule can be located, thereby predicting the active site of the ligand/molecule based on the electron density maps. The energy was calculated to visualize molecular orbital potential using a BLYP function in the absence of a solvent for a fine quality output. The loosely bound electrons in the HOMO shown in Fig. 10.3B are located in the carbonyl group of the guaiacol ring, in contrast to those displayed in the LOMO.

Exploring many structural alternatives and their corresponding spectroscopic and thermodynamic properties in this way is an important step in cross-validating theory and experiment, forming the basis for further elaboration toward more realistic models. Other molecular properties can be evaluated using a potentially more accurate hybrid functional. In this work, we use the obvious traditional choice, the BLYP functional.

10.2.3.5 Docking

Docking was performed on the input site surrounded with a sphere depicting the binding site. CDOCKER is an implementation of a CHARMm-based docking tool using a rigid receptor [29]. As the full forcefield final minimization has proven to be beneficial for both reducing the grid's energetic discrepancies and improving the docking accuracy, it has been adopted as the default docking strategy in CDOCKER. A maximum of 10 random confirmations at dynamic target temperature of 1000. A set of ligand conformations are generated using high-temperature molecular dynamics with different random seeds, producing random orientations of the conformations. The heating and cooling temperatures were set at 700 and 300 with 2000 and 5000 steps, respectively. This docking simulation was performed with CHARMm forcefield. The interaction energy including ligand strain and the interaction energy alone were calculated, and the most favorable (more negative) was used for further analysis. Finally, binding affinities predicted based on the scoring function [31].

10.2.3.6 Score ligand poses and analysis

The structure of CD was used as a receptor and checked for the availability of hydrogens and complete valence shells. The scoring option has several output parameters, but for this work, the interest was on LigScore [31] and CDOCKER energy [29]. Hence they are discussed further. The final ranking of the ligand's docking poses is based on the total docking energy (including the intramolecular energy for ligands and the ligand–protein interactions). A ligand–receptor docking is considered successful if the RMSD between the top ranking (lowest energy) docking pose and the ligand's X-ray position is less than 2.0 Å [29]. Furthermore, the scores were analyzed by the calculation of receptor–ligand interactions. The non-bonded interactions were calculated between receptor and ligand poses. Although there are several interaction categories, our focus was on hydrogen bond, hydrophobic and charge-based interactions.

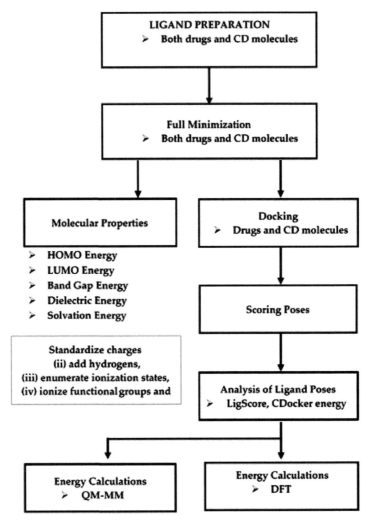

Scheme 10.1 Flowchart of molecular simulation protocol.

10.3 Results and Discussion

10.3.1 Experimental Enantioseparation

The authors did evaluate the influence of concentration of the chiral selector, but eventually choose 20 mM as it proved to provide the best resolution. In order to maintain positive charge on the analytes, measurements were carried out at relatively strong acidic conditions (at pH 2.5 or 3.0). The mobility of electroosmotic flow was very slow, and the adsorption of the basic analytes to the silica capillary wall could be neglected since the deprotonation of silanol groups is depressed at a pH as low as 2.5. Application of CM-β-CD provided excellent chiral resolution, as can be seen in Fig. 10.3, exceeding the selectivity of other CD derivatives reported previously.

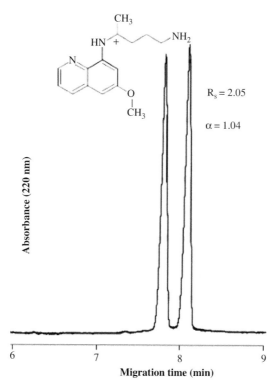

Figure 10.3 Simultaneous separation of primaquine from quinocide and their enantiomeric resolution by 20 mM CM-β-CD at pH 3.0; for experimental details. Reprinted from Ref. [25], Copyright 2011, with permission from Elsevier.

The best separation was achieved by CM-β-CD. The author reported that the presence of carboxyl or sulfate functional groups on CD derivatives is advantageous for the resolution of CLQ enantiomers. We could not correlate the ring size of CDs with the efficiency of resolutions. The excellent resolution properties of CM-β-CD (Fig. 10.2) can be attributed to its good enantioselectivity for both substances, resulting in four peaks in a single run.

The characteristic structural feature of MFQ is the additional piperidyl ring (see Fig. 10.2) with a size that may match the cavity of β-CD derivatives. However, it was not the largest in terms of molecular volume with 266.5 and 264.8. The remaining parts of the molecule might prefer complexation with CD selectors of other ring sizes. Accordingly, the order of efficiency in enantiorecognition by different CDs in the corresponding series could vary.

10.3.2   Score Ligand Poses and Analysis

The final functional form of Lig Score 1 is straightforward and contains only three descriptors:

$$pK_i = C - \beta_1 \langle E_{vdw} \rangle + \beta_2 C\text{pos\_tot} - \beta_3 TotPol^2 \qquad (10.1)$$

where $\langle \, \rangle$ indicates the summation of all interactions between the ligand and protein atoms. $C\text{pos\_tot}$ is the total surface area of the ligand involved in attractive polar interactions with the receptor, and $TotPol^2$ is equal to $C\text{pos\_tot}^2 + C\text{neg\_tot}^2$, where $C\text{neg\_tot}$ is the total surface area of the ligand involved in repulsive polar interactions with the protein.

LigScore 2 is derived by enhancing Eq. 10.1 to give

$$pK_i = C - \beta_1 \langle E_{vdw} \rangle + \beta_2 C\text{pos\_tot} - \\ \beta_3 (SolvPlty\_lig^2 + SolvPlty\_prot^2) \qquad (10.2)$$

Table 10.2   CDOCKER output parameters of antimalarial drugs-CD complexes

Enantiomers	LigScore		Interaction Energy (kcal/mol)
	1	2	
R-CRQ	−999.9	−999.9	−76.9007
S-CRQ	−334.5	−545.2	−2612.06
R-MFQ	−335.5	−548.6	−3009.68
S-MFQ	−290.1	−474.7	−2540.59
D-PMQ	−328.6	−535.5	−2927.74
L-PMQ	−272.9	−446.7	−2343.87

Results and Discussion | 375

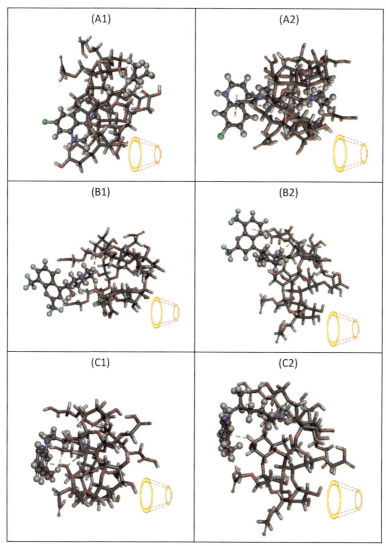

Figure 10.4 Antimalarial grids–CD complexes obtained after CDOCKER calculations using CHARMm forcefield. (A) (*R*)- and (*S*)-Chloroquine, (B) (*R*)- and (*S*)-Mefloquine, (C) D- and L-Primaquine. Corresponding details: type of bonds, lengths are show in Tables 10.4A–10.4C.

The first term represents the overall polar ligand surface that is buried inside the receptor-binding site, while the second term indirectly accounts for the polar receptor surface in contact with the bound ligand molecule. The polarity of all atoms is based on atom types rather than partial atom charges [31].

Table 10.3 Energy calculated with DFT

	Dmol^3 Energies				
QM Energy	Total Energy	Binding Energy	HOMO Energy	LOMO Energy	Band Gap Energy
−4669.62	−1326.19	−7.22	−0.384	−0.322	0.0622
-4675.65	−1326.20	−7.23	−0.397	−0.311	0.0858
−4525.05	−1441.30	−6.87	−0.309	−0.214	0.0941
−4526.08	−1441.30	−6.87	−0.310	−0.215	0.0944
−4199.97	−823.31	−6.52	−0.147	−0.058	0.0883
−3770.91	−824.15	−5.87	−0.535	−0.465	0.0706

QM: quantum mechanics

10.3.3 Analysis of Enantioselective Interaction

In this work, we consider the most influential interactions to understand the interaction mechanism of enantioseparation bearing in mind that molecular simulation conditions do not perfectly mimic the experimental separation conditions. With that in mind, we explore the number of hydrogen bonds, types of bonds, length of bonds, and dihedral angles. With reference to Table 10.4A, for the individual interaction of chloroquine enantiomers with our cyclodextrin, we observe that there was an equal number of conventional hydrogen bonds, but those of S enantiomer were much shorter in distance. This means participation of O23 draws the S enantiomer molecule much closer. However, R enantiomer shows a total of five hydrogen bonds, while (S)-chloroquine shows four bonds and two additional π donar hydrogen bonds. All the π bonds reported in this work are due to CD-donating hydrogen to the corresponding ligand. With regards to mefloquine, there was an equal number of hydrogen bonds for the two enantiomers as shown in Table 10.4B. There are two conventional H bonds in (R)MFQ-CMβCD complex, while there is only one in (S)MFQ-CMβCD complex. Corresponding Fig. 10.4B shows that the molecule was perfectly inserted within CD.

Results and Discussion | 377

Table 10.4A Details of chloroquine-CD complexes

	H Donar → Acceptor	Distance	Bond Type	DHA/θ
(R)-Chloroquine	CLQ:H49 → CMβCD:O5	2.00	Conv. H Bond	157.1
	CLQ:H49 → CMβCD:O19	2.17	Conv. H Bond	116.1
	CLQ:H27 → CMβCD:O9	1.69	C–H Bond	107.8
	CLQ:H28 → CMβCD:O32	2.48	C–H Bond	124.7
	CLQ:H29 → CMβCD:O46	2.24	C–H Bond	96.3
	CLQ:H33 → CMβCD:O46	2.12	C–H Bond	106.5
	CLQ:H36 → CMβCD:O19	2.23	C–H Bond	101.7
(S)-Chloroquine	CLQ:H48 → CMβCD:O5	1.85	Conv. H Bond	127.3
	CLQ:H49 → CMβCD:O23	1.78	Conv. H Bond	130.0
	CLQ:H27 → CMβCD:O46	1.90	C–H Bond	110.9
	CLQ:H34 → CMβCD:O28	2.23	C–H Bond	117.3
	CLQ:H35 → CMβCD:O12	2.08	C–H Bond	170.4
	CLQ:H36 → CMβCD:O8	2.81	C–H Bond	126.4
	CMβCD:H186 → CLQ Pi Orb.	2.68	Pi-Donor H Bond	
	CMβCD:H186 → CLQ Pi Orb.	2.75	Pi-Donor H Bond	

Table 10.4B Details of mefloquine-CD complexes

	H Donar → Acceptor	Distance	Bond Type	DHA/θ
(R)-Mefloquine	CMβCD:H177 → MFQ:O7	2.62	Conv. H Bond	158.1
	CMβCD:H184 → MFQ:F3	2.23	Conv. H Bond	151.4
	MFQ:H27 → CMβCD:O48	2.62	C–H Bond	147.2
	MFQ:H35 → CMβCD:O28	2.52	C–H Bond	134.9
	MFQ:H36 → CMβCD:O48	2.37	C–H Bond	151.8
(S)-Mefloquine	CMβCD:H177 → MFQ:O7	2.37	Conv. H Bond	156.3
	MFQ:H30 → CMβCD O44	2.29	C–H Bond	166.0
	MFQ:H35 → CMβCD:O34	2.50	C–H Bond	146.1
	MFQ:H35 → CMβCD:O48	2.42	C–H Bond	131.9
	CMβCD:H184 → MFQ Pi Orb.	3.36	Pi-Donor H Bond	

With regards to primaquine, the L enantiomer showed more C-H bonds and conventional H bonds as shown in Table 10.4C. There are two conventional H bonds in (D)PRQ-CMβCD complex, while there is three in (L)PRQ-CMβCD complex. Corresponding Fig. 10.4c shows that molecule was partly inserted within CD. This complex also showed a unusual proton transfer where CMβCD:H159 was donated to PRQ:O1.

Table 10.4C Details of primaquine-CD complexes

	H Donar → Acceptor	Distance	Bond Type	DHA/θ
D-Primaquine	PRQ:H35 → CMβCD:O9	2.37	Conv. H Bond	131.5
	PRQ:H35 → CMβCD:O23	2.53	Conv. H Bond	146.8
	PRQ:H29 → CMβCD:O8	2.31	C–H Bond	166.9
	PRQ:H30 → CMβCD:O12	2.40	C–H Bond	120.8
	PRQ:H30 → CMβCD:O34	2.62	C–H Bond	127.2
	PRQ:H40 → CMβCD:O44	2.87	C–H Bond	123.3
	CMβCD:H177 → PRQ Poi Orb.	2.13	Pi-Donor H Bond	
L-Primaquine	PRQ:H40 → CMβCD:O8	2.31	Conv. H Bond	156.8
	PRQ:H41 → CMβCD:O12	2.03	Conv. H Bond	149.0
	PRQ:H42 → CMβCD:O34	2.20	Conv. H Bond	118.2
	CMβCD:H159 → PRQ:O1	2.42	C–H Bond	158.2
	PRQ:H20 → CMβCD:O44	2.35	C–H Bond	123.3
	PRQ:H28 → CMβCD:O23	2.59	C–H Bond	148.5
	PRQ:H29 → CMβCD:O34	2.56	C–H Bond	113.7
	PRQ:H29 → CMβCD:O48	2.58	C–H Bond	166.9
	CMβCD:H177 → PRQ Pi Orb.	2.77	Pi-Donor H Bond	

The significant differences between binding interaction energies reaffirm that structural changes on the ligands are key factors in enantioseparation.

Such inhibitors would be valuable as tools for probing the enantioselective interaction roles of these CDs and possibly for development of novel chiral selector derivatives. The region of

Results and Discussion | 379

ligands that interacts with this portion of the ligand binding pocket is not charged; therefore, no electrostatic repulsion would result binding, leading to a modeled electrostatic repulsion.

10.3.4 Enantioresolution

It was interesting to see that when changing the size of the molecules, the atoms that dominate in hydrogen bonding change. For example, with the sphere of 5 Å used for docking chloroquine and primaquine, it was observed 05, 09, 08, and 23. The size of the binding sites used had a radius of 5 Å, 6.8 Å, 5 Å for chloroquine, mefloquine, and primaquine, respectively. Obviously, the difference was significantly high, and it was a good indication of the ability of the ligands to penetrate the chiral selector, which can be correlated to the geometry of the ligands.

The efficiency of enantiodiscrimination by CDs depended on the ring size and was different for PRQ and CLQ likewise the docking dependent on the size of the binding site, as we observed in the two aminoquinone derivative. Chloroquoniline and primaquine fitted very well in the binding site of 5 Å, while mefloquine did not; therefore, the size was adjusted to 6.8 Å.

On comparing the overall bond lengths, we observe that chloroquine had the most of the shortest hydrogen bonds, which is sensible if we consider the molecular shape and volume.

With the extensive application of cyclodextrins in improving drug solubility and oral absorption, there is an urgent need for a QSAR model to predict the binding constants for the drug molecules. While the separation/analytical aspects related with enantioselective protein binding estimations have been extensively investigated, it was reported based on the experimental results that carboxyalkyl-CD derivatives provided improvement on the resolution, which is interesting from the simulation point of view. A similar conclusion can be made considering interaction energy of the ligand-CD complexes [25].

The more negative the binding energy, the stronger the interaction between (R/S)-enantiomers and β-CD. It was proved that the complexation process is energetically favorable as observed by other authors in similar studies [15, 32], which is an advantage of having numerous rotatable bonds in the penetration of the ligand through CD. Aromaticity may reduce chances of penetration, hence hindering enantioseparation due to limited flexibility of the ligands. These observations are qualitatively correlated with ligand flexibility expressed as the number of rotatable bonds and not with the size or potency of the ligands (data from Table 10.2). The docking success frequencies detailed on a per-ligand–protein complex basis.

10.4 Conclusion and Future Prospect

The molecular modeling techniques based on docking and absorption further enable us to elucidate the chiral recognition mechanisms, which can only be imagined based on the outcome of the CE separation results. The information described herein is obtained from a molecular level, and it explains the nature of the chiral selectivity. These results provide an excellent platform for understanding enantioseparation and its mechanism of separation within the capillary. Furthermore, the observations confirm that inclusion into the hydrophobic cavity occurred through the phenyl ring, which is a prerequisite, and that this phenomenon (associated to other interactions with the CD rims) is essential for enantioselectivity. Having observed that it is possible to obtain a relatively similar trend on the separation, it is fair to say that simulation results have to be treated with caution as the simulation protocols do not consider some of the electrophoretic critical parameters, in particular the high voltage supplied through the capillary. The results obtained are very close to the reality of the experiment as the mobility of electroosmotic flow was very slow and the adsorption of the basic analytes to the silica capillary. This means there was minimal interaction with surface since deprotonation of silanol groups is depressed at a pH as low as 2.5. Participation of the carboxyl or sulfate functional groups on the CMβCD proved to improve the enantioseparation.

The overall molecular simulation offers greater perspective, which can also help to mitigate for cost implications by narrowing experimental research. This chapter seeks to provide a broader perspective of the enantioseparation that occurs within the capillary of the CE. We believe that the results obtained are in good agreement with the experimental results; therefore, it would be great to simulate the enantioseparation prior to the experiments.

A large difference in the interaction energies observed between the two enantiomers represents significant enantiodifferentiation. Our results also suggest that the host–guest interactions between the quinoline ring of the selected antimalarial drugs and the open cavity of the CM-β-CD are mainly due to hydrophobic interactions. When direct correlations are made to experimental results, caution is to be exercised because as the ligands become even more complex, there is no clear correlation between docking success frequency and properties of the ligands.

Acknowledgments

Our grateful acknowledgment goes to Durban University of Technology and National Research Foundation of South Africa for the financial support. We also thank the Centre for High Performance (CHPC), Cape Town, South Africa, for the access to the Accelrys BIOVIA | DISCOVERY STUDIO 2016 license.

References

1. Li, W., Tan, G., Zhao, L., Chen, X., Zhang, X., Zhu, Z., and Chai, Y. (2012). Computer-aided molecular modeling study of enantioseparation of iodiconazole and structurally related triadimenol analogues by capillary electrophoresis: Chiral recognition mechanism and mathematical model for predicting chiral separation, *Anal. Chim. Acta.*, **718**, pp. 138–147.

2. Choi, Y.-H., Yang, C.-H., Kim, H.-W., and Jung, S. (2000). Monte Carlo simulations of the chiral recognition of fenoprofen enantiomers by cyclomaltoheptaose (β-cyclodextrin), *Carbohydr. Res.*, **328**, pp. 393–397.

3. Bikádi, Z., Fodor, G., Hazai, I., Hári, P., Szemán, J., Szente, L., Fülöp, F., Péter, A., and Hazai, E. (2010). Molecular modeling of enantioseparation of phenylazetidin derivatives by cyclodextrins, *Chromatographia*, **71**, pp. 21–28.

4. Servais, A.-C., Rousseau, A., Dive, G., Frederich, M., Crommen, J., and Fillet, M. (2012). Combination of capillary electrophoresis, molecular modelling and nuclear magnetic resonance to study the interaction mechanisms between single-isomer anionic cyclodextrin derivatives and basic drug enantiomers in a methanolic background electrolyte, *J. Chromatogr. A.*, **1232**, pp. 59–64.

5. Elbashir, A. A. and Suliman, F. O. (2011). Computational modeling of capillary electrophoretic behavior of primary amines using dual system of 18-crown-6 and β-cyclodextrin, *J. Chromatogr. A*, **1218**, pp. 5344–5351.

6. Suliman, F. O. and Elbashir, A. A. (2012). Enantiodifferentiation of chiral baclofen by β-cyclodextrin using capillary electrophoresis: A molecular modeling approach, *J. Mol. Struct.*, **1019**, pp. 43–49.

7. Khedkar, J. K., Gobre, V. V., Pinjari, R. V., and Gejji, S. P. (2010). Electronic structure and normal vibrations in (+)-catechin and (–)-epicatechin encapsulated β-cyclodextrin, *J. Phys. Chem. A*, **114**, pp. 7725–7732.

8. Shi, J.-H., Hu, Y., and Ding, Z.-J. (2011). Theoretical study on chiral recognition mechanism of ethyl-3-hydroxybutyrate with permethylated β-cyclodextrin, *Comp. Theor. Chem.*, **973**, pp. 62–68.

9. Rafferty, J. L., Siepmann, J. I., and Schure, M. R. (2012). A molecular simulation study of the effects of stationary phase and solute chain length in reversed-phase liquid chromatography, *J. Chrom. A*, **1223**, pp. 24–34.

10. Lippa, K. A., Sander, L. C., and Mountain, R. D. (2005). Molecular dynamics simulations of alkylsilane stationary-phase order and disorder. 1. Effects of surface coverage and bonding chemistry, *Anal. Chem.*, **77**, pp. 7852–7861.

11. Lippa, K. A., Sander, L. C., and Mountain, R. D. (2005). Molecular dynamics simulations of alkylsilane stationary-phase order and disorder. 2. Effects of temperature and chain length, *Anal. Chem.*, **77**, pp. 7862–7871.

12. Zhang, G., Sun, Q., Hou, Y., Hong, Z., Zhang, J., Zhao, L., Zhang, H., and Chai, Y. (2009). New mathematic model for predicting chiral separation using molecular docking: Mechanism of chiral recognition of triadimenol analogues, *J. Sep. Sci.*, **32**, pp. 2401–2407.

13. Shi, J.-H., Ding, Z.-J., and Hu, Y. (2012). Theoretical study on chiral recognition mechanism of methyl mandelate enantiomers on permethylated β-cyclodextrin, *J. Mol. Mod.*, **18**, pp. 803–813.

14. Shi, J.-H., Ding, Z.-J. and Hu, Y. (2011). Experimental and theoretical studies on the enantioseparation and chiral recognition of mandelate and cyclohexylmandelate on permethylated β-cyclodextrin chiral stationary phase, *Chromatographia*, **74**, pp. 319–325.

15. Li, W.-S. Chung, W.-S., and Chao, I. (2003). A computational study of regioselectivity in a cyclodextrin-mediated Diels–Alder reaction: Revelation of site selectivity and the importance of shallow binding and multiple binding modes, *Chemistry*, **9**, pp. 951–962.

16. Li, W., Liu, C., Tan, G., Zhang, X., Zhu, Z., and Chai, Y. (2012). Molecular modeling study of chiral separation and recognition mechanism of β-adrenergic antagonists by capillary electrophoresis, *Int. J. Mol. Sci.*, **13**, pp. 710–725.

17. Felton, L. A., Popescu, C., Wiley, C., Esposito, E. X., Lefevre, P., and Hopfinger, A. J. (2014). Experimental and computational studies of physicochemical properties influence NSAID-cyclodextrin complexation, *AAPS Pharm. Sci. Tech.*, **15**, pp. 872–881.

18. Kim, H., Jeong, K., Lee, S., and Jung, S. (2013). Molecular modeling of the chiral recognition of propranolol enantiomers by a β-cyclodextrin, *Bull. Korean Chem. Soc.*, **24**, pp.

19. Onodera, R., Hayashi, T., Nakamura, T., Aibe, K., Tahara, K., and Takeuchi, H. (2016). Preparation of silymarin nanocrystals using a novel high pressure crystallization technique and evaluation of its dissolution and absorption properties, *Asian J. Pharmacol.*, **11**, pp. 211–212.

20. Li, H., Sun, J., Wang, Y., Sui, X., Sun, L., Zhang, J., and He, Z. (2011). Structure-based in silico model profiles the binding constant of poorly soluble drugs with β-cyclodextrin, *Eur J. Pharm. Sci.*, **42**, pp. 55–64.

21. Fasinu, P. S., Tekwani, B. L., Nanayakkara, N. D., Avula, B., Herath, H. B., Wang, Y.-H., Adelli, V. R., Elsohly, M. A., Khan, S. I., Khan, I. A., Pybus, B. S., Marcsisin, S. R., Reichard, G. A., McChesney, J. D., and Walker, L. A. (2014). Enantioselective metabolism of primaquine by human CYP2D6, *Malar. J*, **13**, pp. 1–12.

22. Su, C., Li, H., Shi, Y., Wang, G., Liu, L., Zhao, L., and Su, R. (2014). Carboxymethyl-β-cyclodextrin conjugated nanoparticles facilitate therapy for folate receptor-positive tumor with the mediation of folic acid, *Int. J. Pharm.*, **474**, pp. 202–211.

23. Tan, H., Qin, F., Chen, D., Han, S., Lu, W., and Yao, X. (2013). Study of glycol chitosan-carboxymethyl β-cyclodextrins as anticancer drugs carrier, *Carbohydr. Polym.*, **93**, pp. 679–685.

24. Chari, R., Qureshi, F., Moschera, J., Tarantino, R., and Kalonia, D. (2009). Development of improved empirical models for estimating the binding constant of a β-cyclodextrin inclusion complex, *Pharm. Res.*, **26**, pp. 161–171.

25. Németh, K., Tárkányi, G., Varga, E., Imre, T., Mizsei, R., Iványi, R., Visy, J., Szemán, J., Jicsinszky, L., Szente, L., and Simonyi, M. (2011). Enantiomeric separation of antimalarial drugs by capillary electrophoresis using neutral and negatively charged cyclodextrins, *J. Pharm. Biomed. Anal.*, **4**, pp. 475–481.

26. Boudhar, A., Ng, X. W., Loh, C. Y., Chia, W. N., Tan, Z. M., Nosten, F., Dymock, B. W., and Tan, K.S.W. (2016). Overcoming chloroquine resistance in malaria: Design, synthesis and structure–activity relationships of novel chemoreversal agents, *Eur. J. Med. Chem.,* **119**, pp. 231–249.

27. Shimizu, S., Kikuchi, T., Koga, M., Kato, Y., Matsuoka, H., Maruyama, H., and Kimura, M. (2015). Optimal primaquine use for radical cure of *Plasmodium vivax* and *Plasmodium ovale* malaria in Japanese travelers: A retrospective analysis, *Travel Med. Infect. Dis.,* **13**, pp. 235–240.

28. Grigg, M. J., William, T., Menon, J., Dhanaraj, P., Barber, B. E., Wilkes, C. S., von Seidlein, L., Rajahram, G. S., Pasay, C.,. McCarthy, J. S, Price, R. N., Anstey, N. M., and Yeo, T. W. (2016). Artesunate–mefloquine versus chloroquine for treatment of uncomplicated *Plasmodium knowlesi* malaria in Malaysia (ACT KNOW): An open-label, randomised controlled trial, *Lancet Infect. Dis.,* **16**, pp. 180–188.

29. Wu, G., Robertson, D. H., Brooks, C. L., and Vieth, M. (2003). Detailed analysis of grid-based molecular docking: A case study of CDOCKER—A CHARMm-based MD docking algorithm, *J. Comp. Chem.,* **24**, pp. 1549–1562.

30. Orio, M., Pantazis, D. A., and Neese, F. (2009). Density functional theory, *Photosynth. Res.,* **102**, pp. 443–453.

31. Krammer, A., Kirchhoff, P. D., Jiang, X., Venkatachalam, C. M., and Waldman, M. (2005). LigScore: A novel scoring function for predicting binding affinities, *J. Mol. Graph. Model.,* **23**, pp. 395–407.

32. Yap, K., Liu, X., Thenmozhiyal, J., and Ho, P. (2005). Characterization of the 13-cis-retinoic acid/cyclodextrin inclusion complexes by phase solubility, photostability, physicochemical and computational analysis, *Eur. J. Pharm. Sci.,* **25**, pp. 49–56.

Index

AAS *see* atomic absorption spectrometry
ACE *see* affinity capillary electrophoresis
AD *see* Alzheimer's disease
affinity capillary electrophoresis (ACE) 6, 230, 232, 233, 247
albumin 227, 236, 248
Alzheimer's disease (AD) 158
amino acid 18, 24, 28, 29, 38, 119, 128, 195, 197, 228, 260, 310, 321
  inhibitory 264
  neutral 264, 265
amino group 81, 82, 96, 156, 202, 203
  primary 334
  protonated 202
  secondary 202
  tertiary 84
amlodipine 41, 77–79, 82, 84, 85, 94, 246, 248, 249
amlodipine enantiomer 77, 102
amperometric detection 14, 15, 130
amphetamine 37, 43, 45, 74, 96, 97, 185, 187, 188, 191
  beta-keto 96
analyte 2–7, 13–15, 72–75, 78–81, 85, 86, 126, 127, 130, 150–153, 169, 174–177, 179, 180, 182, 184–186, 193, 194, 197
  basic 373, 380
  diverse 1
  enantiomeric 86, 364
  hydrophobic 81
  non-labeled 209
  separation capillary 172

antibody 6, 305, 307, 311, 313–316, 318, 319, 320, 323, 327, 329, 335
antigen 6, 229, 338, 340
antimalarial drug 74, 366, 368, 381
atenolol 36, 41, 78, 81, 82, 86, 88, 99, 195, 207
atomic absorption spectrometry (AAS) 16, 17, 46
atom 17, 82, 133, 376, 379
  ground-state 17
  hydrogen 133
  metal ion 17
  oxygen 133
azithromycin 84

background electrolyte (BGE) 10, 11, 13, 14, 19–22, 24, 72, 73, 79–81, 172, 173, 175, 177–179, 181, 182, 184–186, 190–192, 232, 293–295, 328–330
baclofen 39, 190, 364
bacteria 86, 87, 93, 322
basal ganglia 260–262
baseline separation 85, 130, 156
BGE *see* background electrolyte
BGE pH 11, 184, 250, 326
binding constant 236, 249, 250, 365, 379
binding energy 132, 133, 136, 137, 370, 376, 379
binding equilibrium 230, 232, 249
binding site 227, 228, 236, 237, 247, 249, 371, 379
biomarker 158, 271, 272
biomolecule 24, 25, 152, 154, 259, 336

386 | *Index*

bisoprolol 78, 82, 86
blocker 24, 34, 39, 166, 181, 197, 207, 228, 292
bovine serum albumin (BSA) 120, 181, 242, 244, 246, 248
BR *see* buffer reservoir
brain 32, 89, 158, 261, 263
  equine 104
  human 261
BSA *see* bovine serum albumin
buffer 6, 7, 11, 21, 22, 27, 35–37, 40, 44, 80, 105, 122, 127, 153, 154, 156–158, 331, 333
  acetate 32, 34
  aqueous 81, 154
  basic 11
  borax 83
  boric acid 29
  carbonate 264
  citrate 30
  gel 330
  low-pH 11
  Tris-citrate 121
  Tris-phosphate 95, 290
  volatile 328
buffer reservoir (BR) 2, 129
bupivacaine 74, 78, 79, 85, 94, 99, 187, 247, 249, 364

capillary electrochromatography (CEC) 6, 7, 72, 128, 130, 156, 159, 168, 201
capillary electrophoresis (CE) 1–8, 10–26, 46, 117–120, 147–150, 152–158, 167–172, 177–179, 225–228, 230–234, 242–250, 279, 305–308, 320–322, 346–349
capillary electrophoresis–mass spectrometry (CE-MS) 165, 169, 170, 181, 200, 201, 203, 208, 209, 348

capillary gel electrophoresis (CGE) 5, 6, 148, 307, 322, 324, 329, 331, 347–349
capillary isoelectric focusing (CIEF) 6, 148, 307, 324, 336, 337, 340–344, 347–349
capillary zone electrophoresis (CZE) 5, 6, 148, 168, 259, 260, 262–268, 270, 272, 307, 324–328, 343, 345, 347–349
carbon nanotube (CNT) 151–153, 158, 159
carvedilol 77, 79–81, 84, 88, 94, 102
cathinone derivative 74, 78, 80, 96, 191, 192
cavity 73, 74, 76, 77, 130, 133, 137, 183, 184, 200, 228, 374, 381
CCE *see* chiral capillary electrophoresis
CD *see* conductivity detection
CD cavity 76, 80, 84, 130, 133, 136
CD derivative 73, 74, 77, 184, 185, 373
CDR *see* complementarity determining region
CE *see* capillary electrophoresis
CEC *see* capillary electrochromatography
CEKC *see* capillary electrokinetic chromatography
cell culture 309, 313, 314, 318
CE-MS *see* capillary electrophoresis–mass spectrometry
central nervous system (CNS) 263, 264, 268, 271
cerebrospinal fluid 89, 98, 104, 106, 265
CE separation 97, 157, 172, 177, 178, 181

CE technique 118, 154, 159, 347, 349
cetirizine 35, 74, 78, 79, 84, 89, 99, 103, 105
cetuximab 319, 343, 344
CGE *see* capillary gel electrophoresis
Chinese hamster ovary (CHO) 316, 318
chiral analysis 71, 99, 230, 252
chiral capillary electrophoresis (CCE) 71–85, 87–106
chiral compounds 24, 72, 130, 170, 194, 199, 229
chiral discrimination 81, 128, 250
chiral drug 72–77, 79–81, 83, 85, 86, 93, 99, 103, 165, 205, 226, 227, 235, 242–244, 249, 252
chiral polymer 193, 194, 197, 198, 200
chiral recognition 73, 84, 86, 123, 130, 363
chiral selector (CS) 72–95, 99, 102–105, 120, 121, 123–125, 130, 131, 167, 168, 181–191, 197, 198, 201–207, 232–235, 243–250, 363–366, 368, 378–380
chiral separation 22, 73, 76, 80, 99, 105, 117–121, 168, 186, 187, 198, 205, 234, 235, 242, 247, 286
chiral stationary phase (CSP) 72, 119, 166, 206, 207
chiral surfactant 180, 193, 194, 198, 199
chloroquine 74, 82, 94, 366, 367, 369, 370, 375–377, 379
chlorpheniramine 77, 186, 193, 243, 247
chlorpromazine 36
CHO *see* Chinese hamster ovary

chromatography 12, 20, 24
  gas 119
  size-exclusion 348
  thin-layer 119
CIEF *see* capillary isoelectric focusing
cimetidine 282, 284–286
circulatory system 226, 227
CL *see* clarithromycin lactobionate
clarithromycin lactobionate (CL) 82, 83, 123, 152, 282, 283
clenbuterol 38, 41, 43, 74, 78, 86, 188
CNS *see* central nervous system
CNT *see* carbon nanotube
coated capillary 179, 188, 190, 191, 203, 323, 336, 338
coating 10, 155, 324, 326, 328, 336, 344
  anionic surface 127
  dynamic 325, 328
  inner surface 127
  metal 173
  polybrene cationic 343
complementarity determining region (CDR) 310, 312
complex 24, 73, 79, 127, 132, 133, 135, 136, 138
  chloroquine-CD 377
  diastereomeric 167, 187, 190, 200
  enantiomer-CD 78
  enantiomer– chiral selector 72
  host–guest 138
  ligand-CD 379
  ligand-exchange 72
  mefloquine-CD 377
  metal ion 93
  primaquine-CD 378
  protein-SDS 334
  solvated 134, 135
conductivity 11–14, 81, 93, 154

conductivity detection (CD)
   38–45, 73–82, 88–92, 95–97,
   99, 102–105, 120, 121, 123,
   124, 130–139, 181–193,
   204–207, 242–245, 248–250,
   282–284, 295, 296
contamination 172, 178, 183, 185,
   193, 205, 206, 232
   bacterial 309
   detector 181
counter-current migration
   179–181, 186, 189, 190, 193,
   205, 209
coupling 17, 18, 307, 328
covalent bond 167, 310, 313, 330
crown ether 72, 180, 200
CS *see* chiral selector
CSP *see* chiral stationary phase
cyclodextrin 27, 35, 40, 41, 43–45,
   72, 73, 120, 121, 133, 134,
   183–185, 187, 205, 210, 211,
   244–246, 249, 365, 368, 369
CZE *see* capillary zone
   electrophoresis

deamidation 310–313, 324, 327,
   340, 347
degradation 34, 94, 294, 310, 312,
   318, 323, 324
   chemical 310, 349
   physical 306, 316
   protein clipping 333
demethylation 104, 278, 288
density functional theory (DFT)
   366, 367, 370, 376
detection limit 2, 16–18, 20, 46,
   123, 234
detection method 15, 16, 336
detection reservoir (DR) 129, 130
detection sensitivity 72, 151, 169,
   191, 194, 208, 331
detection system 87, 93, 99, 105,
   230, 334
detection window 10, 125, 181

detector 2, 3, 7, 8, 11–13, 15, 16,
   18–21, 46, 87, 120, 123, 159,
   181, 331, 336, 341, 364
   absorbance 157
   conductometry 120
   electrochemical 120
   evaporative light scattering 172
   fluorescent 16
   lamp-based 16
   laser-induced fluorescence 158
   optical 19, 120
   photodiode array 12, 13
   photometric 12, 13
   potential gradient 13
   radio 19
   radioactive 46
   refractive index 18
   spectrophotometric 178
dextrofloxacin 118, 121, 123–125
DFT *see* density functional theory
diazepam 246, 248
dipeptide 29, 39, 40, 201
disease 260, 261, 271, 272, 306
   infectious 365
   neurodegenerative 158
   neurological 260
   parasitic 365
   peptic ulcer 285
disopyramide 250
disulfide bond 307, 311, 314, 315,
   319, 320, 331
   inter-chain 320
   intra-chain 314, 315
docking simulation 136, 367, 371
DR *see* detection reservoir
drug 33, 34, 77, 85–94, 97–99,
   104–106, 117–119, 138, 166,
   167, 225–233, 235–237, 247,
   248, 251, 252, 278, 279, 307,
   308, 365, 366
drug enantiomer 74, 105, 106,
   225, 226, 247
drug interaction 196, 278
drug metabolism 72, 278, 281

duloxetine 74, 78, 82, 84, 94, 191, 192

electric field 2, 80, 279, 280, 329, 336
electrochemiluminescence 25, 29, 34, 36, 37, 92, 93
electrolyte 8, 11, 98, 174, 178, 180, 182
  basic 197
  carrier 6
  non-aqueous 204
electromigration 5, 7, 21, 128
electroosmotic flow 3, 8, 9, 139, 171, 323
electropherogram 20, 23, 75, 83, 87, 95, 97, 98, 100, 101, 121–123, 125, 126, 234, 265, 290, 295, 297, 298
electrophoretically mediated microanalysis (EMMA) 279, 281–283, 285, 287, 289, 291, 293, 295, 297, 299
electrophoretic mobility 2, 4, 9, 20, 72, 78, 79, 168, 181, 200, 264, 279, 343
electrophoretic principle 93, 98, 99, 105
electrophoretic technique 127, 148, 232, 242
electrospray 174, 181
  discontinuous 176
  stable 171, 175, 176
electrospray ionization (ESI) 170–172, 181, 188, 189, 193, 195, 196, 199, 206, 209, 328, 342
electrostatic interaction 85, 155, 186, 203, 326
ELSD see evaporative light scattering detector
EMMA see electrophoretically mediated microanalysis

enantiomeric separation 72, 73, 75, 77–79, 82, 85, 97, 118, 120, 122–124, 126–130, 132, 134, 136, 138, 249
enantiomer 71, 77–80, 85, 86, 93, 94, 99, 101–105, 117–119, 130–133, 136, 137, 166–168, 204–209, 231, 232, 234, 235, 239–250, 296, 297
  catechin 248
  charged 288
  chlorinated alkyl phenoxy propanoate 207
  chloroquine 376
  citalopram 103
  desmethylvenlafaxine 198
  hexobarbital 206
  methoxytolterodine 78, 79
  nomifensine 245
  norketamine 76, 104, 284
  norverapamil 292, 295
  ornidazole 131, 133
  pheniramine 192
  salbutamol 205
  sotalol 75
  terbutaline 184, 185
  tolterodine 102
  trimipramine 76, 102
  tropic acid 189
enantioresolution 75, 78, 80, 84, 85, 183, 198, 379
enantioselective analysis 120, 139, 232
enantioselective binding 235, 242, 247–249, 251
enantioselective metabolism 281, 283, 285, 287, 289, 291–293, 295, 297
enantioselective protein binding 230, 242
enantioselectivity 128, 207, 226, 240, 246, 247, 250, 296, 297, 380

Index

enantioseparation 72–85, 96, 125, 127, 129, 131, 137, 139, 166, 167, 180–185, 187–195, 197–209, 294, 364, 365, 380
  benzoin 199
  salbutamol 204
  simulated 364
enzymatic reaction 226, 279, 281, 284, 286, 288, 296, 297, 313
enzyme 6, 24, 103, 104, 225, 278–284, 286–289, 291–294, 299, 309
  carboxypeptidase 345
  catabolic 272
  cytochrome P450 103, 289
  metabolic 284
ephedrine 31, 43, 74, 96, 184, 185, 195–197
epinephrine 32, 86
ESI *see* electrospray ionization
eszopiclone 89, 95, 242
evaporative light scattering detector (ELSD) 29, 172
excretion 103, 126, 226, 278

FA *see* frontal analysis
FASI *see* field-amplified sample injection
FESI *see* field-enhanced sample injection
field-amplified sample injection (FASI) 88, 91–93, 99, 102, 105
field-enhanced sample injection (FESI) 193
fluorescence 12, 15, 19, 25–29, 31, 33, 37, 38, 270
  laser-induced 120, 159, 331
fluoxetine 80, 85, 190, 237, 240, 241, 283, 292, 296, 297
frontal analysis (FA) 230, 232–234, 244, 248, 251

gas chromatography (GC) 119

GC *see* gas chromatography
globus pallidum (GP) 260–263
glycation 310, 312, 314, 334, 340
glycosylation 314, 318, 343
GP *see* globus pallidum

HDL *see* high-density lipoprotein
HFEP *see* high-frequency electrical pulses
high-density lipoprotein (HDL) 229, 251
highest occupied molecular orbitals (HOMO) 370, 376
high-frequency electrical pulses (HFEP) 266–268
high-performance liquid chromatography (HPLC) 7, 119, 120, 139, 230, 263, 308
high-throughput analysis 327, 334, 335, 342, 344
Hill model 282, 283
HOMO *see* highest occupied molecular orbitals
HPLC *see* high-performance liquid chromatography
human plasma 32, 44, 88, 101, 102, 227, 237, 242, 246, 247
human urine 26, 29, 30, 35, 45, 81, 88–92, 99, 120, 126, 195
hydrogen bond 203, 228, 371, 376, 379
hydrophobic cavity 73, 138, 183, 184, 365, 380
hydroxyzine 74, 84, 85, 89, 99, 103, 105, 243, 247

ICP *see* Inductively coupled plasma
IL *see* ionic liquid
illicit drug 93, 96, 97, 106
incubation 104, 234, 286, 288, 290–292, 295, 310
Inductively coupled plasma (ICP) 11, 12, 17, 46, 170

interaction 72, 73, 75, 128, 130, 137, 138, 151, 153, 154, 156, 157, 168, 183, 184, 228, 229, 368, 374, 376, 379, 380
  charge-based 371
  drug–AGP 229
  drug–protein 233, 236
  enantiomer–enantiomer 235
  enzyme–xenobiotic 278
  host–guest 381
  hydrophobic 81, 137, 138, 381
  inclusion–complexation 85
  ion-pair 86
  ligand–protein 371
  low-affinity 232
  non-bonded 371
  receptor–ligand 369, 371
ionic liquid (IL) 41, 72, 77, 82, 85
isoform 278, 279, 281, 285, 306, 308, 340, 346
  C-terminal 316
  disulfide bond 310
isomerization 310–312
isomer 121, 138, 159, 166, 249, 250, 369
  geometric 166, 364
  optical 99, 191
  positional 198
isradipine 74, 78, 94

ketamine 40, 44, 76, 89, 90, 103, 104, 185, 190, 282, 284, 287–289
  hepatic 103
  racemic 104, 288, 290–292
ketoconazole 40, 77, 80, 103, 104, 291, 292
ketoprofen 82, 203, 205, 249

Lamarckian genetic algorithm (LGA) 134
laser-induced fluorescence (LIF) 27, 32, 120, 158, 159, 264, 331, 342

LC *see* liquid chromatography
LDL *see* low-density lipoprotein
levocetirizine 45, 94, 105
levofloxacin 118, 121, 123–125, 128
LGA *see* Lamarckian genetic algorithm
LIF *see* laser-induced fluorescence
LIF detector 120, 334
ligand 127, 128, 155, 231, 368, 369, 371, 374, 378–381
  chemical 368
  low-molecular-weight 231
  marker 246
  resin 309
limit of detection (LOD) 87–93, 97, 99, 101–103, 105, 121, 130, 131, 178, 191–193, 198, 207, 324, 327, 331
limit of quantification (LOQ) 97, 99, 101–103, 131, 324, 327
lipoprotein 227, 229, 242, 249, 251
liquid chromatography (LC) 169, 172, 201, 230, 314, 319, 332, 333, 347
LOD *see* limit of detection
LOQ *see* limit of quantification
lorazepam 195, 197
low-density lipoprotein (LDL) 229, 251
lowest unoccupied molecular orbital (LUMO) 370
LUMO *see* lowest unoccupied molecular orbital

macrocyclic antibiotic 72, 81–83, 93, 95, 105, 201, 203
MALDI *see* matrix-assisted laser desorption ionization
matrix-assisted laser desorption ionization (MALDI) 170, 328
MCE *see* microchip capillary electrophoresis

392 | *Index*

mefloquine 74, 367, 369, 370, 375–377, 379
MEKC *see* micellar electrokinetic chromatography (MEKC)
meptazinol 41, 42, 74, 77, 79, 80, 94
metabolism 103, 104, 226, 278, 285, 296, 297
  cell 271
  clozapine 281
  dextromethorphan 281
  polyamine 272
  xenobiotic 277
metabolite 35, 36, 38, 75, 102, 103, 126, 165, 187, 188, 190, 197, 198, 203, 279, 281, 284, 294, 299
methadone 188, 190, 191
methamphetamine 43, 90, 96, 97, 99, 185, 187, 188, 191
methanol 28, 32, 35, 42, 44, 80, 83, 84, 86, 102, 204, 205, 342, 368
methylephedrine 31, 96, 97
metoprolol 36, 41, 78, 81, 82, 86, 88, 99, 195, 207
micellar electrokinetic chromatography (MEKC) 6, 7, 86, 148, 153, 168, 259, 264
micelle 6, 7, 153, 193, 194
Michaelis–Menten model 282, 284, 287, 288, 290
microchip capillary electrophoresis (MCE) 120, 121
microchip zone electrophoresis (MZE) 327, 348
microdialysis 259, 263, 264, 266
migration 3, 19, 101, 177, 179, 181, 183, 188, 209
migration time 3, 4, 20, 23, 75–77, 79–81, 125, 151, 154, 157, 198, 200, 295, 325, 326, 338, 340

MZE *see* microchip zone electrophoresis

NACE *see* non-aqueous capillary electrophoresis
naproxen 77, 80, 82, 95, 203
nebulizing gas 175, 176, 188
nefopam 78, 82, 83, 197
non-aqueous capillary electrophoresis (NACE) 81, 83, 84, 204, 205
norephedrine 43, 85, 96, 195, 196
norfluoxetine 296, 297
norketamine 89, 103, 104, 288, 291, 292
norverapamil 294, 295

ofloxacin 37, 74, 77, 86, 91, 94, 95, 117–139
ofloxacin enantiomer 119–121, 123–126, 128, 130–133, 137
ornidazole 74, 78, 79, 91, 94, 131–133
orphenadrine 243, 247
oxprenolol 77, 91, 99, 195, 207

Parkinson's disease (PD) 259–262, 266, 268, 269, 272
partial filling technique (PFT) 155, 166, 178–182, 187–193, 200, 201, 204, 210, 232, 233, 281, 295, 299
PD *see* Parkinson's disease
PD patient 266, 267, 269–272
penicillin 42, 158, 309
peptide 7, 18, 24, 29, 152, 153, 158, 159, 337
PFT *see* partial filling technique
PGD *see* potential gradient detector
piperazine 22, 35, 133
plasma 17, 88, 89, 91, 92, 98, 102, 106, 188, 191, 196, 198, 226, 227, 229, 231, 269, 271
  human blood 29

rabbit 36
plasma protein 225–229, 232, 233, 242, 246, 247, 249, 251, 252
plug 155, 233, 280, 281, 286, 288–290, 293
  ferrofluid 100
  nondiffused 280
  sandwich 293
  standard injection 280
polyamine 25, 26, 260, 268, 271, 272
polymeric surfactant 93, 193, 194, 198
polysaccharide 72, 73, 78, 84, 85, 93, 105
post-translational modification (PTM) 306, 307, 312, 318
potassium cyanide 331
potassium sorbate 27
potential gradient detector (PGD) 13
primaquine 74, 366, 367, 370, 373, 378, 379
propranolol 41, 77, 78, 81, 82, 86, 91, 99, 195, 207, 244, 249–251, 364
propranolol enantiomer 76, 102, 250, 251
protein 6, 7, 18, 29, 30, 152, 153, 226–237, 240, 242, 306, 309, 311, 313, 314, 316–318, 324, 326, 329–331
  glycosylated 332
  high-affinity drug-binding 229
  host cell 331
  immobilized 231
  microheterogeneous 306
  plasmatic 228, 229
  tau 158
pseudoephedrine 31, 43, 96, 195, 196, 206
PTM, *see* post-translational modification

putrescine 25, 259, 260, 268–272

rapid polarity switching (RPS) 280, 281
reaction 86, 123, 156, 277–280, 293, 311, 315
  condensation 135
  electrochemical 14
  enzymatic CYP 294
  in-capillary 290
  non-enzymatic 314
  redox 370
reagent 119, 120, 252, 279, 280, 284, 293, 309
relative standard deviation (RSD) 121, 325, 326, 335, 337, 341, 342, 345
residue 311, 313, 315, 318
  aspartate 313
  charged 343
  cysteine 319
  end-terminal galactose 250
  lysine 314
  mannose 319
  oxidized Trp 310
  sialic acid 250
RMSD *see* root mean square deviation
root mean square deviation (RMSD) 134, 135, 138, 371
ropivacaine 182, 185, 187
RPS *see* rapid polarity switching
RSD *see* relative standard deviation

SDS *see* sodium dodecyl sulfate
selected ion monitoring (SIM) 208
separation 3–7, 22–24, 76–78, 129–131, 151–154, 167–169, 175–181, 183–187, 189–191, 203–205, 207, 208, 295, 296, 324–327, 329–332, 342, 343
  chemical 149
  chiral HPLC 204

counter-current 233
electrophoretic 186, 204, 334
homogenous 156
liquid-phase 154
proteins 157
shape-selective 364
stereoselective 75
separation buffer 97, 158, 327, 335, 343
separation capillary 1, 13–15, 18, 104, 169, 170, 175, 177, 178, 181, 182, 191, 209
separation condition 10, 286, 290, 297
separation efficiency 10, 120, 139, 148, 153, 157, 167, 183, 230, 233, 327, 329, 342
separation mechanism 6, 72, 119, 131, 150, 153, 348
separation medium 24, 127, 139, 330
separation science 18, 19, 23, 46, 147
separation technique 2, 46, 125, 154, 157, 178, 230, 231
separation voltage 100, 121, 122, 129, 130, 245, 294, 295
separation zone 183, 293, 294
serum 26, 98, 165, 309
    human 26, 30, 31, 158, 246
    rabbit 88, 99
serum albumin 120, 181, 242, 248
sibutramine 74, 77, 94
SIM see selected ion monitoring
SLM see supported liquid membrane
sodium dodecyl sulfate (SDS) 25, 29, 30, 33, 36, 45, 80, 158, 265, 267, 308, 329, 330, 332, 334
sotalol 36, 41, 78, 80, 81, 86, 91, 94
species 9, 11, 12, 15, 17, 19, 21–23, 103, 341

animal 288
microbial 309
nanostructured 149
radioactive 19
reactive oxygen 310
spermidine 25, 268, 269, 272
spermine 25, 268, 269
spraying capillary 170, 171, 174–177
stereoselectivity 103, 104, 197, 242, 248, 251, 278, 285, 288
substrate 6, 278–285, 287, 289, 290, 293, 294, 299
supported liquid membrane (SLM) 101, 102
surfactants 72, 80, 153, 155, 158, 172, 177–179, 193, 197, 264, 265
    low-molecular 194
    polydipeptide 199
    polymerized 197
    unpolymerized 197

technique 2, 8, 103, 105, 139, 148, 154, 158, 243, 244, 252, 263, 268, 307, 329, 337
    atmospheric-pressure ionization 170, 206
    conventional 148, 233
    diffusion-based 289
    electroosmotic–electromigration 8
    in-capillary preconcentration 93
    ionization 169
    microseparation 174
    molecular modeling 132, 380
    soft ionization 170
thalamus 261–263, 266–268, 272
    ventrolateral 262, 267
therapeutic proteins 307, 308, 311, 316, 318, 322, 341, 342, 347

thin-layer chromatography (TLC) 119

TLC *see* thin-layer chromatography

tolterodine 42, 74, 78, 79, 84, 85, 92, 94

toxicity 103, 118, 119, 139, 226

tramadol 84, 85, 94, 190, 281

trastuzumab 326, 343–345

urine 26, 28, 32, 92, 98, 102, 106, 124, 126, 165, 186, 187, 192, 193, 205, 207, 208

vancomycin 39, 82, 120, 182, 202–204

venlafaxine 77, 188, 190, 196, 198, 199

verapamil 244, 249–251, 283, 292, 294–296

very low-density lipoprotein (VLDL) 229

vitamin 18, 24, 33, 229

water-soluble 159

VLDL *see* very low-density lipoprotein

xenobiotics 24, 277–279, 281, 284, 285

zopiclone 37, 74, 75, 77, 79, 94, 95, 102, 104, 242

PGSTL 08/14/2017